W9-BEI-165

# ANALYTICAL MECHANICS

## GRANT R. FOWLES

*University of Utah*

# ANALYTICAL MECHANICS

## THIRD EDITION

HOLT, RINEHART AND WINSTON
*New York Chicago San Francisco Atlanta*
*Dallas Montreal Toronto London Sydney*

# Preface

This textbook is intended primarily for the undergraduate course in mechanics for students majoring in physics, physical science, or engineering science. It is assumed that the reader has taken general physics and has a mathematical background which includes some familiarity with matrix algebra and a working knowledge of differential and integral calculus. In addition, it is recommended that an introductory course in ordinary differential equations or a course in advanced mathematics including differential equations be taken prior to or concurrently with this course in mechanics.

This third edition is basically the same in outline as the previous two editions. Some new material has been added and there is more extensive use of matrix notation. At the request of several users of the previous edition, the chapter on special relativity has been deleted. The present trend is to teach relativity as a separate course.

The first chapter (formerly the first two) presents a brief preparation in vector algebra and vector differentiation. Newton's laws of motion are introduced in Chapter 2, which also includes the rectilinear motion of a single particle.

The general motion of a particle in space is studied in Chapter 3, followed by the study of noninertial reference systems in Chapter 4.

The motion of a system of many particles is treated in Chapter 6, which also includes such subjects as collisions and rocket motion. The motion of rigid bodies is presented in the next two chapters. Some theorems on static equilibrium of rigid bodies are briefly mentioned in Chapter 7. It is assumed that the student has already had experience in solving problems in statics. Elasticity and hydrodynamics are not included because, in the opinion of the author, these subjects are best postponed until the senior or graduate years.

Lagrangian mechanics is introduced in Chapter 9. Chapter 9 also includes a brief discussion of Hamilton's equations. The method of Lagrange is employed in the study of oscillating systems in the final chapter.

Drill exercises and problems are given at the end of each chapter. Some problems are important theorems for the student to prove, perhaps with a hint from the instructor. The author feels that students should participate in the development of the subject, rather than merely substitute numbers into equations already developed in the text. Answers to selected odd-numbered problems are given at the end of the text.

There is a collection of useful mathematical formulas in the appendixes. Also included is a brief outline of matrix algebra.

The author wishes to express his thanks to all who have offered constructive criticism of the previous edition and to the editorial staff for expert assistance in the preparation of this edition.

Grant R. Fowles

# Contents

## 3. GENERAL MOTION OF A PARTICLE IN THREE DIMENSIONS     77

## 4. NONINERTIAL REFERENCE SYSTEMS     117

## 5. CENTRAL FORCES AND CELESTIAL MECHANICS     137

## 6.  DYNAMICS OF SYSTEMS OF MANY PARTICLES      168

## 7.  MECHANICS OF RIGID BODIES. PLANAR MOTION      187

## 8.  MOTION OF RIGID BODIES IN THREE DIMENSIONS      217

## 9.  LAGRANGIAN MECHANICS      251

# ANALYTICAL MECHANICS

# 1. Fundamental Concepts. Vectors

In any scientific theory, and in mechanics in particular, it is necessary to begin with certain primitive concepts. It is also necessary to make a certain number of reasonable assumptions. Two of the most basic concepts are *space* and *time*. In our initial study of the science of motion, mechanics, we shall assume that the physical space of ordinary experience is adequately described by the three-dimensional mathematical space of Euclidean geometry. And with regard to the concept of time, we shall assume that an ordered sequence of events can be measured on a uniform absolute time scale. We shall further assume that space and time are distinct and independent entities. Later, when we study the theory of relativity, we shall reexamine the concepts of space and time and we shall find that they are not absolute and independent. However, this is a matter to which we shall return after we study the classical foundations of mechanics.

In order to define the position of a body in space, it is necessary to have a reference system. In mechanics we use a *coordinate system*. The basic type of coordinate system for our purpose is the *Cartesian* or *rectangular* coordinate system, a set of three mutually perpendicular straight lines or *axes*. The position of a point in such a coordinate system is specified by three numbers or coordinates, $x$, $y$, and $z$. The coordinates of a *moving* point change with time; that is, they are functions of the quantity $t$ as measured on our time scale.

A very useful concept in mechanics is the *particle* or mass point, an entity that has mass[1] but does not have spatial extension. Strictly speaking

---

[1] The concept of mass will be discussed in Chapter 2.

1

the particle is an idealization that does not exist—even an electron has a
finite size—but the idea is useful as an approximation of a small body, or
rather, one whose size is relatively unimportant in a particular discussion.
The earth, for example, might be treated as a particle in celestial mechanics.

### 1.1. Physical Quantities and Units

The observational data of physics are expressed in terms of certain
fundamental entities called *physical quantities*—for example, length, time,
force, and so forth.  A physical quantity is something that can be measured
quantitatively in relation to some chosen unit.  When we say that the length
of a certain object is, say 7 in., we mean that the quantitative measure 7 is
the relation (ratio) of the length of that object to the length of the unit (1 in.).
It has been found that it is possible to define all of the *unit* physical quantities
of mechanics in terms of just three basic ones, namely *length, mass,* and *time.*

#### The Unit of Length

The standard unit of length is the *meter*. The meter  was formerly the
distance between two scratches on a platinum bar kept at the International
Bureau of Metric Standards, Sevres, France.  The meter is now defined as
the distance occupied by exactly 1,650,763.73 wavelengths of light of the
orange spectrum line of the isotope krypton 86.

#### The Unit of Mass

The standard unit of length is the *meter*.  The meter was formerly the
platinum iridium also kept at the International Bureau.

#### The Unit of Time

The basic unit for measurement of time, the *second*, was formerly defined
in terms of the earth's rotation.  But, like the meter, the second is now
defined in terms of a specific atomic standard.  The second is, by definition,
the amount of time required for exactly 9,192,631,770 oscillations of a particular atomic transition of the cesium isotope of mass number 133.

The above system of units is called the mks system.[2]  The modern

---

    [2] In this system there is a fourth unit, the coulomb, which is used to define electrical
units.

atomic standards of length and time in this system are not only more precise than the former standards, but they are also universally reproducible and indestructible.  Unfortunately, it is not at present technically feasible to employ an atomic standard of mass.

Actually, there is nothing particularly sacred about the physical quantities length, mass, and time as a basic set to define units.  Other sets of physical quantities may be used.  The so-called gravitational systems use length, *force*, and time.

In addition to the mks system, there are other systems in common use, namely, the cgs, or centimeter-gram-second, system, and the fps, or foot-pound-second, system.  These latter two systems may be regarded as secondary to the mks system because their units are specifically defined fractions of the mks units:

$$1 \text{ cm} = 10^{-2} \text{ m}$$
$$1 \text{ g} = 10^{-3} \text{ kg}$$
$$1 \text{ ft} = 0.3048 \text{ m}$$
$$1 \text{ lb} = 0.4536 \text{ kg}$$

## 1.2.  Scalar and Vector Quantities

A physical quantity that is completely specified by a single magnitude is called a *scalar*.  Familiar examples of scalars are density, volume, and temperature.  Mathematically, scalars are treated as ordinary real numbers. They obey all the regular rules of algebraic addition, subtraction, multiplication, division, and so on.

There are certain physical quantities that possess a directional characteristic, such as a displacement from one point in space to another.  Such quantities require a direction *and* a magnitude for their complete specification. These quantities are called *vectors* if they combine with each other according to the parallelogram rule of addition as discussed in Section 1.7.[3]  Besides displacement in space, other familiar examples of vectors are velocity, acceleration, and force.  The vector concept and the development of a whole mathematics of vector quantities have proved indispensible to the development of the science of mechanics.  The remainder of this chapter will be largely devoted to a study of the mathematics of vectors.

---

[3] An example of a directed quantity that does not obey the rule for addition is a finite rotation of an object about a given axis.  The reader can readily verify that two successive rotations about *different* axes do not produce the same effect as a single rotation determined by the parallelogram rule.  For the present we shall not be concerned with such non vector directed quantities, however.

## 1.3.  Notation

Vector quantities are denoted in print by boldface type, for example, **A**, whereas ordinary italic type represents scalar quantities.  In written work it is customary to use a distinguishing mark, such as an arrow, $\vec{A}$, to designate a vector.

A given vector **A** is specified by stating its magnitude and its direction relative to some chosen reference system.  A vector is represented diagrammatically by a directed line segment, as shown in Figure 1.1.  A vector can

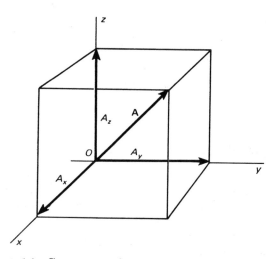

**FIGURE 1.1**   Components of a vector in Cartesian coordinates.

also be specified by listing its *components* or projections along the coordinate axes.   The component symbol $[A_x, A_y, A_z]$ will be used as an alternate designation of a vector.   The equation

$$\mathbf{A} = [A_x, A_y, A_z]$$

means that the vector **A** is expressed on the right in terms of its components in a particular coordinate system.   (It will be assumed that a Cartesian coordinate system is meant, unless stated otherwise.)   For example, if the vector **A** represents a *displacement* from a point $P_1(x_1, y_1, z_1)$ to the point $P_2(x_2, y_2, z_2)$, then $A_x = x_2 - x_1$, $A_y = y_2 - y_1$, $A_z = z_2 - z_1$.   If **A** represents a *force*, then $A_x$ is the $x$ component of the force, and so on.   Clearly, the numerical values of the scalar components of a given vector depend on the choice of the coordinate axes.

If a particular discussion is limited to vectors in a plane, only two components are necessary.   On the other hand, one can define a mathematical space of any number of dimensions. Thus the symbol $[A_1,A_2,A_3, . . .,A_n]$ denotes an $n$-dimensional vector. In this abstract sense a vector is an ordered set of numbers.

## 1.4. Formal Definitions and Rules

We begin the study of vector algebra with some formal statements concerning vectors.

*1. Equality of Vectors*

The equation

$$\mathbf{A} = \mathbf{B}$$

or

$$[A_x,A_y,A_z] = [B_x,B_y,B_z]$$

is equivalent to the three equations

$$A_x = B_x \qquad A_y = B_y \qquad A_z = B_z$$

That is, two vectors are equal if, and only if, their respective components are equal.

*2. Vector Addition*

The addition of two vectors is defined by the equation

$$\mathbf{A} + \mathbf{B} = [A_x,A_y,A_z] + [B_x,B_y,B_z] = [A_x + B_x,A_y + B_y,A_z + B_z]$$

The sum of two vectors is a vector whose components are sums of the components of the given vectors.

*3. Multiplication by a Scalar*

If $c$ is a scalar and $\mathbf{A}$ a vector,

$$c\mathbf{A} = c[A_x,A_y,A_z] = [cA_x,cA_y,cA_z] = \mathbf{A}c$$

The product $c\mathbf{A}$ is a vector whose components are $c$ times those of $\mathbf{A}$.

*4. Vector Subtraction*

Subtraction is defined as follows:

$$\mathbf{A} - \mathbf{B} = \mathbf{A} + (-1)\mathbf{B} = [A_x - B_x,A_y - B_y,A_z - B_z]$$

### 5. The Null Vector

The vector $\mathbf{O} = [0,0,0]$ is called the *null* vector. The direction of the null vector is undefined. From (4) it follows that $\mathbf{A} - \mathbf{A} = \mathbf{O}$. Since there can be no confusion when the null vector is denoted by a "zero," we shall hereafter use the notation: $\mathbf{O} = 0$.

### 6. The Commutative Law of Addition

This law holds for vectors; that is,

$$\mathbf{A} + \mathbf{B} = \mathbf{B} + \mathbf{A}$$

since $A_x + B_x = B_x + A_x$, and similarly for the $y$ and $z$ components.

### 7. The Associative Law

The associative law is also true, because

$$\begin{aligned}
\mathbf{A} + (\mathbf{B} + \mathbf{C}) &= [A_x + (B_x + C_x), A_y + (B_y + C_y), A_z + (B_z + C_z)]\\
&= [(A_x + B_x) + C_x, (A_y + B_y) + C_y, (A_z + B_z) + C_z]\\
&= (\mathbf{A} + \mathbf{B}) + \mathbf{C}
\end{aligned}$$

### 8. The Distributive Law

Under multiplication by a scalar the distributive law is valid, because, from (2) and (3),

$$\begin{aligned}
c(\mathbf{A} + \mathbf{B}) &= c[A_x + B_x, A_y + B_y, A_z + B_z]\\
&= [c(A_x + B_x), c(A_y + B_y), c(A_z + B_z)]\\
&= [cA_x + cB_x, cA_y + cB_y, cA_z + cB_z]\\
&= c\mathbf{A} + c\mathbf{B}
\end{aligned}$$

Thus vectors obey the rules of ordinary algebra as far as the above operations are concerned.

## 1.5.  Magnitude of a Vector

The magnitude of a vector $\mathbf{A}$, denoted by $|\mathbf{A}|$ or by $A$, is defined as the square root of the sum of the squares of the components, namely,

$$A = |\mathbf{A}| = (A_x{}^2 + A_y{}^2 + A_z{}^2)^{1/2} \tag{1.1}$$

where the positive root is understood. Geometrically, the magnitude of a vector is its length, that is, the length of the diagonal of the rectangular parallelepiped whose sides are $A_x$, $A_y$, and $A_z$.

## 1.6. Unit Coordinate Vectors

A *unit vector* is a vector whose magnitude is unity. Unit vectors are often designated by the symbol **e** from the German word *einheit*. The three unit vectors

$$\mathbf{e}_x = [1,0,0] \qquad \mathbf{e}_y = [0,1,0] \qquad \mathbf{e}_z = [0,0,1] \tag{1.2}$$

are called *unit coordinate vectors* or *basis vectors*. In terms of basis vectors, any vector can be expressed as a vector sum of components as follows:

$$\begin{aligned} \mathbf{A} = [A_x,A_y,A_z] &= [A_x,0,0] + [0,A_y,0] + [0,0,A_z] \\ &= A_x[1,0,0] + A_y[0,1,0] + A_z[0,0,1] \\ &= \mathbf{e}_x A_x + \mathbf{e}_y A_y + \mathbf{e}_z A_z \end{aligned} \tag{1.3}$$

A widely used notation for Cartesian unit vectors are the letters **i**, **j**, and **k**, namely

$$\mathbf{i} = \mathbf{e}_x \qquad \mathbf{j} = \mathbf{e}_y \qquad \mathbf{k} = \mathbf{e}_z$$

We shall usually employ this notation hereafter.

The directions of the unit coordinate vectors are defined by the coordinate axes (Figure 1.2). They form a right-handed or a left-handed triad, depending on which type of coordinate system is used. It is customary to use right-handed coordinate systems. The system shown in Figure 1.2 is right-handed.

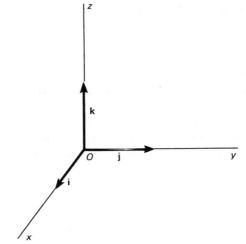

FIGURE 1.2   The unit coordinate vectors **ijk**.

## 1.7. Geometric Meaning of Vector Operations

If we consider a vector to be represented by a directed line segment, it is easily verified that the definitions stated above have the following simple interpretations:

### 1. Equality of Vectors

If two vectors are equal then the vectors are parallel and have the same length, but they do not necessarily have the same position.   Equal vectors are shown in Figure 1.3, where only two components are drawn for clarity.

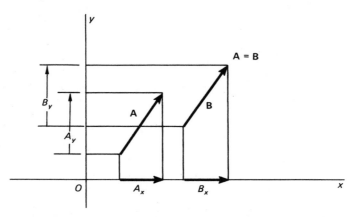

FIGURE 1.3   Illustrating equal vectors.

Notice that the vectors form opposite sides of a parallelogram.   (Equal vectors are not necessarily equivalent in all respects.   Thus two vectorially equal forces acting at *different* points on an object may produce different mechanical effects.)

### 2. Vector Addition

The vector sum of two vectors is equal to the third side of a triangle, two sides of which are the given vectors.   The vector sum is illustrated in Figure 1.4.   The sum is also given by the parallelogram rule, as shown in the figure.   [The vector sum is defined, however, according to definition 1.4(2) even if the vectors do not have a common point.]

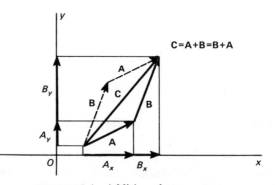

FIGURE 1.4   Addition of two vectors.

### 3. Multiplication of a Vector by a Scalar

The vector $c\mathbf{A}$ is parallel to $\mathbf{A}$ and is $c$ times the length of $\mathbf{A}$. When $c = -1$, the vector $-\mathbf{A}$ is one whose direction is the reverse of that of $\mathbf{A}$, as shown in Figure 1.5.

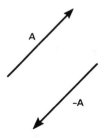

FIGURE 1.5   The negative of a vector.

## 1.8.   The Scalar Product

Given two vectors $\mathbf{A}$ and $\mathbf{B}$, the scalar product or "dot" product, $\mathbf{A}\cdot\mathbf{B}$, is the scalar defined by the equation

$$\mathbf{A}\cdot\mathbf{B} = A_xB_x + A_yB_y + A_zB_z \qquad (1.4)$$

It follows from the above definition that

$$\mathbf{A}\cdot\mathbf{B} = \mathbf{B}\cdot\mathbf{A} \qquad (1.5)$$

since $A_xB_x = B_xA_x$, and so on.  It also follows that

$$\mathbf{A}\cdot(\mathbf{B} + \mathbf{C}) = \mathbf{A}\cdot\mathbf{B} + \mathbf{A}\cdot\mathbf{C} \qquad (1.6)$$

because if we apply the definition [(1.4)] in detail

$$\begin{aligned}
\mathbf{A}\cdot(\mathbf{B} + \mathbf{C}) &= A_x(B_x + C_x) + A_y(B_y + C_y) + A_z(B_z + C_z) \\
&= A_xB_x + A_yB_y + A_zB_z + A_xC_x + A_yC_y + A_zC_z \\
&= \mathbf{A}\cdot\mathbf{B} + \mathbf{A}\cdot\mathbf{C}
\end{aligned}$$

From analytical geometry we recall the formula for the cosine of the angle between two line segments

$$\cos\theta = \frac{A_xB_x + A_yB_y + A_zB_z}{(A_x{}^2 + A_y{}^2 + A_z{}^2)^{1/2}(B_x{}^2 + B_y{}^2 + B_z{}^2)^{1/2}}$$

Using Equations (1.1) and (1.4), the above formula may be written

$$\cos\theta = \frac{\mathbf{A}\cdot\mathbf{B}}{AB}$$

or

$$\mathbf{A}\cdot\mathbf{B} = AB\cos\theta \qquad (1.7)$$

The above equation may be regarded as an alternate definition of the dot product.   Geometrically, $\mathbf{A} \cdot \mathbf{B}$ is equal to the length of the projection of $\mathbf{A}$ on $\mathbf{B}$, times the length of $\mathbf{B}$.

If the dot product $\mathbf{A} \cdot \mathbf{B}$ is equal to zero, then $\mathbf{A}$ is perpendicular to $\mathbf{B}$, provided neither $\mathbf{A}$ nor $\mathbf{B}$ is null.

The square of the magnitude of a vector $\mathbf{A}$ is given by the dot product of $\mathbf{A}$ with itself,

$$A^2 = |\mathbf{A}|^2 = \mathbf{A} \cdot \mathbf{A}$$

From the definitions of the unit coordinate vectors $\mathbf{i}$, $\mathbf{j}$, and $\mathbf{k}$, it is clear that the following relations hold:

$$\mathbf{i} \cdot \mathbf{i} = \mathbf{j} \cdot \mathbf{j} = \mathbf{k} \cdot \mathbf{k} = 1$$

$$\mathbf{i} \cdot \mathbf{j} = \mathbf{i} \cdot \mathbf{k} = \mathbf{j} \cdot \mathbf{k} = 0 \tag{1.8}$$

### 1.9.   Some Examples of the Scalar Product

*1. Component of a Vector.   Work*

As an example of the dot product, suppose that an object under the action of a constant force[4] undergoes a linear displacement $\Delta \mathbf{s}$, as shown in Figure 1.6.   By definition, the *work* $\Delta W$ done by the force is given by the product of the component of the force $\mathbf{F}$ in the direction of $\Delta \mathbf{s}$, multiplied by the magnitude $\Delta s$ of the displacement, that is,

$$\Delta W = (F \cos \theta)\, \Delta s$$

where $\theta$ is the angle between $\mathbf{F}$ and $\Delta \mathbf{s}$.   But the expression on the right is just the dot product of $\mathbf{F}$ and $\Delta \mathbf{s}$, that is,

$$\Delta W = \mathbf{F} \cdot \Delta \mathbf{s}$$

---

[4] The concept of force will be discussed later in Chapter 3.

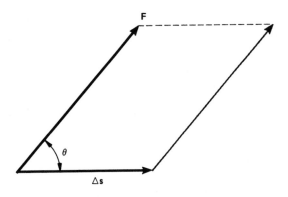

FIGURE 1.6   A force undergoing a displacement.

### 2. *Law of Cosines*

Given the triangle whose sides are **A**, **B**, and **C**, as shown in Figure 1.7.
Then **C** = **A** + **B**.   Take the dot product of **C** with itself

$$\mathbf{C \cdot C} = (\mathbf{A} + \mathbf{B}) \cdot (\mathbf{A} + \mathbf{B})$$
$$= \mathbf{A \cdot A} + 2\mathbf{A \cdot B} + \mathbf{B \cdot B}$$

The second step follows from the application of the rules in Equations (1.5) and
(1.6).   Replace **A·B** by $AB \cos \theta$ to obtain

$$C^2 = A^2 + 2AB \cos \theta + B^2$$

which is the familiar law of cosines. This is just one example of the use of
vector algebra to prove theorems in geometry.

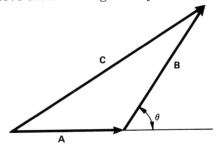

FIGURE 1.7   The law of cosines.

### 1.10.   The Vector Product

Given two vectors **A** and **B**, the vector product or "cross product,"
**A ✕ B**, is defined as the vector whose components are given by the equation

$$\mathbf{A \times B} = [A_yB_z - A_zB_y, A_zB_x - A_xB_z, A_xB_y - A_yB_x] \qquad (1.9)$$

The geometric interpretation of the cross product is given in Section 1.12. It can be shown that the following rules hold for cross multiplication.

$$\mathbf{A} \times \mathbf{B} = -\mathbf{B} \times \mathbf{A} \tag{1.10}$$
$$\mathbf{A} \times (\mathbf{B} + \mathbf{C}) = \mathbf{A} \times \mathbf{B} + \mathbf{A} \times \mathbf{C} \tag{1.11}$$
$$n(\mathbf{A} \times \mathbf{B}) = (n\mathbf{A}) \times \mathbf{B} = \mathbf{A} \times (n\mathbf{B}) \tag{1.12}$$

The proofs of these follow directly from the definition and are left as an exercise.

According to the algebraic definitions of the unit coordinate vectors, Equation (1.2), it readily follows that the following relations for the cross product are true:

$$\begin{aligned}
\mathbf{i} \times \mathbf{i} = \mathbf{j} \times \mathbf{j} &= \mathbf{k} \times \mathbf{k} = 0 \\
\mathbf{j} \times \mathbf{k} = \mathbf{i} &= -\mathbf{k} \times \mathbf{j} \\
\mathbf{i} \times \mathbf{j} = \mathbf{k} &= -\mathbf{j} \times \mathbf{i} \\
\mathbf{k} \times \mathbf{i} = \mathbf{j} &= -\mathbf{i} \times \mathbf{k}
\end{aligned} \tag{1.13}$$

For example,

$$\mathbf{i} \times \mathbf{j} = [0 - 0, 0 - 0, 1 - 0] = [0,0,1] = \mathbf{k}$$

The remaining equations are easily proved in a similar manner.

## 1.11.   Geometric Interpretation of the Cross Product

The cross product expressed in **ijk** form is

$$\mathbf{A} \times \mathbf{B} = \mathbf{i}(A_y B_z - A_z B_y) + \mathbf{j}(A_z B_x - A_x B_z) + \mathbf{k}(A_x B_y - A_y B_x)$$

Each term in parentheses is equal to a determinant

$$\mathbf{A} \cdot \mathbf{B} = \mathbf{i} \begin{vmatrix} A_y A_z \\ B_y B_z \end{vmatrix} + \mathbf{j} \begin{vmatrix} A_z A_x \\ B_z B_x \end{vmatrix} + \mathbf{k} \begin{vmatrix} A_x A_y \\ B_x B_y \end{vmatrix}$$

and finally

$$\mathbf{A} \times \mathbf{B} = \begin{vmatrix} \mathbf{i} & \mathbf{j} & \mathbf{k} \\ A_x A_y A_z \\ B_x B_y B_z \end{vmatrix} \tag{1.14}$$

which is readily verified by expansion.   The determinant form is a convenient aid for remembering the definition of the cross product.   From the properties of determinants, it can be seen at once that if **A** is parallel to **B**, that is, if $\mathbf{A} = c\mathbf{B}$, then the two lower rows of the determinant are proportional and so the determinant is null.   Thus the cross product of two parallel vectors is null.

Let us calculate the magnitude of the cross product.   We have

$$|\mathbf{A} \times \mathbf{B}|^2 = (A_y B_z - A_z B_y)^2 + (A_z B_x - A_x B_z)^2 + (A_x B_y - A_y B_x)^2$$

With a little patience this can be reduced to

$$|\mathbf{A} \times \mathbf{B}|^2 = (A_x^2 + A_y^2 + A_z^2)(B_x^2 + B_y^2 + B_z^2) - (A_xB_x + A_yB_y + A_zB_z)^2$$

or, from the definition of the dot product, the above equation may be written in the form

$$|\mathbf{A} \times \mathbf{B}|^2 = A^2B^2 - (\mathbf{A} \cdot \mathbf{B})^2$$

Taking the square root of both sides of the above equation and using Equation (1.7), we can express the magnitude of the cross product as

$$|\mathbf{A} \times \mathbf{B}| = AB(1 - \cos^2 \theta)^{1/2} = AB \sin \theta \qquad (1.15)$$

where $\theta$ is the angle between **A** and **B**.

To interpret the cross product geometrically, we observe that the vector $\mathbf{C} = \mathbf{A} \times \mathbf{B}$ is perpendicular to both **A** and to **B**, because

$$
\begin{aligned}
\mathbf{A} \cdot \mathbf{C} &= A_xC_x + A_yC_y + A_zC_z \\
&= A_x(A_yB_z - A_zB_y) + A_y(A_zB_x - A_xB_z) + A_z(A_xB_y - A_yB_x) \\
&= 0
\end{aligned}
$$

Similarly, $\mathbf{B} \cdot \mathbf{C} = 0$. Thus the vector **C** is perpendicular to the plane containing the vectors **A** and **B**.

The sense of the vector $\mathbf{C} = \mathbf{A} \times \mathbf{B}$ is determined from the requirement that the three vectors **A**, **B**, and **C** form a right-handed triad, as shown in Figure 1.8. (This is consistent with the previously established result that in

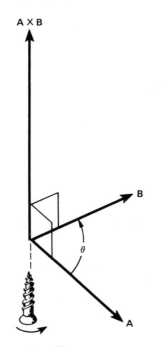

FIGURE 1.8   The vector or cross product.

the right-handed triad **ijk** we have $\mathbf{i} \times \mathbf{j} = \mathbf{k}$.)   Therefore, from Equation (1.15) we see that we can write

$$\mathbf{A} \times \mathbf{B} = (AB \sin \theta)\mathbf{n} \tag{1.16}$$

where **n** is a unit vector normal to the plane of the two vectors **A** and **B**. The sense of **n** is given by the *right-hand rule*, that is, the direction of advancement of a right-handed screw rotated from the positive direction of **A** to that of **B** through the smallest angle between them, as illustrated in Figure 1.8.   Equation (1.16) may be regarded as an alternate definition of the cross product.

## EXAMPLES

1. Given the two vectors $\mathbf{A} = 2\mathbf{i} + \mathbf{j} - \mathbf{k}$, $\mathbf{B} = \mathbf{i} - \mathbf{j} + 2\mathbf{k}$, find $\mathbf{A} \cdot \mathbf{B}$ and $\mathbf{A} \times \mathbf{B}$.

$$\mathbf{A} \cdot \mathbf{B} = (2)(1) + (1)(-1) + (-1)(2) = 2 - 1 - 2 = -1$$

$$\mathbf{A} \times \mathbf{B} = \begin{vmatrix} \mathbf{i} & \mathbf{j} & \mathbf{k} \\ 2 & 1 & -1 \\ 1 & -1 & 2 \end{vmatrix} = \mathbf{i}(2 - 1) + \mathbf{j}(-1 - 4) + \mathbf{k}(-2 - 1)$$

$$= \mathbf{i} - 5\mathbf{j} - 3\mathbf{k}$$

2. Find the angle between **A** and **B**.   We have, from the definition of the dot product,

$$\cos \theta = \frac{\mathbf{A} \cdot \mathbf{B}}{AB} = \frac{-1}{[2^2 + 1^2 + (-1)^2]^{1/2}[1^2 + (-1)^2 + 2^2]^{1/2}}$$

$$= \frac{-1}{6^{1/2}6^{1/2}} = -\frac{1}{6}$$

Hence

$$\theta = \cos^{-1}\left(-\tfrac{1}{6}\right) = 99.6°$$

### 1.12.   An Example of the Cross Product.   Moment of a Force

A particularly useful application of the cross product is the representation of moments.   Let a force **F** act at a point $P(x,y,z)$, as shown in Figure 1.9, and let the vector $\overrightarrow{OP}$ be designated by **r**, that is,

$$\overrightarrow{OP} = \mathbf{r} = \mathbf{i}x + \mathbf{j}y + \mathbf{k}z$$

The moment **N**, or the *torque*, about a given point $O$ is defined as the cross product

$$\mathbf{N} = \mathbf{r} \times \mathbf{F} \tag{1.17}$$

Thus the moment of a force about a point is a vector quantity having a magnitude and a direction.   If a single force is applied at a point $P$ on a body that is free to turn about a fixed point $O$ as a pivot, then the body tends to rotate.   The axis of this rotation is perpendicular to the force $\mathbf{F}$, and it is also perpendicular to the line $OP$.   Hence the direction of the torque $\mathbf{N}$ is along the axis of rotation.

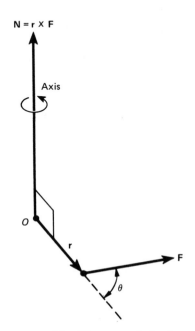

FIGURE 1.9   The moment of a force.

The magnitude of the torque is given by

$$|\mathbf{N}| = |\mathbf{r} \times \mathbf{F}| = rF \sin \theta \tag{1.18}$$

in which $\theta$ is the angle between $\mathbf{r}$ and $\mathbf{F}$.   Thus $|\mathbf{N}|$ can be regarded as the product of the magnitude of the force and the quantity $r \sin \theta$ which is just the perpendicular distance from the line of action of the force to the point $O$.

When several forces are applied to a single body at different points, the moments add vectorially.   This follows from the distributive law of vector multiplication, Equation (1.11).   The condition for rotational equilibrium is that the vector sum of all the moments is zero:

$$\sum_i (\mathbf{r}_i \times \mathbf{F}_i) = \sum_i \mathbf{N}_i = 0$$

A more complete discussion of this will be given later in Chapter 7.

### 1.13. Representation of a Given Vector as the Product of a Scalar and a Single Unit Vector

Consider the equation

$$\mathbf{A} = \mathbf{i}A_x + \mathbf{j}A_y + \mathbf{k}A_z$$

Multiply and divide on the right by the magnitude of $\mathbf{A}$

$$\mathbf{A} = A\left(\mathbf{i}\frac{A_x}{A} + \mathbf{j}\frac{A_y}{A} + \mathbf{k}\frac{A_z}{A}\right)$$

Now $A_x/A = \cos\alpha$, $A_y/A = \cos\beta$, and $A_z/A = \cos\gamma$ are the *direction cosines* of the vector $\mathbf{A}$, and $\alpha$, $\beta$, and $\gamma$ are the *direction angles*. Thus we can write

$$\mathbf{A} = A(\mathbf{i}\cos\alpha + \mathbf{j}\cos\beta + \mathbf{k}\cos\gamma) = A[\cos\alpha, \cos\beta, \cos\gamma]$$

or

$$\mathbf{A} = A\mathbf{n} \tag{1.19}$$

where $\mathbf{n}$ is a unit vector whose components are $\cos\alpha$, $\cos\beta$, and $\cos\gamma$. Consider any other vector $\mathbf{B}$. Clearly, the projection of $\mathbf{B}$ on $\mathbf{A}$ is just

$$B\cos\theta = \frac{\mathbf{B}\cdot\mathbf{A}}{A} = \mathbf{B}\cdot\mathbf{n} \tag{1.20}$$

where $\theta$ is the angle between $\mathbf{A}$ and $\mathbf{B}$.

### EXAMPLE

Find a unit vector normal to the plane containing the two vectors $\mathbf{A} = 2\mathbf{i} + \mathbf{j} - \mathbf{k}$ and $\mathbf{B} = \mathbf{i} - \mathbf{j} + 2\mathbf{k}$. From Example 1, p. 14, we have $\mathbf{A} \times \mathbf{B} = \mathbf{i} - 5\mathbf{j} - 3\mathbf{k}$. Hence

$$\mathbf{n} = \frac{\mathbf{A}\times\mathbf{B}}{|\mathbf{A}\times\mathbf{B}|} = \frac{\mathbf{i} - 5\mathbf{j} - 3\mathbf{k}}{[1^2 + 5^2 + 3^2]^{1/2}}$$

$$= \frac{\mathbf{i}}{\sqrt{35}} - \frac{5\mathbf{j}}{\sqrt{35}} - \frac{3\mathbf{k}}{\sqrt{35}}$$

### 1.14. Triple Products

The expression

$$\mathbf{A}\cdot(\mathbf{B}\times\mathbf{C})$$

is called the *triple scalar product* of $\mathbf{A}$, $\mathbf{B}$, and $\mathbf{C}$. It is a scalar since it is the dot product of two vectors. Referring to the determinant expression for the cross product, Equation (1.14), we see that the triple scalar product may be written

$$\mathbf{A} \cdot (\mathbf{B} \times \mathbf{C}) = \begin{vmatrix} A_x A_y A_z \\ B_x B_y B_z \\ C_x C_y C_z \end{vmatrix} \tag{1.21}$$

From the well-known property of determinants that the exchange of the terms of two rows or of two columns changes the sign but does not change the absolute value of the determinant, we can easily derive the following useful equation:

$$\mathbf{A} \cdot (\mathbf{B} \times \mathbf{C}) = (\mathbf{A} \times \mathbf{B}) \cdot \mathbf{C} \tag{1.22}$$

Thus the dot and the cross may be interchanged in the triple scalar product.
The expression

$$\mathbf{A} \times (\mathbf{B} \times \mathbf{C})$$

is called the *triple vector product*.  It is left for the student to prove that the following equation holds for the triple vector product:

$$\mathbf{A} \times (\mathbf{B} \times \mathbf{C}) = (\mathbf{A} \cdot \mathbf{C})\mathbf{B} - (\mathbf{A} \cdot \mathbf{B})\mathbf{C} \tag{1.23}$$

### 1.15.  Change of Coordinate System.   The Transformation Matrix

Consider the vector $\mathbf{A}$ expressed relative to the triad $\mathbf{ijk}$

$$\mathbf{A} = \mathbf{i}A_x + \mathbf{j}A_y + \mathbf{k}A_z$$

Relative to a new triad $\mathbf{i'j'k'}$ having a different orientation from that of $\mathbf{ijk}$, the *same* vector $\mathbf{A}$ is expressed as

$$\mathbf{A} = \mathbf{i'}A_{z'} + \mathbf{j'}A_{y'} + \mathbf{k'}A_{z'}$$

Now the dot product $\mathbf{A} \cdot \mathbf{i'}$ is just $A_{x'}$, that is, the projection of $\mathbf{A}$ on the unit vector $\mathbf{i'}$.   Thus we may write

$$\begin{aligned} A_{x'} &= \mathbf{A} \cdot \mathbf{i'} = (\mathbf{i} \cdot \mathbf{i'})A_x + (\mathbf{j} \cdot \mathbf{i'})A_y + (\mathbf{k} \cdot \mathbf{i'})A_z \\ A_{y'} &= \mathbf{A} \cdot \mathbf{j'} = (\mathbf{i} \cdot \mathbf{j'})A_x + (\mathbf{j} \cdot \mathbf{j'})A_y + (\mathbf{k} \cdot \mathbf{j'})A_z \\ A_{z'} &= \mathbf{A} \cdot \mathbf{k'} = (\mathbf{i} \cdot \mathbf{k'})A_x + (\mathbf{j} \cdot \mathbf{k'})A_y + (\mathbf{k} \cdot \mathbf{k'})A_z \end{aligned} \tag{1.24}$$

The scalar products $(\mathbf{i} \cdot \mathbf{i'})$, $(\mathbf{i} \cdot \mathbf{j'})$, and so on, are called the *coefficients of transformation*.   They are equal to the direction cosines of the axes of the primed coordinate system relative to the unprimed system.   The unprimed components are similarly expressed as

$$\begin{aligned} A_x &= \mathbf{A} \cdot \mathbf{i} = (\mathbf{i'} \cdot \mathbf{i})A_{x'} + (\mathbf{j'} \cdot \mathbf{i})A_{y'} + (\mathbf{k'} \cdot \mathbf{i})A_{z'} \\ A_y &= \mathbf{A} \cdot \mathbf{j} = (\mathbf{i'} \cdot \mathbf{j})A_{x'} + (\mathbf{j'} \cdot \mathbf{j})A_{y'} + (\mathbf{k'} \cdot \mathbf{j})A_{z'} \\ A_z &= \mathbf{A} \cdot \mathbf{k} = (\mathbf{i'} \cdot \mathbf{k})A_{x'} + (\mathbf{j'} \cdot \mathbf{k})A_{y'} + (\mathbf{k'} \cdot \mathbf{k})A_{z'} \end{aligned} \tag{1.25}$$

All of the coefficients of transformation in Equation (1.25) also appear in Equation (1.24), because $\mathbf{i}\cdot\mathbf{i}' = \mathbf{i}'\cdot\mathbf{i}$, etc., but those in the rows (equations) of Equation (1.25) appear in the columns of terms in Equation (1.24), and conversely.   The transformation rules expressed in these two sets of equations are a general property of vectors.   As a matter of fact, they constitute an alternative way of defining vectors.[5]

The equations of transformation are conveniently expressed in matrix notation.   Thus Equation (1.24) is written

$$\begin{bmatrix} A_{x'} \\ A_{y'} \\ A_{z'} \end{bmatrix} = \begin{bmatrix} \mathbf{i}\cdot\mathbf{i}' & \mathbf{j}\cdot\mathbf{i}' & \mathbf{k}\cdot\mathbf{i}' \\ \mathbf{i}\cdot\mathbf{j}' & \mathbf{j}\cdot\mathbf{j}' & \mathbf{k}\cdot\mathbf{j}' \\ \mathbf{i}\cdot\mathbf{k}' & \mathbf{j}\cdot\mathbf{k}' & \mathbf{k}\cdot\mathbf{k}' \end{bmatrix} \begin{bmatrix} A_x \\ A_y \\ A_z \end{bmatrix} \qquad (1.26)$$

The 3 by 3 matrix in the above equation is called the *transformation matrix*. One advantage of the matrix notation is that successive transformations are readily handled by means of matrix multiplication.

The reader will observe that the application of a given transformation matrix to some vector $\mathbf{A}$ is also formally equivalent to rotating that vector within the unprimed (fixed) coordinate system, the components of the rotated vector being given by Equation (1.26).   Thus finite rotations can be represented by matrices.   (Note that the sense of rotation of the vector in this context is opposite that of the rotation of the coordinate system in the previous context.)

## EXAMPLES

1. Express the vector $\mathbf{A} = 3\mathbf{i} + 2\mathbf{j} + \mathbf{k}$ in terms of the triad $\mathbf{i}'\mathbf{j}'\mathbf{k}'$ where the $x'y'$ axes are rotated $45°$ around the $z$ axis, the $z$ and the $z'$ axes coinciding, as shown in Figure 1.10.   Referring to the figure, we have for the coefficients of transformation,

$$\mathbf{i}\cdot\mathbf{i}' = 1/\sqrt{2} \qquad \mathbf{j}\cdot\mathbf{i}' = 1/\sqrt{2} \qquad \mathbf{k}\cdot\mathbf{i}' = 0$$
$$\mathbf{i}\cdot\mathbf{j}' = -1/\sqrt{2} \qquad \mathbf{j}\cdot\mathbf{j}' = 1/\sqrt{2} \qquad \mathbf{k}\cdot\mathbf{j}' = 0$$
$$\mathbf{i}\cdot\mathbf{k}' = 0 \qquad \mathbf{j}\cdot\mathbf{k}' = 0 \qquad \mathbf{k}\cdot\mathbf{k}' = 1$$

These give

$$A_{x'} = \frac{3}{\sqrt{2}} + \frac{2}{\sqrt{2}} = \frac{5}{\sqrt{2}} \qquad A_{y'} = \frac{-3}{\sqrt{2}} + \frac{2}{\sqrt{2}} = \frac{-1}{\sqrt{2}} \qquad A_{z'} = 1$$

[5] See, for example, L. P. Smith, *Mathematical Methods for Scientists and Engineers*, Prentice-Hall, Englewood Cliffs, N.J., 1953.

so that, in the primed system, the vector **A** is given by

$$\mathbf{A} = \frac{5}{\sqrt{2}}\,\mathbf{i}' - \frac{1}{\sqrt{2}}\,\mathbf{j}' + \mathbf{k}'$$

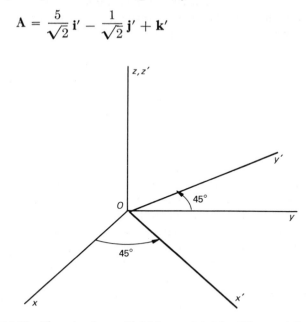

FIGURE 1.10  The primed axes $O'x'y'z'$ are rotated by 45° around the $z$ axis.

2. Find the transformation matrix for a rotation of the primed coordinate system through an angle $\phi$ about the $z$ axis. (The previous example is a special case of this.)  We have

$$\mathbf{i}\cdot\mathbf{i}' = \mathbf{j}\cdot\mathbf{j}' = \cos\phi$$
$$\mathbf{j}\cdot\mathbf{i}' = -\mathbf{i}\cdot\mathbf{j}' = \sin\phi$$
$$\mathbf{k}\cdot\mathbf{k}' = 1$$

and all other dot products are zero.  Hence the transformation matrix is

$$\begin{bmatrix} \cos\phi & \sin\phi & 0 \\ -\sin\phi & \cos\phi & 0 \\ 0 & 0 & 1 \end{bmatrix}$$

It is clear from the above example that the transformation matrix for a rotation about a different coordinate axis, say the $y$ axis through an angle $\theta$, will be given by the matrix

$$\begin{bmatrix} \cos\theta & 0 & -\sin\theta \\ 0 & 1 & 0 \\ \sin\theta & 0 & \cos\theta \end{bmatrix}$$

Consequently the matrix for the combination of two rotations, the first being about the $z$ axis (angle $\phi$) and the second being about the new $y'$ axis (angle $\theta$) is given by the matrix product

$$\begin{bmatrix} \cos\theta & 0 & -\sin\theta \\ 0 & 1 & 0 \\ \sin\theta & 0 & \cos\theta \end{bmatrix} \begin{bmatrix} \cos\phi & \sin\phi & 0 \\ -\sin\phi & \cos\phi & 0 \\ 0 & 0 & 1 \end{bmatrix} = \begin{bmatrix} \cos\theta\cos\phi & \cos\theta\sin\phi & -\sin\theta \\ -\sin\phi & \cos\phi & 0 \\ \sin\theta\cos\phi & \sin\theta\sin\phi & \cos\theta \end{bmatrix}$$

Now matrix multiplication is, in general, noncommutative. Hence we might expect that if the order of the rotations were reversed, and therefore the order of the matrix multiplication on the left, the final result would be different. This turns out to be the case, which the reader can verify. This is in keeping with a remark made earlier, namely that finite rotations do not obey the law of vector addition and hence are not vectors even though a single rotation has a direction (the axis) and a magnitude (the angle of rotation). However, we shall show later that infinitesimal rotations do obey the law of vector addition, and can be represented by vectors.

## 1.16.  Derivative of a Vector

Consider a vector $\mathbf{A}$, the components of which are functions of a single variable $u$. The vector may represent position, velocity, and so on. The parameter $u$ is usually the time $t$, but it can be any quantity which determines the components of $\mathbf{A}$:

$$\mathbf{A}(u) = \mathbf{i}A_x(u) + \mathbf{j}A_y(u) + \mathbf{k}A_z(u)$$

The derivative of $\mathbf{A}$ with respect to $u$ is defined, quite analogously to the ordinary derivative of a scalar function, by the limit

$$\frac{d\mathbf{A}}{du} = \lim_{\Delta u \to 0} \frac{\Delta\mathbf{A}}{\Delta u} = \lim_{\Delta u \to 0} \left( \mathbf{i}\frac{\Delta A_x}{\Delta u} + \mathbf{j}\frac{\Delta A_y}{\Delta u} + \mathbf{k}\frac{\Delta A_z}{\Delta u} \right)$$

where $\Delta A_x = A_x(u + \Delta u) - A_x(u)$, and so on. Hence

$$\frac{d\mathbf{A}}{du} = \mathbf{i}\frac{dA_x}{du} + \mathbf{j}\frac{dA_y}{du} + \mathbf{k}\frac{dA_z}{du} \tag{1.27}$$

The derivative of a vector, therefore, is a vector whose components are ordinary derivatives.

It follows from the above equation that the derivative of the sum of two vectors is equal to the sum of the derivatives, namely,

$$\frac{d}{du}(\mathbf{A} + \mathbf{B}) = \frac{d\mathbf{A}}{du} + \frac{d\mathbf{B}}{du} \tag{1.28}$$

Rules for differentiating vector products will be treated later in Section 1.22.

### 1.17.   Position Vector of a Particle

In a given reference system the position of a particle can be specified by a single vector, namely, the displacement of the particle relative to the origin of the coordinate system.   This vector is called the *position vector* of the particle.   In rectangular coordinates, Figure 1.11, the position vector is simply

$$\mathbf{r} = \mathbf{i}x + \mathbf{j}y + \mathbf{k}z$$

The components of the position vector of a moving particle are functions of the time, namely,

$$x = x(t) \qquad y = y(t) \qquad z = z(t)$$

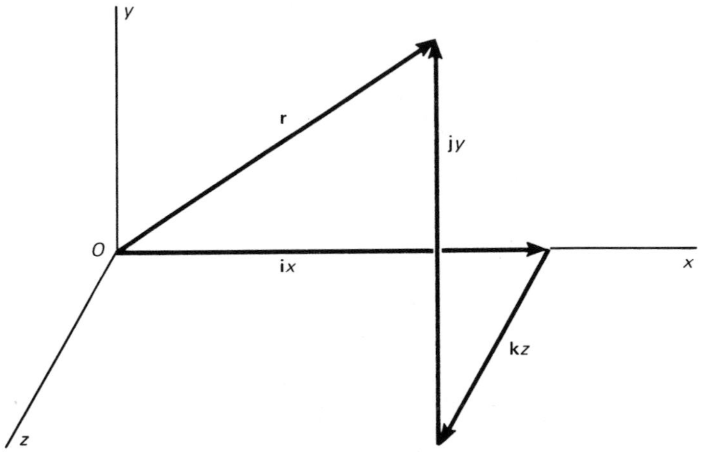

FIGURE 1.11   The position vector.

### 1.18.   The Velocity Vector

In Equation (1.27) we gave the formal definition of the derivative of any vector with respect to some parameter.   In particular, if the vector is the position vector $\mathbf{r}$ of a moving particle and the parameter is the time $t$, the derivative of **r** with respect to $t$ is called the *velocity*, which we shall denote by **v**.   Hence

$$\mathbf{v} = \frac{d\mathbf{r}}{dt} = \mathbf{i}\dot{x} + \mathbf{j}\dot{y} + \mathbf{k}\dot{z} \tag{1.29}$$

where the dots indicate differentiation with respect to $t$.   (This convention is standard and will be used throughout the book.)   Let us examine the geo- metric significance of the velocity vector.   Suppose a particle is at a certain position at time $t$.   At a time $\Delta t$ later, the particle will have moved from the position $\mathbf{r}(t)$ to the position $\mathbf{r}(t + \Delta t)$.   The vector displacement during the time interval $\Delta t$ is

$$\Delta \mathbf{r} = \mathbf{r}(t + \Delta t) - \mathbf{r}(t)$$

so the quotient $\Delta\mathbf{r}/\Delta t$ is a *vector* which is parallel to the displacement.   As we consider smaller and smaller time intervals, the quotient $\Delta\mathbf{r}/\Delta t$ approaches a limit $d\mathbf{r}/dt$ which we call the velocity.   The vector $d\mathbf{r}/dt$ expresses both the direction of motion and the rate.   This is shown graphically in Figure 1.12.   In

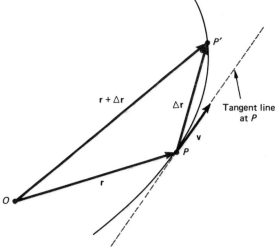

FIGURE 1.12   Displacement vector of a moving particle.

the time interval $\Delta t$ the particle moves along the path from $P$ to $P'$.   As $\Delta t$ approaches zero, the point $P'$ approaches $P$, and the direction of the vector $\Delta\mathbf{r}/\Delta t$ approaches the direction of the tangent to the path at $P$.   The velocity vector, therefore, is always tangent to the path of motion.

The magnitude of the velocity is called the *speed*.   In rectangular com- ponents the speed is just

$$v = |\mathbf{v}| = (\dot{x}^2 + \dot{y}^2 + \dot{z}^2)^{1/2} \tag{1.30}$$

If we denote the scalar distance along the path by $s$, then we can alternately express the speed as

$$v = \frac{ds}{dt} = \lim_{\Delta t \to 0} \frac{\Delta s}{\Delta t} = \lim_{\Delta t \to 0} \frac{[(\Delta x)^2 + (\Delta y)^2 + (\Delta z)^2]^{1/2}}{\Delta t}$$

which reduces to the expression on the right of Equation (1.30).

### 1.19.  Acceleration Vector

The time derivative of the velocity is called the *acceleration*.  Denoting the acceleration by **a**, we have

$$\mathbf{a} = \frac{d\mathbf{v}}{dt} = \frac{d^2\mathbf{r}}{dt^2} \tag{1.31}$$

In rectangular components

$$\mathbf{a} = \mathbf{i}\ddot{x} + \mathbf{j}\ddot{y} + \mathbf{k}\ddot{z} \tag{1.32}$$

Thus acceleration is a vector quantity whose components, in rectangular coordinates, are the second derivatives of the positional coordinates of a moving particle.   The resolution of **a** into tangential and normal components will be discussed in Section 1.23.

## EXAMPLES

1. Let us examine the motion represented by the equation

$$\mathbf{r}(t) = \mathbf{i}bt + \mathbf{j}\left(ct - \frac{gt^2}{2}\right) + \mathbf{k}0$$

This represents motion in the $xy$ plane, since the $z$ component is constant and equal to zero.   The velocity **v** is obtained by differentiating with respect to $t$, namely,

$$\mathbf{v} = \frac{d\mathbf{r}}{dt} = \mathbf{i}b + \mathbf{j}(c - gt)$$

The acceleration, likewise, is given by

$$\mathbf{a} = \frac{d\mathbf{v}}{dt} = -\mathbf{j}g$$

Thus **a** is in the negative $y$ direction and has the constant magnitude $g$.   The path of motion is a parabola, as shown in Figure 1.13.   (This equation actually represents the motion of a projectile.)   The speed $v$ varies with $t$ according to the equation

$$v = [b^2 + (c - gt)^2]^{1/2}$$

2. Suppose the position vector of a particle is given by

$$\mathbf{r} = \mathbf{i}b \sin \omega t + \mathbf{j}b \cos \omega t + \mathbf{k}c$$

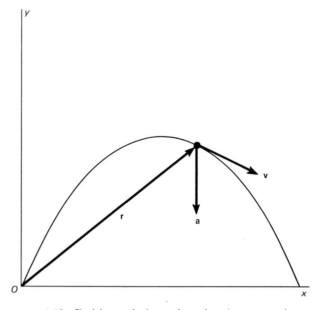

FIGURE 1.13   Position, velocity and acceleration vectors for a particle
moving in a parabolic path.

Let us analyze the motion.   The distance from the origin remains constant

$$|\mathbf{r}| = r = (b^2 \sin^2 \omega t + b^2 \cos^2 \omega t + c^2)^{1/2} = (b^2 + c^2)^{1/2}$$

Differentiating $\mathbf{r}$, we find

$$\mathbf{v} = \frac{d\mathbf{r}}{dt} = \mathbf{i}b\omega \cos \omega t - \mathbf{j}b\omega \sin \omega t + \mathbf{k}0$$

Since the $z$ component of $\mathbf{v}$ is zero, the velocity vector is parallel to the $xy$
plane.   The particle traverses its path with constant speed

$$v = |\mathbf{v}| = (b^2\omega^2 \cos^2 \omega t + b^2\omega^2 \sin^2 \omega t)^{1/2} = b\omega$$

The acceleration

$$\mathbf{a} = \frac{d\mathbf{v}}{dt} = -\mathbf{i}b\omega^2 \sin \omega t - \mathbf{j}b\omega^2 \cos \omega t$$

which is perpendicular to the velocity, since the dot product of $\mathbf{v}$ and $\mathbf{a}$
vanishes, thus

$$\mathbf{v}\cdot\mathbf{a} = (b\omega \cos \omega t)(-b\omega^2 \sin \omega t) + (-b\omega \sin \omega t)(-b\omega^2 \cos \omega t) = 0$$

Further, the acceleration is perpendicular to the $z$ axis, as shown in the
figure, because $\mathbf{a}\cdot\mathbf{k} = 0$.   The actual path is a circle of radius $b$, the plane of
the circle being in the plane $z = c$.   The motion is illustrated in Figure 1.14.

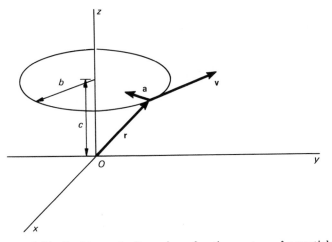

**FIGURE 1.14** Position, velocity and acceleration vectors of a particle moving in a circle.

## 1.20. Vector Integration

Suppose that the time derivative of a vector $\mathbf{r}$ is given in rectangular coordinates where each component is known as a function of time, namely,

$$\frac{d\mathbf{r}}{dt} = \mathbf{i} f_1(t) + \mathbf{j} f_2(t) + \mathbf{k} f_3(t)$$

It is possible to integrate with respect to $t$ to obtain

$$\mathbf{r} = \mathbf{i} \int f_1(t)\, dt + \mathbf{j} \int f_2(t)\, dt + \mathbf{k} \int f_3(t)\, dt \tag{1.33}$$

This is, of course, just the inverse of the process of finding the velocity vector when the position vector is given as a function of time. The same applies to the case in which the acceleration is given as a function of time and an integration yields the velocity.

### EXAMPLE

The velocity vector of a moving particle is given by

$$\mathbf{v} = \mathbf{i} A + \mathbf{j} Bt + \mathbf{k} C t^{-1}$$

in which $A$, $B$, and $C$ are constants. Find $\mathbf{r}$. By integrating, we get

$$\mathbf{r} = \mathbf{i} \int A\, dt + \mathbf{j} \int Bt\, dt + \mathbf{k} \int C t^{-1}\, dt = \mathbf{i} At + \mathbf{j} B \frac{t^2}{2} + \mathbf{k} C \ln t + \mathbf{r}_0$$

The vector $\mathbf{r}_0$ is the constant of integration.

### 1.21.   Relative Velocity

Consider two particles whose position vectors are $\mathbf{r}_1$ and $\mathbf{r}_2$, respectively, as shown in Figure 1.15.   The displacement of the second particle with respect to the first is the difference $\mathbf{r}_2 - \mathbf{r}_1$ which we shall call $\mathbf{r}_{12}$.   The velocity of the

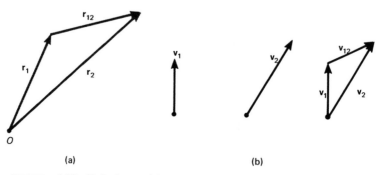

(a)                                                              (b)

FIGURE 1.15   Relative position vector (a) and relative velocity vector (b) of two particles.

second particle relative to the first is therefore

$$\mathbf{v}_{12} = \frac{d\mathbf{r}_{12}}{dt} = \frac{d(\mathbf{r}_2 - \mathbf{r}_1)}{dt} = \mathbf{v}_2 - \mathbf{v}_1$$

which we shall call the *relative velocity*.   By transposing $\mathbf{v}_1$, we have

$$\mathbf{v}_2 = \mathbf{v}_1 + \mathbf{v}_{12}$$

for the actual velocity of particle 2 in terms of the velocity of particle 1 and the relative velocity of the two particles.

It should be noted that the magnitude of the relative velocity of two particles is *not* the same as the time rate of change of the distance between them.   This latter quantity is

$$\frac{d}{dt} |\mathbf{r}_{12}| = \frac{d}{dt} |\mathbf{r}_2 - \mathbf{r}_1|$$

which is different, in general, from $|\mathbf{v}_{12}|$.

## EXAMPLES

1. A particle moves along the $x$ axis with speed $v$, so its position vector is given by $\mathbf{r}_1 = \mathbf{i}(a + vt)$ where $a$ is a constant.   A second particle moves

along the $y$ axis with the same speed, so its position is $\mathbf{r}_2 = \mathbf{j}(b + vt)$. Thus the relative velocity of the second particle with respect to the first is $\mathbf{v}_{12} = \mathbf{v}_2 - \mathbf{v}_1 = \mathbf{j}v - \mathbf{i}v = v(\mathbf{j} - \mathbf{i})$. The relative speed is then $v_{12} = v\sqrt{2}$. What is the value of $d|\mathbf{r}_{12}|/dt$?

2. **A wheel of radius $b$ rolls along the ground with a forward speed $v_0$.** Find the velocity of any point $P$ on the rim relative to the ground. First, consider the expression

$$\mathbf{r}_{OP} = \mathbf{i}b \cos \theta - \mathbf{j}b \sin \theta$$

where

$$\theta = \omega t$$

This represents clockwise circular motion about the origin, the center of the wheel, in this case. The time derivative then gives the velocity of $P$ relative to the center of the wheel as

$$\mathbf{v}_{\text{rel}} = -\mathbf{i}b\omega \sin \theta - \mathbf{j}b\omega \cos \theta$$

But the angular velocity $\omega = v_0/b$, and since the velocity of the center of the wheel relative to the ground is $\mathbf{i}v_0$, then the true velocity of $P$ relative to the ground is

$$\begin{aligned} \mathbf{v} &= \mathbf{i}v_0 - \mathbf{i}b\omega \sin \theta - \mathbf{j}b\omega \cos \theta \\ &= \mathbf{i}v_0(1 - \sin \theta) - \mathbf{j}v_0 \cos \theta \end{aligned}$$

A diagram showing the velocity vectors for various values of $\theta$ is shown in Figure 1.16.

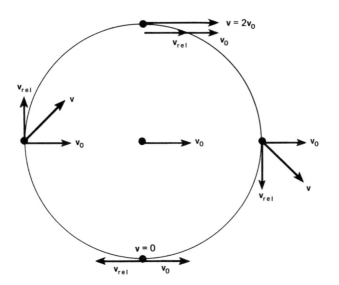

**FIGURE 1.16**   Velocity vectors for various points on a rolling wheel.

### 1.22.   Derivatives of Products of Vectors

It is often necessary to deal with derivatives of the products $n\mathbf{A}$, $\mathbf{A} \cdot \mathbf{B}$, and $\mathbf{A} \times \mathbf{B}$ where the scalar $n$ and the vectors $\mathbf{A}$ and $\mathbf{B}$ are functions of a single parameter $u$, as in Section 1.16.   From the general definition of the derivative, we have

$$\frac{d(n\mathbf{A})}{du} = \lim_{\Delta u \to 0} \frac{n(u + \Delta u)\mathbf{A}(u + \Delta u) - n(u)\mathbf{A}(u)}{\Delta u}$$

$$\frac{d(\mathbf{A} \cdot \mathbf{B})}{du} = \lim_{\Delta u \to 0} \frac{\mathbf{A}(u + \Delta u) \cdot \mathbf{B}(u + \Delta u) - \mathbf{A}(u) \cdot \mathbf{B}(u)}{\Delta u}$$

$$\frac{d(\mathbf{A} \times \mathbf{B})}{du} = \lim_{\Delta u \to 0} \frac{\mathbf{A}(u + \Delta u) \times \mathbf{B}(u + \Delta u) - \mathbf{A}(u) \times \mathbf{B}(u)}{\Delta u}$$

By adding and subtracting expressions like $n(u + \Delta u)\mathbf{A}(u)$ in the numerators, we obtain the following rules:

$$\frac{d(n\mathbf{A})}{du} = \frac{dn}{du}\mathbf{A} + n\frac{d\mathbf{A}}{du} \tag{1.34}$$

$$\frac{d(\mathbf{A} \cdot \mathbf{B})}{du} = \frac{d\mathbf{A}}{du} \cdot \mathbf{B} + \mathbf{A} \cdot \frac{d\mathbf{B}}{du} \tag{1.35}$$

$$\frac{d(\mathbf{A} \times \mathbf{B})}{du} = \frac{d\mathbf{A}}{du} \times \mathbf{B} + \mathbf{A} \times \frac{d\mathbf{B}}{du} \tag{1.36}$$

Notice that it is necessary to preserve the order of the terms in the derivative of the cross product.   The steps are left as an exercise for the student.

### 1.23.   Tangential and Normal Components of Acceleration

In Section 1.13, it was shown that any vector can be expressed as the product of its magnitude and a unit vector giving its direction.   Accordingly, the velocity vector of a moving particle can be written as the product of the particle's speed $v$ and a unit vector $\boldsymbol{\tau}$ that gives the direction of the particle's motion.   Thus

$$\mathbf{v} = v\boldsymbol{\tau} \tag{1.37}$$

The vector $\boldsymbol{\tau}$ is called the *unit tangent vector*.   As the particle moves the speed $v$ may change and the direction of $\boldsymbol{\tau}$ may change.   Let us use the rule for differentiation of the product of a scalar and a vector to obtain the acceleration.   The result is

$$\mathbf{a} = \frac{d\mathbf{v}}{dt} = \frac{d(v\boldsymbol{\tau})}{dt} = \dot{v}\boldsymbol{\tau} + v\frac{d\boldsymbol{\tau}}{dt} \tag{1.38}$$

The unit vector $\boldsymbol{\tau}$, being of constant magnitude, has a derivative $d\boldsymbol{\tau}/dt$ that must necessarily express the change in the direction of $\boldsymbol{\tau}$ with respect to time.

This is illustrated in Figure 1.17(a).   The particle is initially at some point $P$ on its path of motion.   In a time interval $\Delta t$ the particle moves to another point $P'$ a certain distance $\Delta s$ along the path.   Let us denote the unit tangent vectors at $P$ and $P'$ by $\tau$ and $\tau'$, respectively, as shown.   The directions of these two unit vectors differ by a certain angle $\Delta \psi$ as shown in Figure 1.17(b).

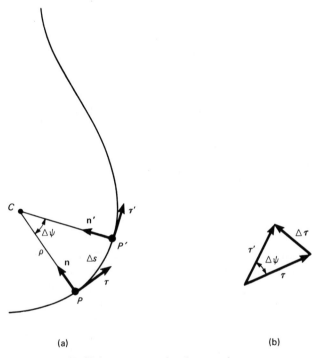

(a)                                (b)

**FIGURE 1.17**   Unit tangent and unit normal vectors.

It is apparent that for small values of $\Delta \psi$, the difference $\Delta \tau$ approaches $\Delta \psi$ in magnitude.   Also, the direction of $\Delta \tau$ becomes perpendicular to the direction of $\tau$ in the limit as $\Delta \psi$ and $\Delta s$ approach zero.   It follows that the derivative $d\tau/d\psi$ is of magnitude unity and is perpendicular to $\tau$.   We shall therefore call it the *unit normal vector* and denote it by $\mathbf{n}$:

$$\frac{d\tau}{d\psi} = \mathbf{n} \tag{1.39}$$

Next, in order to find the time derivative $d\tau/dt$, we use the chain rule as follows

$$\frac{d\tau}{dt} = \frac{d\tau}{d\psi}\frac{d\psi}{dt} = \mathbf{n}\,\frac{d\psi}{ds}\frac{ds}{dt} = \mathbf{n}\,\frac{v}{\rho}$$

in which

$$\rho = \frac{ds}{d\psi}$$

is the radius of curvature of the path of the moving particle at the point $P$. The above value for $d\tau/dt$ is now inserted into Equation (1.38) to yield the final result

$$\mathbf{a} = \dot{v}\boldsymbol{\tau} + \frac{v^2}{\rho}\mathbf{n} \tag{1.40}$$

Thus the acceleration of a moving particle has a component of magnitude

$$a_\tau = \dot{v} = \ddot{s}$$

in the direction of motion.   This is the *tangential acceleration*.   The other component of magnitude

$$a_n = \frac{v^2}{\rho}$$

is the normal component.   This component is always directed toward the center of curvature on the concave side of the path of motion.   Hence the normal component is also called the *centripetal acceleration*.

From the above considerations we see that the time derivative of the speed is only the tangential component of the acceleration.   The magnitude of the total acceleration is given by

$$|\mathbf{a}| = \left|\frac{d\mathbf{v}}{dt}\right| = \left(\dot{v}^2 + \frac{v^4}{\rho^2}\right)^{1/2} \tag{1.41}$$

For example, if a particle moves on a circle with constant speed $v$, the acceleration vector is of magnitude $v^2/R_0$ where $R_0$ is the radius of the circle. The acceleration vector always points to the center of the circle in this case. However, if the speed is not constant but increases at a certain rate $\dot{v}$, then the acceleration has a forward component of this amount and is slanted away from the center of the circle towards the forward direction as illustrated in Figure 1.18.   If the particle is slowing down, then the acceleration vector is slanted in the opposite direction.

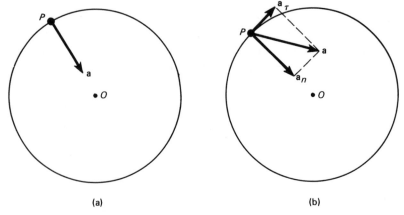

(a)                                              (b)

FIGURE 1.18   Acceleration vectors for a particle moving in a circular path.
(a) Constant speed; (b) increasing speed.

## 1.24.   Velocity and Acceleration in Plane Polar Coordinates

It is often convenient to employ polar coordinates $r$, $\theta$ to express the position of a particle moving in a plane. Vectorially, the position of the particle can be written as the product of the radial distance $r$ by a unit radial vector $\mathbf{e}_r$:

$$\mathbf{r} = r\mathbf{e}_r \qquad (1.42)$$

As the particle moves, both $r$ and $\mathbf{e}_r$ vary, thus they are both functions of the time. Hence, if we differentiate with respect to $t$, we have

$$\mathbf{v} = \frac{d\mathbf{r}}{dt} = \dot{r}\mathbf{e}_r + r\frac{d\mathbf{e}_r}{dt} \qquad (1.43)$$

In order to calculate the derivative $d\mathbf{e}_r/dt$, let us consider the vector diagram shown in Figure 1.19. A study of the figure shows that when the direction

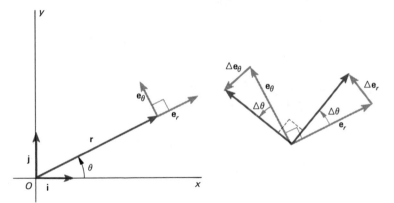

FIGURE 1.19   Unit vectors for plane polar coordinates.

of $\mathbf{r}$ changes by an amount $\Delta\theta$, the corresponding change $\Delta\mathbf{e}_r$ of the unit radial vector is as follows: The magnitude $|\Delta\mathbf{e}_r|$ is approximately equal to $\Delta\theta$, and the direction of $\Delta\mathbf{e}_r$ is very nearly perpendicular to $\mathbf{e}_r$. Let us introduce another unit vector $\mathbf{e}_\theta$ whose direction is perpendicular to $\mathbf{e}_r$. Then we have

$$\Delta\mathbf{e}_r \simeq \mathbf{e}_\theta\Delta\theta$$

If we divide by $\Delta t$ and take the limit, we get

$$\frac{d\mathbf{e}_r}{dt} = \mathbf{e}_r\frac{d\theta}{dt} \qquad (1.44)$$

for the time derivative of the unit radial vector. In a precisely similar way, we can argue that the change in the unit vector $\mathbf{e}_\theta$ is given by the approximation

$$\Delta\mathbf{e}_\theta \simeq -\mathbf{e}_r\Delta\theta$$

Here the minus sign is inserted to indicate that the direction of the change $\Delta \mathbf{e}_\theta$ is opposite to the direction of $\mathbf{e}_r$ as can be seen from the figure.   Consequently, the time derivative is given by

$$\frac{d\mathbf{e}_\theta}{dt} = -\mathbf{e}_r \frac{d\theta}{dt} \tag{1.45}$$

By using Equation (1.44) for the derivative of the unit radial vector, we can finally write the equation for the velocity as

$$\mathbf{v} = \dot{r}\mathbf{e}_r + r\dot{\theta}\mathbf{e}_\theta \tag{1.46}$$

Thus $\dot{r}$ is the magnitude of the radial component of the velocity vector, and $r\dot{\theta}$ is the magnitude of the transverse component.

In order to find the acceleration vector, we take the derivative of the velocity with respect to time.   This gives

$$\mathbf{a} = \frac{d\mathbf{v}}{dt} = \ddot{r}\mathbf{e}_r + \dot{r}\frac{d\mathbf{e}_r}{dt} + (\dot{r}\dot{\theta} + r\ddot{\theta})\mathbf{e}_\theta + r\dot{\theta}\frac{d\mathbf{e}_\theta}{dt}$$

The values of $d\mathbf{e}_r/dt$ and $d\mathbf{e}_\theta/dt$ are given by Equations (1.44) and (1.45) and yield the following equation for the acceleration vector in plane polar coordinates:

$$\mathbf{a} = (\ddot{r} - r\dot{\theta}^2)\mathbf{e}_r + (r\ddot{\theta} + 2\dot{r}\dot{\theta})\mathbf{e}_\theta \tag{1.47}$$

Thus the magnitude of the radial component of the acceleration vector is

$$a_r = \ddot{r} - r\dot{\theta}^2 \tag{1.48}$$

and that of the transverse component is

$$a_\theta = r\ddot{\theta} + 2\dot{r}\dot{\theta} = \frac{1}{r}\frac{d}{dt}(r^2\dot{\theta}) \tag{1.49}$$

The above results show, for instance, that if a particle moves on a circle of constant radius $b$, so that $\dot{r} = 0$, then the radial component of the acceleration is of magnitude $b\dot{\theta}^2$ and is directed inward toward the center of the circular path.   The magnitude of the transverse component in this case is $b\ddot{\theta}$.   On the other hand, if the particle moves along a fixed radial line, that is, if $\theta$ is constant, then the radial component is just $\ddot{r}$ and the transverse component is zero.   If $r$ and $\theta$ both vary, then the general expression (1.47) gives the acceleration.

## EXAMPLE

A particle moves on a spiral path such that the position in polar coordinates is given by

$$r = bt^2 \qquad \theta = ct$$

where $b$ and $c$ are constants. Find the velocity and acceleration as functions of $t$. From Equation (1.46), we find

$$\mathbf{v} = \mathbf{e}_r \frac{d}{dt}(bt^2) + \mathbf{e}_\theta(bt^2)\frac{d}{dt}(ct)$$
$$= (2bt)\mathbf{e}_r + (bct^2)\mathbf{e}_\theta$$

Similarly, from Equation (1.47), we have

$$\mathbf{a} = \mathbf{e}_r(2b - bt^2c^2) + \mathbf{e}_\theta[0 + 2(2bt)c]$$
$$= b(2 - t^2c^2)\mathbf{e}_r + 4bct\mathbf{e}_\theta$$

It is interesting to note that the radial component of the acceleration becomes negative for large $t$, in this example, although the radius is always increasing monotonically with time.

## 1.25. Velocity and Acceleration in Cylindrical and Spherical Coordinates

*Cylindrical Coordinates*

In the case of three-dimensional motion, the position of a particle can be described in cylindrical coordinates $R$, $\varphi$, $z$. The position vector is then written as

$$\mathbf{r} = R\mathbf{e}_R + z\mathbf{e}_z \tag{1.50}$$

where $\mathbf{e}_R$ is a unit radial vector in the $xy$ plane and $\mathbf{e}_z$ is the unit vector in the $z$ direction. A third unit vector $\mathbf{e}_\varphi$ is needed so that the three vectors $\mathbf{e}_R\mathbf{e}_\varphi\mathbf{e}_z$ constitute a right-handed triad as illustrated in Figure 1.20. We note that $\mathbf{k} = \mathbf{e}_z$.

The velocity and acceleration vectors are found by differentiating, as before. This will again involve derivatives of the unit vectors. An argument similar to that used for the plane case shows that $d\mathbf{e}_R/dt = \mathbf{e}_\varphi\dot\varphi$ and $d\mathbf{e}_\varphi/dt = -\mathbf{e}_R\dot\varphi$. The unit vector $\mathbf{e}_z$ does not change in direction, so its time derivative is zero.

In view of these facts, the velocity and acceleration vectors are easily seen to be given by the following equations:

$$\mathbf{v} = \dot R\mathbf{e}_R + R\dot\varphi\mathbf{e}_\varphi + \dot z\mathbf{e}_z \tag{1.51}$$

$$\mathbf{a} = (\ddot R - R\dot\varphi^2)\mathbf{e}_R + (2\dot R\dot\varphi + R\ddot\varphi)\mathbf{e}_\varphi + \ddot z\mathbf{e}_z \tag{1.52}$$

These give the values of $\mathbf{v}$ and $\mathbf{a}$ in terms of their components in the *rotated* triad $\mathbf{e}_R\mathbf{e}_\varphi\mathbf{e}_z$.

An alternative way of obtaining the derivatives of the unit vectors is to differentiate the following equations which are the relationships between the fixed unit triad **ijk** and the rotated triad:

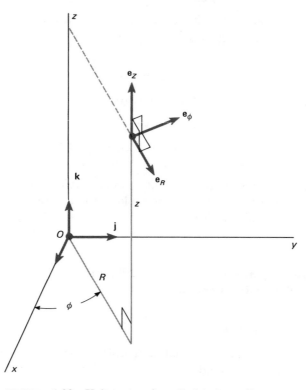

FIGURE 1.20    Unit vectors for cylindrical coordinates.

$$\mathbf{e}_R = \mathbf{i} \cos \varphi + \mathbf{j} \sin \varphi$$
$$\mathbf{e}\varphi = -\mathbf{i} \sin \varphi + \mathbf{j} \cos \varphi \qquad (1.53)$$
$$\mathbf{e}_z = \mathbf{k}$$

The steps are left as an exercise.    The result can also be found by use of the rotation matrix as given in Example 2, Section 1.15.

*Spherical Coordinates*

When spherical coordinates $r$, $\theta$, $\varphi$ are employed to describe the position of a particle, the position vector is written as the product of the radial distance $r$ and the unit radial vector $\mathbf{e}_r$, as with plane polar coordinates.    Thus

$$\mathbf{r} = r\mathbf{e}_r$$

The direction of $\mathbf{e}_r$ is now specified by the two angles $\varphi$ and $\theta$.    We introduce two more unit vectors $\mathbf{e}_\varphi$ and $\mathbf{e}_\theta$ as shown in Figure 1.21.

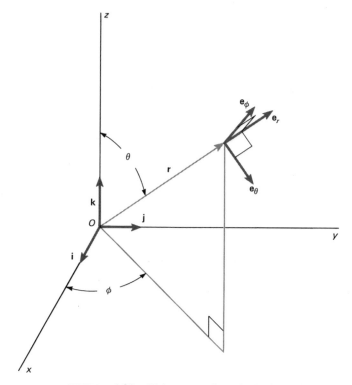

**FIGURE 1.21**   Unit vectors for spherical coordinates.

The velocity is

$$\mathbf{v} = \frac{d\mathbf{r}}{dt} = \dot{r}\mathbf{e}_r + r\frac{d\mathbf{e}_r}{dt} \tag{1.54}$$

Our next problem is how to express the derivative $d\mathbf{e}_r/dt$ in terms of the unit vectors in the rotated triad.

Referring to the figure, we see that the following relationships hold between the two triads

$$\begin{aligned}
\mathbf{e}_r &= \mathbf{i} \sin\theta \cos\varphi + \mathbf{j} \sin\theta \sin\varphi + \mathbf{k} \cos\theta \\
\mathbf{e}_\theta &= \mathbf{i} \cos\theta \cos\varphi + \mathbf{j} \cos\theta \sin\varphi - \mathbf{k} \sin\theta \\
\mathbf{e}_\varphi &= -\mathbf{i} \sin\varphi + \mathbf{j} \cos\varphi
\end{aligned} \tag{1.55}$$

which express the unit vectors of the rotated triad in terms of the fixed triad **ijk**. We note the similarity between this transformation and that of the second part of Example 2 in Section 1.15. The two are, in fact, identical if the correct identification of rotations is made. Let us differentiate the first equation with respect to time. The result is.

$$\frac{d\mathbf{e}_r}{dt} = \mathbf{i}(\dot{\theta}\cos\theta\cos\varphi - \dot{\varphi}\sin\theta\sin\varphi)$$

$$+ \mathbf{j}(\dot{\theta}\cos\theta\sin\varphi + \dot{\varphi}\sin\theta\cos\varphi) - \mathbf{k}\dot{\theta}\sin\theta$$

Next, by using the expressions for $\mathbf{e}_\varphi$ and $\mathbf{e}_\theta$ in Equation (1.55), we find that the above equation reduces to

$$\frac{d\mathbf{e}_r}{dt} = \dot{\varphi}\mathbf{e}_\varphi\sin\theta + \dot{\theta}\mathbf{e}_\theta \tag{1.56}$$

The other two derivatives are found by a similar procedure.   The results are

$$\frac{d\mathbf{e}_\theta}{dt} = -\dot{\theta}\mathbf{e}_r + \dot{\varphi}\mathbf{e}_\varphi\cos\theta \tag{1.57}$$

$$\frac{d\mathbf{e}_\varphi}{dt} = -\dot{\varphi}\mathbf{i}_r\sin\theta - \dot{\varphi}\mathbf{e}_\theta\cos\theta \tag{1.58}$$

The steps are left as an exercise.   Returning now to the problem of finding $\mathbf{v}$, we insert the expression for $d\mathbf{e}_r/dt$ given by Equation (1.56) into Equation (1.54).   The final result is

$$\mathbf{v} = \mathbf{e}_r\dot{r} + \mathbf{e}_\varphi r\dot{\varphi}\sin\theta + \mathbf{e}_\theta r\dot{\theta} \tag{1.59}$$

giving the velocity vector in terms of its components in the rotated triad.

To find the acceleration, we differentiate the above expression with respect to time.   This gives

$$\mathbf{a} = \frac{d\mathbf{v}}{dt} = \mathbf{e}_r\ddot{r} + \dot{r}\frac{d\mathbf{e}_r}{dt} + \mathbf{e}_\varphi\frac{d(r\dot{\varphi}\sin\theta)}{dt} + r\dot{\varphi}\sin\theta\frac{d\mathbf{e}_\varphi}{dt} + \mathbf{e}_\theta\frac{d(r\dot{\theta})}{dt} + r\dot{\theta}\frac{d\mathbf{e}_\theta}{dt}$$

Upon using the previous formulas for the derivatives of the unit vectors, it is readily found that the above expression for the acceleration reduces to

$$\mathbf{a} = (\ddot{r} - r\dot{\varphi}^2\sin^2\theta - r\dot{\theta}^2)\mathbf{e}_r + (r\ddot{\theta} + 2\dot{r}\dot{\theta} - r\dot{\varphi}^2\sin\theta\cos\theta)\mathbf{e}_\theta$$
$$+ (r\ddot{\varphi}\sin\theta + 2\dot{r}\dot{\varphi}\sin\theta + 2r\dot{\theta}\dot{\varphi}\cos\theta)\mathbf{e}_\varphi \tag{1.60}$$

giving the acceleration vector in terms of its components in the triad $\mathbf{e}_r\mathbf{e}_\theta\mathbf{e}_\varphi$.

### 1.26.  Angular Velocity

Let a particle, whose position vector is initially $\mathbf{r}$, undergo a displacement produced by a rotation through an angle $\delta\phi$ about an axis whose direction is defined by a unit vector $\mathbf{e}$, Figure 1.22.   Thus the particle will move along an arc of a circle of radius $r \sin \phi$ in which $\theta$ is the angle between **r** and **e**.   The magnitude of the particle's displacement is thus $|\delta \mathbf{r}| = r \sin \theta \, \delta \phi$, and the direction of the displacement is perpendicular to both **r** and **e**.   Thus we can express the displacement vectorially as a cross product, namely

$$\delta \mathbf{r} = \delta \phi \, \mathbf{e} \times \mathbf{r}$$

Accordingly, the velocity of the particle is given by

$$\dot{r} = \lim_{\delta t \to 0} \frac{\delta \mathbf{r}}{\delta t} = \dot{\phi} \mathbf{e} \times \mathbf{r} \qquad (1.61)$$

We now introduce the vector **ω** defined as the product

$$\boldsymbol{\omega} = \dot{\phi} \mathbf{e}$$

called the angular velocity.   The velocity of the particle is thus expressed as

$$\dot{\mathbf{r}} = \boldsymbol{\omega} \times \mathbf{r} \qquad (1.62)$$

We now proceed to show that angular velocities obey the rule of vector addition.   Consider the displacement caused by an infinitesimal rotation $\delta \phi_1$ about an axis $\mathbf{e}_1$ followed by a second such rotation $\delta \phi_2$ about a different axis $\mathbf{e}_2$.   The first rotation changes the position vector **r** to a new position vector $\mathbf{r} + \delta \phi_1 \mathbf{e}_1 \times \mathbf{r}_1$.   Hence the net displacement due to the two rotations is

$$\delta \mathbf{r}_{12} = \delta \phi_1 \mathbf{e}_1 \times \mathbf{r} + \delta \phi_2 \mathbf{e}_2 \times (\mathbf{r} + \delta \phi_1 \mathbf{e}_1 \times \mathbf{r})$$

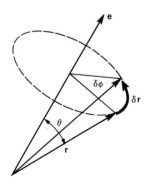

**FIGURE 1.22**   Displacement produced by rotation. The radius of the circular path is $r \sin \theta$.

If the angular rotations are both small enough so that we can neglect the product $\delta\phi_1\,\delta\phi_2$, then we find upon expansion that

$$\delta\mathbf{r}_{12} = (\delta\phi_1\,\mathbf{e}_1 + \delta\phi_2\,\mathbf{e}_2) \times \mathbf{r}$$

If the order of rotations is reversed, we indeed find the same result, that is $\delta\mathbf{r}_{12} = \delta\mathbf{r}_{21}$. In other words, the two infinitesimal rotations are commutative. Finally, let us divide by $\delta t$ and take the limit, as in Equation (1.61) above. We can then write

$$\dot{\mathbf{r}} = (\omega_1 + \omega_2) \times \mathbf{r}$$

for the velocity of the particle where $\omega_1 = \dot{\phi}_1\mathbf{e}_1$ and $\omega_2 = \dot{\phi}_2\mathbf{e}_2$. Thus the motion of the particle can be considered to be described by a single angular velocity

$$\omega = \omega_1 + \omega_2$$

given by the regular rule of vector addition.

## DRILL EXERCISES

1.1 Given the two vectors $\mathbf{A} = \mathbf{i} + \mathbf{j}$ and $\mathbf{B} = \mathbf{j} - \mathbf{k}$. Find
(a) $\mathbf{A} + \mathbf{B}$ and $|\mathbf{A} + \mathbf{B}|$
(b) $\mathbf{A} - \mathbf{B}$ and $|\mathbf{A} - \mathbf{B}|$
(c) $\mathbf{A} \cdot \mathbf{B}$
(d) $\mathbf{A} \times \mathbf{B}$ and $|\mathbf{A} \times \mathbf{B}|$
(e) $(\mathbf{A} + 2\mathbf{B}) \cdot (2\mathbf{A} - \mathbf{B})$
(f) $(\mathbf{A} + \mathbf{B}) \times (\mathbf{A} - \mathbf{B})$

1.2 Given $\mathbf{A} = \mathbf{i} + \mathbf{j} + \mathbf{k}$, $\mathbf{B} = \mathbf{i} + 2\mathbf{j}$, $\mathbf{C} = 2\mathbf{j} - \mathbf{k}$. Find
(a) $\mathbf{A} + \mathbf{B} - \mathbf{C}$
(b) $\mathbf{A} \cdot (\mathbf{B} + \mathbf{C})$ and $(\mathbf{A} + \mathbf{B}) \cdot \mathbf{C}$
(c) $\mathbf{A} \cdot (\mathbf{B} \times \mathbf{C})$ and $(\mathbf{A} \times \mathbf{B}) \cdot \mathbf{C}$
(d) $\mathbf{A} \times (\mathbf{B} \times \mathbf{C})$ and $(\mathbf{A} \times \mathbf{B}) \times \mathbf{C}$

1.3 Find the angle between the vectors $\mathbf{A} = \mathbf{i} + \mathbf{j} + \mathbf{k}$ and $\mathbf{B} = \mathbf{i} + \mathbf{j}$. (Note: These vectors define a body diagonal and a face diagonal of a cube.)

1.4 Given the time-varying vectors $\mathbf{A} = \mathbf{i} \cos \omega t + \mathbf{j} \sin \omega t$ and $\mathbf{B} = t\mathbf{i} + t^2\mathbf{j} + t^3\mathbf{k}$. Find
(a) $d\mathbf{A}/dt$ and $|d\mathbf{A}/dt|$
(b) $d^2\mathbf{B}/dt^2$ and $|d^2\mathbf{B}/dt^2|$
(c) $d(\mathbf{A} \cdot \mathbf{B})/dt$ and $d(\mathbf{A} \times \mathbf{B})/dt$

## PROBLEMS

**1.5** For what values of $q$ are the two vectors $\mathbf{A} = \mathbf{i} + \mathbf{j} + \mathbf{k}q$ and $\mathbf{B} = \mathbf{i}q - 2\mathbf{j} + 2\mathbf{k}q$ perpendicular to each other?

**1.6** Prove the vector identity $\mathbf{A} \times (\mathbf{B} \times \mathbf{C}) = (\mathbf{A} \cdot \mathbf{C})\mathbf{B} - (\mathbf{A} \cdot \mathbf{B})\mathbf{C}$.

**1.7** Two vectors $\mathbf{A}$ and $\mathbf{B}$ represent concurrent sides of a parallelogram. Prove that the area of the parallelogram is $|\mathbf{A} \times \mathbf{B}|$.

**1.8** Prove the trigonometric law of sines using vector methods.

**1.9** Three vectors $\mathbf{A}$, $\mathbf{B}$, and $\mathbf{C}$ represent concurrent sides of a parallelepiped. Show that the volume of the parallelepiped is $|\mathbf{A} \cdot (\mathbf{B} \times \mathbf{C})|$.

**1.10** Express the vector $\mathbf{i} + \mathbf{j}$ in terms of the triad $\mathbf{i}'\mathbf{j}'\mathbf{k}'$ where the $x'z'$ axes are rotated about the $y$ axis (which coincides with the $y'$ axis) through an angle of 60°.

**1.11** Show that the magnitude of a vector is unchanged by a rotation. Use the matrix

$$\begin{bmatrix} \cos\theta & \sin\theta & 0 \\ -\sin\theta & \cos\theta & 0 \\ 0 & 0 & 1 \end{bmatrix}$$

for a rotation about the $z$ axis through an angle $\theta$.

**1.12** Find the transformation matrix for a rotation about the $z$ axis through an angle $\theta$ followed by a rotation about the $y'$ axis through an angle $\varphi$.

**1.13** The two sets of vectors $\mathbf{a}$, $\mathbf{b}$, $\mathbf{c}$, and $\mathbf{a}'$, $\mathbf{b}'$, $\mathbf{c}'$ are said to be *reciprocal* if $\mathbf{a} \cdot \mathbf{a}' = \mathbf{b} \cdot \mathbf{b}' = \mathbf{c} \cdot \mathbf{c}' = 1$ and all other mixed dot products like $\mathbf{a} \cdot \mathbf{b}' = 0$. Show that $\mathbf{c}' = (\mathbf{a} \times \mathbf{b})/Q$, $\mathbf{a}' = (\mathbf{b} \times \mathbf{c})/Q$, $\mathbf{b}' = (\mathbf{c} \times \mathbf{a})/Q$ where $Q = \mathbf{a} \cdot (\mathbf{b} \times \mathbf{c})$.

**1.14** Find a set of vectors that are reciprocal to the set $\mathbf{i}$, $\mathbf{j}$, and $\mathbf{i} + \mathbf{j} + \mathbf{k}$.

**1.15** A particle moves in an elliptical path given by the equation

$$\mathbf{r} = \mathbf{i}b \cos \omega t + \mathbf{j}2b \sin \omega t$$

Find the speed as a function of $t$.

**1.16** In the above problem, find the angle between the velocity vector and the acceleration vector at time $t = \pi/4\omega$.

**1.17** The position of a particle is given in plane polar coordinates by $\mathbf{r} = be^{kt}$, $\theta = ct$. Show that the angle between the velocity vector and the acceleration vector remains constant as the particle spirals outward.

**1.18** A particle moves on a circle of constant radius $b$. If the speed of the particle varies with the time $t$ according to the equation

$$v = At^2$$

for what value, or values, of $t$ does the acceleration vector make an angle of 45° with the velocity vector?

1.19   A particle moves on a helical path such that its position, in cylindrical coordinates, is given by

$$R = b \qquad \varphi = \omega t \qquad z = ct^2$$

Find the speed and the magnitude of the acceleration as functions of $t$.

1.20   Show that the magnitude of tangential component of the acceleration is given by the expression

$$a_\tau = \frac{\mathbf{a} \cdot \mathbf{v}}{|\mathbf{v}|}$$

and that of the normal component is

$$a_n = (a^2 - a_\tau{}^2)^{1/2}$$

1.21   Use the above result to find the tangential and normal components of the acceleration as functions of time in Problem 1.19.

1.22   Prove that $\mathbf{v} \cdot \mathbf{a} = v\dot{v}$, and hence that for a moving particle $\mathbf{v}$ and $\mathbf{a}$ are perpendicular to each other if the speed $v$ is constant.   [Hint: Differentiate both sides of the equation $\mathbf{v} \cdot \mathbf{v} = v^2$ with respect to $t$. Remember that $\dot{v}$ is not the same as $|\mathbf{a}|$.]

1.23   Prove that

$$\frac{d}{dt}[\mathbf{r} \cdot (\mathbf{v} \times \mathbf{a})] = \mathbf{r} \cdot (\mathbf{v} \times \dot{\mathbf{a}})$$

1.24   Prove that $|\mathbf{v} \times \mathbf{a}| = v^3/\rho$, where $\rho$ is the radius of curvature of the path of a moving particle.

1.25   By using the fact that the unit tangent vector $\boldsymbol{\tau}$ can be expressed as'

$$\boldsymbol{\tau} = \frac{\mathbf{v}}{v}$$

find an expression for the unit normal vector $\mathbf{n}$ in terms $\mathbf{a}$, $a$, $\mathbf{v}$, $v$, and $\dot{v}$.

1.26   A wheel of radius $b$ is placed in a gimbal mount and is made to rotate as follows:   The wheel spins with constant angular speed $\omega_1$ about its own axis which, in turn rotates with constant angular speed $\omega_2$ about a vertical axis in such a way that the axis of the wheel stays in a horizontal plane and the center of the wheel is motionless.   Use spherical coordinates to find the acceleration of any point on the rim of the wheel.   In particular, find the acceleration of the highest point on the wheel.   [Hint: Use the fact that spherical coordinates can be chosen such that $r = b$, $\theta = \omega_1 t$, and $\varphi = \omega_2 t$.]

# 2. Newtonian Mechanics. Rectilinear Motion of a Particle

As stated in the introduction, dynamics is that branch of mechanics which deals with the physical laws governing the actual motion of material bodies. One of the fundamental tasks of dynamics is to predict, out of all possible ways a material system can move, which particular motion will occur in any given situation. Our study of dynamics at this point will be based on the laws of motion as they were first formulated by Newton. In a later chapter we shall study alternative ways of expressing the laws of motion in the more advanced equations of Lagrange and Hamilton. These are not different theories, however, for they can be derived from Newton's laws.

## 2.1. Newton's Laws of Motion

The reader is undoubtedly already familiar with Newton's laws of motion. They are as follows:

I. Every body continues in its state of rest or of uniform motion in a straight line, unless it is compelled by a force to change that state.

II. Change of motion is proportional to the applied force and takes place in the direction of the force.

III. To every action there is always an equal and opposite reaction, or, the mutual actions of two bodies are always equal and oppositely directed.

Let us now examine these laws in some detail.

### 2.2.  Newton's First Law.   Inertial Reference Systems

The first law describes a common property shared by all matter, namely *inertia*.   The law states that a moving body travels in a straight line with constant speed unless some influence called *force* prevents the body from doing so.   Whether or not a body moves in a straight line with constant speed depends not only upon external influences (forces) but also upon the particular reference system that is used to describe the motion.   The first law actually amounts to a definition of a particular kind of reference system called a Newtonian or *inertial* reference system.   Such a system is one in which Newton's first law holds.   Rotating or accelerating systems are not inertial.   These will be studied in Chapter 4.

The question naturally arises as to how it is possible to determine whether or not a given coordinate system constitutes an inertial system.   The answer is not simple.   In order to eliminate *all* forces on a body it would be necessary to isolate the body completely.   This is impossible, of course, since there are always at least some gravitational forces acting unless the body was removed to an infinite distance from all other matter.

For many practical purposes not requiring high precision, a coordinate system fixed to the earth is approximately inertial.   Thus, for example, a billiard ball seems to move in a straight line with constant speed as long as it does not collide with other balls or hit the cushion.   If the motion of a billiard ball were measured with very high precision, however, it would be discovered that the path is slightly curved.   This is due to the fact that the earth is rotating and so a coordinate system fixed to the earth is not actually an inertial system.   A better system would be one using the center of the earth, the center of the sun, and a distant star as reference points.   But even this system would not be strictly inertial because of the earth's orbital motion around the sun.   The next best approximation would be to take the center of the sun and two distant stars as reference points, for example.   It is generally agreed that the ultimate inertial system, in the sense of Newtonian mechanics, would be one based on the average background of all the matter in the universe.

### 2.3.  Mass and Force.   Newton's Second and Third Laws

We are all familiar with the fact that a big stone is not only hard to lift, but that such an object is more difficult to set in motion (or to stop) than, say, a small piece of wood.   We say that the stone has more inertia than the wood.   The quantitative measure of inertia is called *mass*.   Suppose we have two bodies $A$ and $B$.   How do we determine the measure of inertia of one relative to the other?   There are many experiments that can be devised to answer this question.   If the two bodies can be made to interact directly with one another, say by a spring connecting them, then it is found, by careful experiments, that the accelerations of the two bodies are always opposite in direction and have a *constant ratio*.   (It is assumed that the accelerations are given in an inertial reference system and that only the *mutual* influence of the two bodies $A$ and $B$ is under consideration.)   We can express this very important and fundamental fact by the equation

$$\frac{d\mathbf{v}_A}{dt} = - \frac{d\mathbf{v}_B}{dt} \mu_{BA} \tag{2.1}$$

The constant $\mu_{BA}$ is, in fact, the measure of relative inertia of $B$ with respect to $A$.   From Equation (2.1) it follows that $\mu_{BA} = 1/\mu_{AB}$.   Thus we might express $\mu_{BA}$ as a ratio

$$\mu_{BA} = \frac{m_B}{m_A}$$

and use some standard body as a unit of inertia.   Now the ratio $m_B/m_A$ ought to be independent of the choice of the unit.   This will be the case if, for any third body $C$,

$$\frac{\mu_{BC}}{\mu_{AC}} = \mu_{BA}$$

This is indeed found to be true.   We call the quantity $m$ the *mass*.

Strictly speaking, $m$ should be called the *inertial mass*, for its definition is based on the properties of inertia.   In actual practice mass ratios are usually determined by weighing.   The weight or gravitational force is proportional to what may be called the *gravitational mass* of a body.   All experience thus far, however, indicates that inertial mass and gravitational mass are strictly proportional to one another.   Hence for our purpose we need not distinguish between the two kinds of mass.

The fundamental fact expressed by Equation (2.1) can now be written in the form

$$m_A \frac{d\mathbf{v}_A}{dt} = - m_B \frac{d\mathbf{v}_B}{dt} \tag{2.2}$$

The product of mass and acceleration in the above equation is the "change of motion" of Newton's second law and, according to that law, is proportional to the *force*.   In other words, we can write the second law as

$$\mathbf{F} = km \frac{d\mathbf{v}}{dt} \tag{2.3}$$

where $\mathbf{F}$ is the force and $k$ is a constant of proportionality.   It is customary to take $k = 1$ and write[1]

---

[1] In the mks system the unit of force, defined by Equation (2.4), is called the *newton*. Thus a force of 1 newton imparts acceleration of 1 m per sec² to an object of 1 kg mass. The cgs unit of force (1 g × 1 cm per sec²) is called the *dyne*.   In engineering, a common unit of force is the *pound force* which imports an acceleration of 1 ft per sec² to an object of 1 *slug* mass.   (1 slug = 32 pounds mass.)

$$\mathbf{F} = m\frac{d\mathbf{v}}{dt} \tag{2.4}$$

The above equation is equivalent to

$$\mathbf{F} = \frac{d(m\mathbf{v})}{dt} \tag{2.5}$$

if the mass is constant. According to the theory of relativity, the mass of a moving body is not constant but is a function of the speed of the body, so that Equations (2.4) and (2.5) are not strictly equivalent. However, for speeds that are small compared to the speed of light, $3 \times 10^8$ m/sec, the change of mass is negligible.

According to Equation (2.4) we can now interpret the fundamental fact expressed by Equation (2.2) as a statement that two directly interacting bodies exert equal and opposite forces on one another:

$$\mathbf{F}_A = -\mathbf{F}_B$$

This is embodied in the statement of the third law. The forces are called *action* and *reaction*.

There are situations in which the third law fails. If the two bodies are separated by a large distance and interact with one another through a force field which propagates with a finite velocity, such as the interaction between moving electric charges, then the forces of action and reaction are not always equal and opposite.[2]

One great advantage of the force concept is that it enables us to restrict our attention to a single body. The physical significance of the idea of force is that, in a given situation, there can usually be found some relatively simple function of the coordinates, called the force function, which when set equal to the product of mass and acceleration correctly describes the motion of a body. This is the essence of Newtonian mechanics.

### 2.4.  Linear Momentum

The product of mass and velocity is called *linear momentum* and is denoted by the symbol $\mathbf{p}$. Thus

$$\mathbf{p} = m\mathbf{v} \tag{2.6}$$

The mathematical statement of Newton's second law, Equation (2.5), may then be written as

---

[2] However, it is possible in such cases to regard the force field as a third "body" with its own action and reaction. The third law thus need not be discarded. See Section 6.1 and reference cited therein.

$$\mathbf{F} = \frac{d\mathbf{p}}{dt} \tag{2.7}$$

In words, *force is equal to the time rate of change of linear momentum.*

The third law, the law of action and reaction, can be expressed conveniently in terms of linear momentum. Thus for two mutually interacting bodies $A$ and $B$, we have

$$\frac{d\mathbf{p}_A}{dt} = -\frac{d\mathbf{p}_B}{dt}$$

or

$$\frac{d}{dt}(\mathbf{p}_A + \mathbf{p}_B) = 0$$

Accordingly

$$\mathbf{p}_A + \mathbf{p}_B = \text{constant}$$

Thus the third law implies that the total linear momentum of two interacting bodies always remains constant.

The constancy of the combined linear momentum of two mutually interacting bodies is a special case of a more general rule that we shall discuss in detail later, namely that *the total linear momentum of any isolated system remains constant in time.* This fundamental statement is known as the *law of conservation of linear momentum* and is one of the most basic rules of physics. It is assumed to be valid even in those cases in which Newtonian mechanics fails to hold.

## 2.5.   Motion of a Particle

The fundamental equation of motion of a particle is given by the analytical statement of Newton's second law, Equation (2.4). When a particle is under the influence of more than one force, it may be regarded as an experimental fact that these forces add vectorially, namely,

$$\mathbf{F} = \Sigma \, \mathbf{F}_i = m\frac{d^2\mathbf{r}}{dt^2} = m\mathbf{a} \tag{2.8}$$

If the acceleration of a particle is known, then the equation of motion [Equation (2.8)] gives the force that acts on the particle. The usual problems of particle dynamics, however, are those in which the forces are certain known functions of the coordinates including the time, and the task is to find the position of the particle as a function of time. This involves the solution of a set of differential equations. In some problems it turns out to be impossible to obtain solutions of the differential equations of motion in terms of known analytic functions, in which case one must use some method of approximation.

In many practical applications, such as ballistics, satellite motion, and so on, the differential equations are so complicated that it is necessary to resort to numerical integration, often done on high-speed electronic computers, to predict the motion.

### 2.6.   Rectilinear Motion.   Uniform Acceleration

When a moving particle remains on a single straight line, the motion is said to be *rectilinear*.   In this case, without loss of generality we can choose the $x$ axis as the line of motion.   The general equation of motion is then written

$$F(x,\dot{x},t) = m\ddot{x}$$

Let us consider some special cases in which the equation can be integrated by elementary methods.

The simplest situation is that in which the force is constant.   In this case we have constant acceleration

$$\frac{dv}{dt} = \frac{F}{m} = \text{constant} = a$$

and the solution is readily obtained by direct integration with respect to time:

$$v = at + v_0 \tag{2.9}$$

$$x = \tfrac{1}{2}at^2 + v_0 t + x_0 \tag{2.10}$$

where $v_0$ is the initial velocity and $x_0$ is the initial position.   By eliminating the time $t$ between Equations (2.9) and (2.10), we obtain

$$2a(x - x_0) = v^2 - v_0^2 \tag{2.11}$$

The student will recall the above familiar equations of uniformly accelerated motion.   There are a number of fundamental applications.   For example, in the case of a body falling freely near the surface of the earth, neglecting air resistance, the acceleration is very nearly constant.   We denote the acceleration of a freely falling body by $\mathbf{g}$.   (By measurement, $g = 9.8$ m per sec$^2$ = 32 ft per sec$^2$.)   The downward force of gravity (the *weight*) is, accordingly, equal to $m\mathbf{g}$.   The gravitational force is always present, regardless of the motion of the body and is independent of any other forces that may be acting.[3]   We shall henceforth call it $m\mathbf{g}$.

---

[3] Effects of the earth's rotation will be studied in Chapter 4.

## EXAMPLE

Consider a particle that is sliding down a smooth plane inclined at an angle $\theta$ to the horizontal, as shown in Figure 2.1(a). We choose the positive direction of the $x$ axis to be down the plane, as indicated. The component of

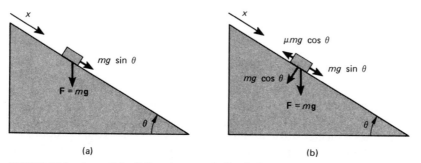

(a)                                                          (b)

FIGURE 2.1   A particle sliding down an inclined plane. (a) Smooth plane; (b) rough plane.

the gravitational force in the $x$ direction is equal to $mg \sin \theta$. This is a constant, hence the motion is given by Equations (2.9), (2.10), and (2.11) where

$$a = \frac{F}{m} = g \sin \theta$$

Suppose that, instead of being smooth, the plane is rough; that is, it exerts a frictional force **f** on the particle. Then the net force in the $x$ direction, as shown in Figure 2.1(b), is equal to $mg \sin \theta - f$. Now for sliding contact it is found that the magnitude of the frictional force is proportional to the magnitude of the normal force $N$, that is,

$$f = \mu N$$

where the constant of proportionality $\mu$ is known as the *coefficient of sliding friction*. In the example under discussion the normal force $N$, as shown in the figure, is equal to $mg \cos \theta$, hence

$$f = \mu mg \cos \theta$$

Consequently, the net force in the $x$ direction is equal to

$$mg \sin \theta - \mu mg \cos \theta$$

Again the force is constant, and Equations (2.9), (2.10), and (2.11) apply, where

$$a = \frac{F}{m} = g(\sin \theta - \mu \cos \theta)$$

The speed of the particle will increase if the expression in parentheses is positive, that is, if $\theta > \tan^{-1} \mu$. The angle $\tan^{-1} \mu$, usually denoted by $\epsilon$, is called the *angle of friction*. If $\theta = \epsilon$, then $a = 0$, and the particle slides down the plane with constant speed. If $\theta < \epsilon$, $a$ is negative, and so the particle will eventually come to rest. It should be noted that for motion *up* the plane the direction of the frictional force is reversed; that is, it is in the positive $x$ direction. The acceleration (actually deceleration) is then $a = g(\sin \theta + \mu \cos \theta)$.

## 2.7.  The Concepts of Kinetic and Potential Energy

It is generally true that the force that a particle experiences depends on the particle's position with respect to other bodies. This is the case, for example, with electrostatic and gravitational forces. It also applies to forces of elastic tension or compression. If the force is independent of velocity or time, then the differential equation for rectilinear motion is simply

$$F(x) = m\ddot{x}$$

It is usually possible to solve this type of differential equation by one of several methods. One useful and significant method of solution is to write the acceleration in the following way:

$$\ddot{x} = \frac{d\dot{x}}{dt} = \frac{dx}{dt}\frac{d\dot{x}}{dx} = v\frac{dv}{dx}$$

so the differential equation of motion may be written

$$F(x) = mv\frac{dv}{dx} = \frac{m}{2}\frac{d(v^2)}{dx} = \frac{dT}{dx} \tag{2.12}$$

The quantity $T = \frac{1}{2}mv^2$ is called the *kinetic energy* of the particle. We can now express Equation (2.12) in integral form

$$\int F(x)\,dx = \int dT = \tfrac{1}{2}m\dot{x}^2 + \text{constant}$$

Now the integral $\int F(x)\,dx$ is the *work* done on the particle by the impressed force $F(x)$. Let us *define* a function $V(x)$ such that

$$-\frac{dV}{dx} = F(x) \tag{2.13}$$

The function $V(x)$ is called the *potential energy*; it is defined only to within an additive (arbitrary) constant. In terms of $V(x)$, the work integral is

$$\int F(x)\,dx = -\int \frac{dV}{dx}\,dx = -V(x) + \text{constant}$$

Consequently we may write

$$T + V = \tfrac{1}{2}mv^2 + V(x) = \text{constant} = E \tag{2.14}$$

We call $E$ the total energy. In words: For one-dimensional motion, if the impressed force is a function of position only, then the sum of the kinetic and potential energies remains constant throughout the motion. The force in this case is said to be *conservative*.[4] Nonconservative forces, that is, those for which no potential function exists, are usually of a dissipational nature, such as friction.

The motion of the particle can be obtained by solving the energy equation [Equation (2.14)] for $v$

$$v = \frac{dx}{dt} = \pm \sqrt{\frac{2}{m}[E - V(x)]} \tag{2.15}$$

which can be written in integral form

$$\int \frac{\pm dx}{\sqrt{\frac{2}{m}[E - V(x)]}} = t \tag{2.16}$$

thus giving $t$ as a function of $x$.

In view of Equation (3.21) we see that the expression for speed is real only for those values of $x$ such that $V(x)$ is less than or equal to the total

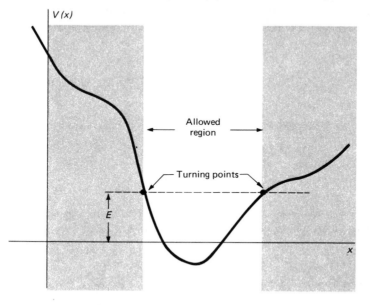

FIGURE 2.2 Graph of a potential energy function $V(x)$ showing allowed region of motion and the turning points for a given value of the total energy $E$.

[4] A more complete discussion of conservative forces will be found in the next chapter.

energy $E$. Physically, this means that the particle is confined to the region, or regions, for which the condition $V(x) \leq E$ is satisfied. Furthermore, the speed goes to zero when $V(x) = E$. This means that the particle must come to rest and reverse its motion at those points for which the equality holds. These points are called the *turning points* of the motion. The above facts are illustrated in Figure 2.2.

## EXAMPLE

The motion of a freely falling body discussed above under the case of a constant force is a special case of conservative motion. If we choose the $x$ direction to be positive upward, then the gravitational force is equal to $-mg$, and the potential energy function is therefore given by $V = mgx + C$. Here $C$ is an arbitrary constant whose value depends merely on the choice of the reference level for $V$. For $C = 0$, the total energy is just

$$E = \tfrac{1}{2}m\dot{x}^2 + mgx$$

Suppose, for example, that a body is projected upward with initial speed $v_0$. Choosing $x = 0$ as the initial point of projection, we have

$$E = \tfrac{1}{2}mv_0^2 = \tfrac{1}{2}m\dot{x}^2 + mgx$$

The turning point is the maximum height attained by the body. It can be found by setting $\dot{x} = 0$. Thus

$$\tfrac{1}{2}mv_0^2 = mgx_{max}$$

or

$$h = x_{max} = \frac{v_0^2}{2g}$$

The motion, as expressed by integrating the energy equation is given by

$$\int_0^x (v_0^2 - 2gx)^{-1/2}\, dx = t$$

$$\frac{v_0}{g} - \frac{1}{g}(v_0^2 - 2gx)^{1/2} = t$$

The student should verify that this reduces to the same relation between $x$ and $t$ as that given by Equation (2.10) when $a$ is set equal to $-g$.

## 2.8. The Force as a Function of Time. The Concept of Impulse

If the force acting on a particle is known explicitly as a function of time, then the equation of motion is

$$F(t) = m\frac{dv}{dt}$$

This can be integrated directly to give the linear momentum (and hence velocity) as a function of the time

$$\int F(t) \, dt = mv(t) + C \tag{2.17}$$

in which $C$ is a constant of integration. The integral $\int F(t) \, dt$ is called the *impulse*.[5] It is equal to the momentum imparted to the particle by the force $F(t)$.

The position of the particle as a function of time can be found by a second integration as follows

$$x = \int v(t) \, dt = \int \left[ \int \frac{F(t')}{m} \, dt' \right] dt \tag{2.18}$$

It should be noted that only in the case of the force being given as a function of $t$, is the solution of the equation of motion expressible as a simple double integral. In all other cases, the various methods of solving second order differential equations must be used to find the position $x$ as a function of $t$.

## EXAMPLE

A block is initially at rest on a smooth horizontal surface. At time $t = 0$ a constantly increasing horizontal force is applied: $F = ct$. Find the velocity and the displacement as functions of time.

We have, for the differential equation of motion,

$$ct = m \frac{dv}{dt}$$

Then

$$v = \frac{1}{m} \int_0^t ct \, dt = \frac{ct^2}{2m}$$

and

$$x = \int_0^t \frac{ct^2}{2m} \, dt = \frac{ct^3}{6m}$$

where the initial position of the block is at the origin ($x = 0$).

### 2.9. Velocity-Dependent Force

It often happens that the force acting on a particle is a function of the particle's velocity. This is true, for example, in the case of viscous resistance exerted on a body moving through a fluid. In the case of fluid resistance, it

[5] The use of the impulse concept will be taken up later in Chapter 6.

is found that, for low velocities, the resistance is approximately proportional to the velocity, whereas, for higher velocities, the resistance is more nearly proportional to the square of $v$. If there are no other forces acting, the differential equation of motion can be expressed as

$$F(v) = m\frac{dv}{dt}$$

A single integration yields $t$ as a function of $v$

$$t = \int \frac{m\,dv}{F(v)} = t(v) \tag{2.19}$$

We can omit the constant of integration, since its value depends only on the choice of the time origin. Assuming that we can solve the above equation for $v$, namely,

$$v = v(t)$$

then a second integration gives the position $x$ as a function of $t$

$$x = \int v(t)\,dt = x(t) \tag{2.20}$$

### EXAMPLE

Suppose a block is projected with initial velocity $v_0$ on a smooth horizontal plane, but that there is air resistance proportional to $v$; that is, $F(v) = -cv$, where $c$ is a constant of proportionality. (The $x$ axis is along the direction of motion.) The differential equation of motion is

$$-cv = m\frac{dv}{dt}$$

which gives, upon integrating,

$$t = \int_{v_0}^{v} -\frac{m\,dv}{cv} = -\frac{m}{c}\ln\left(\frac{v}{v_0}\right)$$

We can easily solve for $v$ as a function of $t$ by multiplying by $-c/m$ and taking the exponent of both sides. The result is

$$v = v_0 e^{-ct/m}$$

Thus the velocity decreases exponentially with time. A second integration gives

$$x = \int_0^t v_0 e^{-ct/m}\,dt$$

$$= \frac{mv_0}{c}(1 - e^{-ct/m})$$

We see, from the above equation, that the block never goes beyond the limiting distance $mv_0/c$.

### 2.10.   Vertical Motion in a Resisting Medium. Terminal Velocity

An object falling vertically through the air or through any fluid is subject to viscous resistance. If the resistance is proportional to the first power of $v$ (the linear case), we can express this force as $-cv$ regardless of the sign of $v$, because the resistance is always opposite to the direction of motion. The constant of proportionality $c$ depends on the size and shape of the object and the viscosity of the fluid. Let us take the $x$ axis to be positive upward. The differential equation of motion is then

$$-mg - cv = m\frac{dv}{dt}$$

If $g$ is a constant, then we have a velocity-dependent force, and we can write

$$t = \int \frac{m\,dv}{F(v)} = \int_{v_0}^{v} \frac{m\,dv}{-mg - cv}$$

$$= -\frac{m}{c}\ln\frac{mg + cv}{mg + cv_0}$$

We can readily solve for $v$

$$v = -\frac{mg}{c} + \left(\frac{mg}{c} + v_0\right)e^{-ct/m} \tag{2.21}$$

The exponential term drops to a negligible value after a sufficient time $(t \gg m/c)$, and the velocity approaches the limiting value $-mg/c$. The limiting velocity of a falling body is called the *terminal velocity*; it is that velocity at which the force of resistance is just equal and opposite to the weight of the body so that the total force is zero. The magnitude of the terminal velocity is called the *terminal speed*. The terminal speed of a falling raindrop, for instance, is roughly 10 to 20 ft per sec, depending on the size.

Equation (3.33) expresses $v$ as a function of $t$, so a second integration will give $x$ as a function of $t$:

$$x - x_0 = \int_0^t v(t)\,dt = -\frac{mg}{c}t + \left(\frac{m^2g}{c^2} + \frac{mv_0}{c}\right)(1 - e^{-ct/m}) \tag{2.22}$$

Let us designate the terminal speed $mg/c$ by $v_t$, and let us write $\tau$ (which we may call the *characteristic time*) for $m/c$. Equation (2.21) may then be written in the more significant form

$$v = -v_t + (v_t + v_0)e^{-t/\tau} \tag{2.23}$$

Thus, an object dropped from rest ($v_0 = 0$) will reach a speed of $1 - e^{-1}$ times the terminal speed in a time $\tau$, $(1 - e^{-2})v_t$ in a time $2\tau$, and so on. After an interval of $10\tau$ the speed is practically equal to the terminal value, namely $0.99995\ v_t$.

If the viscous resistance is proportional to $v^2$ (the quadratic case), the differential equation of motion is, remembering that we are taking the positive direction upward,

$$-mg \pm cv^2 = m\frac{dv}{dt}$$

The minus sign for the resistance term refers to upward motion ($v$ positive), and the plus sign refers to downward motion ($v$ negative). The double sign is necessary for any resistive force that involves an even power of $v$. As in the previous case, the differential equation of motion can be integrated to give $t$ as a function of $v$:

$$t = \int \frac{m\,dv}{-mg - cv^2} = -\tau \tan^{-1}\frac{v}{v_t} + t_0 \qquad (rising)$$

$$t = \int \frac{m\,dv}{-mg + cv^2} = -\tau \tanh^{-1}\frac{v}{v_t} + t_0' \qquad (falling)$$

where

$$\sqrt{\frac{m}{cg}} = \tau \ (the\ characteristic\ time)$$

and

$$\sqrt{\frac{mg}{c}} = v_t \ (the\ terminal\ speed)$$

Solving for $v$,

$$v = v_t \tan\frac{t_0 - t}{\tau} \qquad (rising) \tag{2.24}$$

$$v = -v_t \tanh\frac{t - t_0'}{\tau} \qquad (falling) \tag{2.25}$$

If the body is released from rest at time $t = 0$, then $t_0' = 0$. We have then, from the definition of the hyperbolic tangent,

$$v = -v_t \tanh\frac{t}{\tau} = -v_t\left(\frac{e^{t/\tau} - e^{-t/\tau}}{e^{t/\tau} + e^{-t/\tau}}\right)$$

Again we see that the terminal speed is practically attained after the lapse of a few characteristic times, for example, for $t = 5\tau$, the speed is $0.99991\ v_t$. Graphs of speed versus time of fall for the linear and quadratic laws of resistance are shown in Figure 2.3. It is interesting to note that, in both the linear and the quadratic cases, the characteristic time $\tau$ is equal to $v_t/g$. For instance, if the terminal speed of a parachute is 4 ft per sec, the characteristic time is 4 ft per sec/32 ft per sec$^2$ = $\frac{1}{8}$ sec.

Equations (2.24) and (2.25) can be integrated to give explicit expressions for $x$ as a function of $t$.

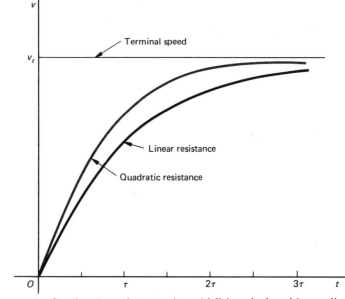

FIGURE 2.3   Graphs of speed versus time of fall for a body subject to linear and quadratic air resistance.

### 2.11.   Variation of Gravity with Height

In the previous section we considered $g$ to be constant. Actually, the gravitational attraction of the earth on a body above the surface falls off as the inverse square of the distance (Newton's law of gravity).[6]   Thus the gravitational force on a body of mass $m$ is

$$F = -\frac{GMm}{r^2}$$

where $G$ is the gravitational constant, $M$ the mass of the earth, and $r$ the distance from the center of the earth to the body.   It can be seen by inspection that this type of force is given by an inverse-first-power potential energy function

$$V(r) = -\frac{GMm}{r}$$

where $F = -\partial V/\partial r$.

If we neglect air resistance, the differential equation for vertical motion is

$$m\ddot{r} = -\frac{GMm}{r^2} \qquad (2.26)$$

Writing $\ddot{r} = \dot{r}\,d\dot{r}/dr$, we can integrate with respect to $r$ to get

$$\frac{1}{2}m\dot{r}^2 - \frac{GMm}{r} = E \qquad (2.27)$$

[6] We shall study Newton's law of gravity in more detail in Chapter 5.

in which $E$ is the constant of integration. This is in fact just the energy equation: the sum of the kinetic energy and the potential energy remains constant throughout the motion of a falling body.

Let us apply the energy equation to the case of a projectile shot upward from the surface of the earth with initial speed $v_0$. The constant $E$ is then given by the initial condition

$$\frac{1}{2} m v_0^2 - \frac{GMm}{r_e} = E \tag{2.28}$$

where $r_e$ is the radius of the earth. The speed at any height $x$ is then found by combining Equations (2.27) and (2.28). The result is

$$v^2 = v_0^2 + 2GM \left( \frac{1}{r_e + x} - \frac{1}{r_e} \right) \tag{2.29}$$

where $r = r_e + x$. Now the acceleration of gravity at the earth's surface is, from Equation (2.26),

$$g = \frac{GM}{r_e^2}$$

The formula for the speed can then be written as

$$v^2 = v_0^2 - 2gx \left( 1 + \frac{x}{r_e} \right)^{-1} \tag{2.30}$$

The above equation reduces to the familiar formula for a uniform gravitational field

$$v^2 = v_0^2 - 2gx$$

if $x$ is very small compared to $r_e$ so that the term $x/r_e$ can be neglected in comparison with unity.

The turning point of the motion of the projectile, that is, the maximum height attained is found by setting $v = 0$ and solving for $x$. The result is

$$x_{max} = h = \frac{v_0^2}{2g} \left( 1 - \frac{v_0^2}{2gr_e} \right)^{-1} \tag{2.31}$$

Again we get the usual formula

$$h = \frac{v_0^2}{2g}$$

if the second term can be ignored.

Finally, let us apply the exact formula (2.31) to find the value of $v_0$ that gives an infinite value of $h$. This is called the *escape speed*, and it is clearly found by setting the quantity in parentheses equal to zero. The result is

$$v_e = (2gr_e)^{1/2}$$

This gives

$$v_e \simeq 7 \text{ mi/sec} \simeq 11 \text{ km/sec}$$

for the numerical value of the escape speed from the surface of the earth.

In the earth's atmosphere, the average speed[7] of air molecules ($O_2$ and $N_2$) is about 0.5 km per sec, which is considerably less than the escape speed,

---

[7] According to kinetic theory, the average speed of a gas molecule is equal to $(3kT/m)^{1/2}$ where $k$ = Boltzmann's constant = $1.38 \times 10^{-16}$ erg per degree, $T$ is the absolute temperature, and $m$ is the mass of the molecule.

so the earth retains its atmosphere. The moon, on the other hand, has no atmosphere, because the escape speed at the moon's surface, owing to the moon's small mass, is considerably smaller than that at the earth's surface; any oxygen or nitrogen would eventually disappear. The earth's atmosphere, however, contains no significant amount of hydrogen, even though hydrogen is the most abundant element in the universe as a whole. A hydrogen atmosphere would have escaped from the earth long ago, because the molecular speed of hydrogen is large enough (owing to the small mass of the hydrogen molecule) so that a significant number of hydrogen molecules would have speeds exceeding the escape speed at any instant.

### 2.12.  Linear Restoring Force.  Harmonic Motion

One of the most important cases of rectilinear motion, from a practical as well as from a theoretical standpoint, is that produced by a *linear restoring force*. This is a force whose magnitude is proportional to the displacement of a particle from some equilibrium position and whose direction is always opposite to that of the displacement. Such a force is exerted by an elastic cord or by a spring obeying Hooke's law

$$F = -k(X - a) = -kx \tag{2.32}$$

where $X$ is the total length, and $a$ is the unstretched (zero load) length of the spring. The variable $x = X - a$ is the displacement of the spring from its equilibrium length. The proportionality constant $k$ is called the *stiffness*. Let a particle of mass $m$ be attached to the spring, as shown in Figure 2.4(a); the force acting on the particle is that given by Equation (2.32). Let the

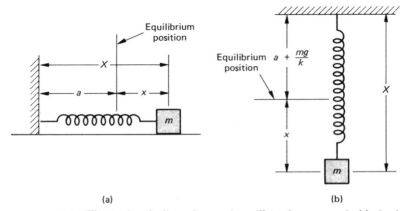

(a)                                                                (b)

FIGURE 2.4  Illustrating the linear harmonic oscillator by means of a block of mass $m$ and a spring.  (a) Horizontal motion; (b) vertical motion.

same spring be held vertically, supporting the same particle, as shown in Figure 2.4(b). The total force now acting on the particle is

$$F = -k(X - a) + mg \qquad (2.33)$$

where the positive direction is downward. Now, in the latter case, let us measure $x$ relative to the new equilibrium position; that is, let $x = X - a - mg/k$. This gives again $F = -kx$, and so the differential equation of motion in either case is

$$m\ddot{x} + kx = 0 \qquad (2.34)$$

The above differential equation of motion is met in a wide variety of physical problems. In the particular example that we are using here, the constants $m$ and $k$ refer to the mass of a body and to the stiffness of a spring, respectively, and the displacement $x$ is a distance. The same equation is encountered, as we shall see later, in the case of a pendulum, where the displacement is an angle, and where the constants involve the acceleration of gravity and the length of the pendulum. Again, in certain types of electrical circuits, this equation is found to apply, where the constants represent the circuit parameters, and the quantity $x$ represents electric current or voltage.

Equation (2.34) can be solved in a number of ways. It is one example of an important class of differential equations known as *linear differential equations with constant coefficients.*[8] Many, if not most, of the differential equations of physics are second-order linear differential equations. To solve Equation (2.34) we shall employ the trial method in which the function $Ae^{qt}$ is the trial solution where $q$ is a constant to be determined. If $x = Ae^{qt}$ is, in fact, a solution, then for all values of $t$ we must have

$$m\frac{d^2}{dt^2}(Ae^{qt}) + k(Ae^{qt}) = 0$$

which reduces, upon canceling the common factors, to the equation[9]

$$mq^2 + k = 0$$

that is

$$q = \pm i\sqrt{\frac{k}{m}} = \pm i\omega_0$$

---

[8] The general $n$th-order equation of this type is

$$c_n\frac{d^n x}{dt^n} + \cdots + c_2\frac{d^2 x}{dt^2} + c_1\frac{dx}{dt} + c_0 = b(t)$$

The equation is called *homogeneous* if $b = 0$.

[9] This equation is called the *auxiliary equation.*

where $i = \sqrt{-1}$, and $\omega_0 = \sqrt{k/m}$.   Now, for linear differential equations, solutions are additive.   (That is, if $f_1$ and $f_2$ are solutions. then the sum $f_1 + f_2$ is also a solution.)   The general solution of Equation (2.34) is then

$$x = A_+ e^{i\omega_0 t} + A_- e^{-i\omega_0 t} \tag{2.35}$$

Since $e^{iu} = \cos u + i \sin u$, alternate forms of the solution are

$$x = a \sin \omega_0 t + b \cos \omega_0 t \tag{2.36}$$

or

$$x = A \cos (\omega_0 t + \theta_0) \tag{2.37}$$

The constants of integration in the above solutions are determined from the initial conditions.   That all three expressions are solutions of Equation (2.34) may be verified by direct substitution.   The motion is a sinusoidal oscillation of the displacement $x$.   For this reason Equation (2.34) is often referred to as the differential equation of the *harmonic oscillator* or the *linear oscillator*.

The coefficient $\omega_0$ is called the *angular frequency*.   The maximum value of $x$ is called the *amplitude* of the oscillation; it is the constant $A$ in Equation (2.37), or $(a^2 + b^2)^{1/2}$ in Equation (2.36).   The period $T_0$ of the oscillation is the time required for one complete cycle, as shown in Figure 2.5; that is,

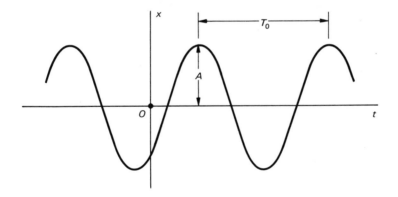

FIGURE 2.5  Graph of displacement versus time for the harmonic oscillator.

the period is the time for which the product $\omega t$ increases by just $2\pi$, thus

$$T_0 = \frac{2\pi}{\omega_0} = 2\pi \sqrt{\frac{m}{k}} \tag{2.38}$$

The *linear frequency* of oscillation $f_0$ is defined as the number of cycles in unit time, therefore

$$\omega_0 = 2\pi f_0$$

and

$$f_0 = \frac{1}{T_0} = \frac{1}{2\pi}\sqrt{\frac{k}{m}} \tag{2.39}$$

It is common usage to employ the word "frequency" for either the angular or the linear frequency; which one is meant is usually clear from context.

## EXAMPLE

A light spring is found to stretch an amount $b$ when it supports a block of mass $m$. If the block is pulled downward a distance $l$ from its equilibrium position and released at time $t = 0$, find the resulting motion as a function of $t$. First, to find the spring stiffness, we note that in the static equilibrium condition

$$F = -kb = -mg$$

so that

$$k = \frac{mg}{b}$$

Hence the angular frequency of oscillation is

$$\omega_0 = \sqrt{\frac{k}{m}} = \sqrt{\frac{g}{b}}$$

In order to find the constants for the equation of motion

$$x = A \cos(\omega_0 t + \theta_0)$$

we have

$$x = l \quad \text{and} \quad \dot{x} = 0$$

at time $t = 0$. But

$$\dot{x} = -A\omega_0 \sin(\omega_0 t + \theta_0)$$

Thus

$$A = l \qquad \theta_0 = 0$$

so

$$x = l \cos\left(\sqrt{\frac{g}{b}}\, t\right)$$

is the required expression.

### 2.13. Energy Considerations in Harmonic Motion

Consider a particle moving under a linear restoring force $F = kx$. Let us calculate the work $W$ done by an external force $F_a$ in moving the particle from the equilibrium position ($x = 0$) to some position $x$. We have $F_a = -F = kx$, and so

$$W = \int F_a \, dx = \int_0^{x} (kx) \, dx = \frac{k}{2} x^2$$

The work $W$ is stored in the spring as potential energy

$$V(x) = W = \frac{k}{2} x^2 \qquad (2.40)$$

Thus $F = -dV/dx = -kx$ as required by the definition of $V$, Equation (2.13). The total energy $E$ is then given by the sum of the kinetic and potential energies as

$$E = \tfrac{1}{2} m \dot{x}^2 + \tfrac{1}{2} k x^2 \qquad (2.41)$$

We can now solve for the velocity as a function of displacement

$$\dot{x} = \left( \frac{2E}{m} - \frac{k}{m} x^2 \right)^{1/2}$$

This can be integrated to give $t$ as a function of $x$ as follows:

$$t = \int \frac{dx}{\sqrt{(2E/m) - (k/m)x^2}} = \sqrt{\frac{m}{k}} \cos^{-1} \left( \frac{x}{A} \right) + C$$

in which

$$A = \sqrt{\frac{2E}{k}}$$

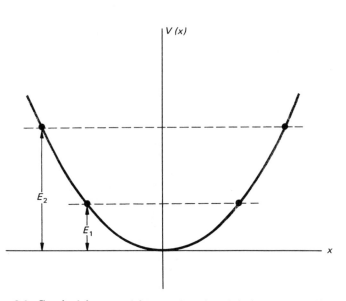

FIGURE 2.6   Graph of the potential energy function of the harmonic oscillator. The turning points defining the amplitude are shown for two values of the total energy.

and $C$ is a constant of integration. Upon solving the integrated equation for $x$ as a function of $t$, we find the very same relationship as that found in the previous section, except that we now obtain an explicit value for the amplitude $A$. We could also have found the amplitude directly from the energy equation (2.41) by noting that $x$ must lie between $\sqrt{2E/k}$ and $-\sqrt{2E/k}$ in order for $\dot{x}$ to be real. This is illustrated in Figure 2.6 which shows the potential energy function and the turning points of the motion for different values of the total energy $E$.

From the energy equation we see that the maximum value of $\dot{x}$, which we shall call $v_{max}$, occurs when $x = 0$, and so we have

$$E = \tfrac{1}{2}mv_{max}^2 = \tfrac{1}{2}kA^2$$

or

$$v_{max} = \sqrt{\frac{k}{m}}\, A = \omega_0 A$$

### 2.14. Damped Harmonic Motion

The above analysis of the harmonic oscillator is somewhat idealized in that we have failed to take into account frictional forces. They are always present in a mechanical system to some extent. Analogously, there is always a certain amount of resistance in an electrical circuit. Let us consider, for example, the motion of an object that is supported by a spring of stiffness $k$. We shall assume that there is a viscous retarding force varying *linearly* with the speed (as in Section 2.8), that is, such as is produced by air resistance. The forces are indicated in Figure 2.7.

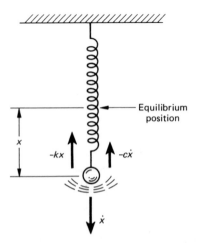

FIGURE 2.7 The damped harmonic oscillator.

If $x$ is the displacement from the equilibrium position, then the restoring force exerted by the spring is $-kx$, and the retarding force is $-c\dot{x}$ where $c$ is a constant of proportionality.   The differential equation of motion $F = m\ddot{x}$ is therefore $-kx - c\dot{x} = m\ddot{x}$ or, by rearranging terms,

$$m\ddot{x} + c\dot{x} + kx = 0 \tag{2.42}$$

Again, as before, we shall use as a trial solution the exponential function $Ae^{qt}$.   This is a solution if

$$m\frac{d^2}{dt^2}(Ae^{qt}) + c\frac{d}{dt}(Ae^{qt}) + k(Ae^{qt}) = 0$$

for all $t$.   This will be the case if $q$ satifies the auxiliary equation

$$mq^2 + cq + k = 0$$

The roots are given by the well-known quadratic formula

$$q = \frac{-c \pm (c^2 - 4mk)^{1/2}}{2m} \tag{2.43}$$

There are three physically distinct cases:

I.   $c^2 > 4mk$     *overdamping*
II.   $c^2 = 4mk$     *critical damping*
III. $c^2 < 4mk$     *underdamping*

I. For the first case, let us call $-\gamma_1$ and $-\gamma_2$ the two real values of $q$ given by Equation (2.43).   The general solution may then be written

$$x = A_1 e^{-\gamma_1 t} + A_2 e^{-\gamma_2 t} \tag{2.44}$$

We see that the motion is nonoscillatory, the displacement $x$ decaying to zero in an exponential manner as time goes on, as shown in Figure 2.8.

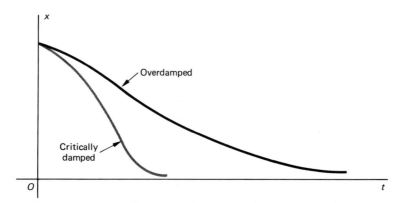

**FIGURE 2.8**   Graphs of displacement versus time for the overdamped and the critically damped cases of the harmonic oscillator.

II. In the case of critical damping the two roots are equal, so Equation (2.44) does not represent a general solution since there is really only one function $e^{-\gamma t}$ and only one constant $A_1 + A_2$, where $\gamma = c/2m$. To find the general solution in this case we can go back to the original differential equation of motion (2.42). For equal roots this equation can be factored as

$$\left(\frac{d}{dt} + \gamma\right)\left(\frac{d}{dt} + \gamma\right) x = 0$$

We now make the substitution $u = \gamma x + dx/dt$ which then gives

$$\left(\frac{d}{dt} + \gamma\right) u = 0$$

This is easily integrated to give $u = A_1 e^{-\gamma t}$. Hence, from the definition of $u$, $\gamma x + dx/dt = A_1 e^{-\gamma t}$ which can also be written

$$A_1 = \left(\gamma x + \frac{dx}{dt}\right) e^{\gamma t} = \frac{d}{dt}(x e^{\gamma t})^{\cdot}$$

A second integration with respect to $t$ then gives $A_1 t = x e^{\gamma t} - A_2$, or finally

$$x = e^{-\gamma t}(A_1 t + A_2) \tag{2.45}$$

This also represents a nonoscillatory motion, the displacement $x$ decaying to zero asymptotically with time, Figure 2.8. Critical damping produces an optimum return to the equilibrium position for applications such as galvanometer suspensions and so on.

III. If the resistance constant $c$ is small enough so that $c^2 < 4mk$, we have the third case: *underdamping*. In this case $q$ is complex. The two roots of the auxiliary equation are conjugate complex numbers, and the motion is given by the general solution

$$x = A_+ e^{(-\gamma + i\omega_1)t} + A_- e^{(-\gamma - i\omega_1)t} \tag{2.46}$$

where $\gamma = c/2m$, and

$$\omega_1 = \sqrt{\frac{k}{m} - \frac{c^2}{4m^2}} = \sqrt{\omega_0^2 - \gamma^2} \tag{2.47}$$

$$x = e^{-\gamma t}(a \sin \omega_1 t + b \cos \omega_1 t) \tag{2.48}$$

where $a = i(A_+ - A_-)$ and $b = A_+ + A_-$. We can also write the solution as

$$x = A e^{-\gamma t} \cos(\omega_1 t + \theta_0) \tag{2.49}$$

where $A = (a^2 + b^2)^{1/2}$ and $\theta_0 = -\tan^{-1}(b/a)$.

The real form of the solution shows that the motion is oscillatory, and that the amplitude $A e^{-\gamma t}$ decays exponentially with time. Further, we note that the angular frequency of oscillation $\omega_1$ is less than that of the undamped oscillator $\omega_0$. The frequency $\omega_1$ is called the *natural frequency*.

In the case of weak damping, that is if $\gamma$ is very small compared to $\omega_0$, we have the approximate relation

$$\omega_1 \simeq \omega_0 - \frac{\gamma^2}{2\omega_0} \tag{2.50}$$

which is obtained by expanding the right side of Equation (2.47) by the binomial theorem and retaining only the first two terms.

A plot of the motion is shown in Figure 2.9.  From Equation (2.49) it follows that the two curves $x = Ae^{-\gamma t}$ and $x = -Ae^{-\gamma t}$ form an envelope of the curve of motion, since the cosine factor takes values between $+1$ and $-1$, including $+1$ and $-1$, at which points the curve of motion touches the envelope.  The points of contact are thus separated by a time interval of one-half period, or $\pi/\omega_1$, but these points are not quite the maxima and minima of the displacement $x$.  It is left to the student to find the values of $t$ at which the displacement does assume its extreme values.

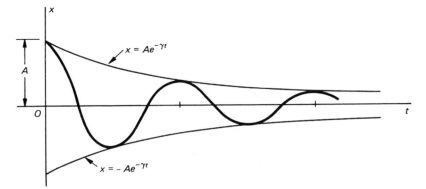

**FIGURE 2.9**   Graph of displacement versus time for the underdamped harmonic oscillator.

*Energy Considerations*

The total energy of the damped harmonic oscillator is, at any instant, equal to the sum of the kinetic energy $\frac{1}{2}m\dot{x}^2$ and the potential energy $\frac{1}{2}kx^2$:

$$E = \tfrac{1}{2}m\dot{x}^2 + \tfrac{1}{2}kx^2$$

We found this to be constant for the undamped oscillator.  Let us differentiate the above equation with respect to $t$ to find the time rate of change of $E$. We have

$$\frac{dE}{dt} = m\ddot{x}\dot{x} + k\dot{x}x = (m\ddot{x} + kx)\dot{x}$$

But, from the differential equation of motion, Equation (2.42),

$$m\ddot{x} + kx = -c\dot{x}$$

Consequently,

$$\frac{dE}{dt} = -c\dot{x}^2 \tag{2.51}$$

This is always negative and represents the rate at which the energy is being dissipated into heat by friction.

## EXAMPLES

1. A particle of mass $m$ is attached to a spring of stiffness $k$. The damping is such that $\gamma = \omega_0/4$. Find the natural frequency. From Equation (2.47), we find

$$\omega_1 = \sqrt{\omega_0{}^2 - \frac{\omega_0{}^2}{16}} = \omega_0 \sqrt{\frac{15}{16}} = \sqrt{\frac{k}{m}} \sqrt{\frac{15}{16}}$$

2. In the above problem, find the ratio of the amplitudes of two successive oscillations. This ratio, from the previous theory, is given by

$$\frac{Ae^{-\gamma T_1}}{A} = e^{-\gamma T_1}$$

where

$$T_1 = \frac{1}{f_1} = \frac{2\pi}{\omega_1}$$

Hence, in our problem

$$T_1 = \frac{2\pi}{\omega_0} \sqrt{\frac{16}{15}} = \frac{2\pi}{4\gamma} \sqrt{\frac{16}{15}}$$

or

$$\gamma T_1 = \frac{\pi}{2} \sqrt{\frac{16}{15}} = 1.56$$

Hence the ratio of two successive swings is $e^{-1.56} = 0.21$.

### 2.15.  Forced Harmonic Motion.  Resonance

In this section we shall study the motion of a damped harmonic oscillator that is driven by an external *harmonic force*, that is, a force that varies sinusoidally with time. Suppose this applied force $F_{ext}$ has an angular frequency $\omega$ and a certain amplitude $F_0$, so that we could write

$$F_{ext} = F_0 \cos (\omega t + \theta)$$

We shall find it convenient, however, to use the exponential form

$$F_{ext} = F_0 e^{i(\omega t + \theta)}$$

rather than the trigonometric, although either can be used.[10]  The total force, then, will be the sum of three forces: the elastic restoring force $-kx$, the viscous damping force $-c\dot{x}$, and the external force $F_{ext}$. The differential equation of motion is therefore

$$-kx - c\dot{x} + F_{ext} = m\ddot{x}$$

or

$$m\ddot{x} + c\dot{x} + kx = F_{ext} = F_0 e^{i(\omega t + \theta)} \tag{2.52}$$

---

[10] The exponential form is equivalent to writing $F_{ext} = F_0 \cos (\omega t + \theta) + iF_0 \times \sin (\omega t + \theta)$. The resulting differential equation is satisfied if the real and the imaginary parts on both sides of the equation are equal.

The solution of the above linear differential equation is given by the sum of two parts, the first being the solution of the homogeneous equation $m\ddot{x} + c\dot{x} + kx = 0$, which we have already solved in the previous section; the second being any particular solution. As we have seen, the solution of the homogeneous equation represents an oscillation which eventually decays to zero—it is called the *transient term*. We are interested in a solution that depends on the nature of the applied force. Since this force is constant in amplitude and varies sinusoidally with time, we can reasonably expect to find a solution for which the displacement $x$ also has a sinusoidal time dependence. Therefore, for the steady-state condition, we shall try a solution of the form

$$x = Ae^{i(\omega t + \theta')}$$

If this "guess" is correct, we must have

$$m\frac{d^2}{dt^2}[Ae^{i(\omega t + \theta')}] + c\frac{d}{dt}[Ae^{(i\omega t + \theta')}] + kAe^{i(\omega t + \theta')} = F_0e^{i(\omega t + \theta)}$$

hold for all values of $t$. This reduces, upon performing the indicated operations and canceling the common factors, to

$$-m\omega^2 A + i\omega cA + kA = F_0e^{i(\theta - \theta')} = F_0[\cos{(\theta - \theta')} + i\sin{(\theta - \theta')}]$$

Equating the real and the imaginary parts, we have

$$A(k - m\omega^2) = F_0 \cos\varphi \qquad (2.53)$$
$$c\omega A = F_0 \sin\varphi \qquad (2.54)$$

where the *phase difference* or phase *angle* $\theta - \theta'$ is denoted by $\varphi$. Upon dividing the second equation by the first and using the identity $\sin\varphi/\cos\varphi = \tan\varphi$, we obtain

$$\tan\varphi = \frac{c\omega}{k - m\omega^2} \qquad (2.55)$$

By squaring both sides of Equations (2.53) and (2.54) and adding and employing the identity $\sin^2\varphi + \cos^2\varphi = 1$, we find

$$A^2(k - m\omega^2)^2 + c^2\omega^2A^2 = F_0^2$$

Solving for $A$, the amplitude of the steady-state oscillation, yields

$$A = \frac{F_0}{\sqrt{(k - m\omega^2)^2 + c^2\omega^2}} \qquad (2.56)$$

In terms of the abbreviations $\omega_0 = \sqrt{k/m}$ and $\gamma = c/2m$, we can write

$$\tan\varphi = \frac{2\gamma\omega}{\omega_0^2 - \omega^2} \qquad (2.57)$$

and

$$A = \frac{F_0/m}{\sqrt{(\omega_0^2 - \omega^2)^2 + 4\gamma^2\omega^2}} \qquad (2.58)$$

The above equation relating the amplitude $A$ to the impressed driving frequency $\omega$ is of fundamental importance. A graph, Figure 2.10, shows that $A$ assumes a maximum value at a certain frequency $\omega_r$, called the *resonant frequency*.

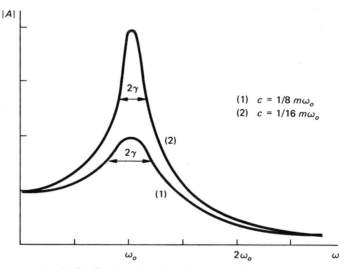

FIGURE 2.10   Graphs of amplitude versus driving frequency.

To find the resonant frequency, we calculate $dA/d\omega$ from Equation (2.58) and set the result equal to zero. Upon solving the resulting equation for $\omega$, we find that the resonant frequency is given by

$$\omega = \omega_r = (\omega_0{}^2 - 2\gamma^2)^{1/2} \tag{2.59}$$

In the case of weak damping, that is, when the damping constant $c$ is very small, $c \ll 2\sqrt{mk}$, or, equivalently, if $\gamma \ll \omega_0$, then we see that the resonant frequency $\omega_r$ is very nearly equal to the frequency of the freely running oscillator with no damping $\omega_0$. If we expand the right side of Equation (2.59) by the binomial theorem and retain only the first two terms, we get

$$\omega_r \simeq \omega_0 - \frac{\gamma^2}{\omega_0} \tag{2.60}$$

Equations (2.59) and (2.60) should be compared to Equations (2.47) and (2.50), which give the frequency of oscillation $\omega_1$ of the freely running oscillator *with* damping. Let $\epsilon$ denote the quantity $\gamma^2/\omega_0$. Then we may write

$$\omega_1 \simeq \omega_0 - \tfrac{1}{2}\epsilon \tag{2.61}$$

for the approximate value of the natural frequency, and

$$\omega_r \simeq \omega_0 - \epsilon \tag{2.62}$$

for the approximate value of the resonant frequency

The steady-state amplitude at the resonant frequency, which we shall call $A_{max}$, is obtained from Equations (2.58) and (2.59). The result is

$$A_{max} = \frac{F_0/m}{2\gamma \sqrt{\omega_0{}^2 - \gamma^2}} = \frac{F_0}{c \sqrt{\omega_0{}^2 - \gamma^2}}$$

In the case of weak damping, we can neglect $\gamma^2$ and write

$$A_{max} \simeq \frac{F_0}{2\gamma m\omega_0} = \frac{F_0}{c\omega_0}$$

Thus the amplitude of the induced oscillation at the resonant condition becomes very large if the damping constant $c$ is very small, and conversely. In mechanical systems it may, or may not, be desirable to have large resonant amplitudes. In the case of electric motors, for example, rubber or spring mounts are used to minimize the transmission of vibration. The stiffness of these mounts is chosen so as to ensure that the resulting resonant frequency is far from the running frequency of the motor.

The sharpness of the resonance peak is frequently of interest. Let us consider the case of weak damping $\gamma \ll \omega_0$. Then in the expression for steady-state amplitude, Equation (2.58), we can make the following substitutions:

$$\omega_0{}^2 - \omega^2 = (\omega_0 + \omega)(\omega_0 - \omega)$$
$$\simeq 2\omega_0(\omega_0 - \omega)$$
$$\gamma\omega \simeq \gamma\omega_0$$

These, together with the expression for $A_{max}$, allow us to write the amplitude equation in the form

$$A = \frac{A_{max}\gamma}{\sqrt{(\omega_0 - \omega)^2 + \gamma^2}} \tag{2.63}$$

The above equation shows that when $|\omega_0 - \omega| = \gamma$, or equivalently, if

$$\omega = \omega_0 \pm \gamma$$

then

$$A^2 = \tfrac{1}{2}A_{max}{}^2$$

This means that $\gamma$ is a measure of the width of the resonance curve. Thus $2\gamma$ is the frequency difference between the points for which the energy is down by a factor of $\frac{1}{2}$ from the energy at resonance, because the energy is proportional to $A^2$. This is illustrated in Figure 2.10.

Another way of designating the sharpness of the resonance peak is in terms of a parameter $Q$ called the *quality factor* of a resonant system. It is defined as

$$Q = \frac{\omega_r}{2\gamma} \tag{2.64}$$

or, for weak damping

$$Q \simeq \frac{\omega_0}{2\gamma} \tag{2.65}$$

Thus the width $\Delta\omega$ at the half-energy points is approximately

$$\Delta\omega = 2\gamma \simeq \frac{\omega_0}{Q}$$

or, since $\omega = 2\pi f$,

$$\frac{\Delta\omega}{\omega_0} = \frac{\Delta f}{f_0} \simeq \frac{1}{Q} \tag{2.66}$$

giving the fractional width of the resonance peak.

Electrically driven quartz crystal oscillators are used to control the frequency of radio broadcasting stations. The $Q$ of the quartz crystals in such applications is of the order of $10^4$. Such high values of $Q$ ensures that the frequency of oscillation remains accurately at the resonance frequency.

The phase difference $\varphi$ between the applied driving force and the response is given by Equation (2.57). This equation is plotted in Figure 2.11 showing

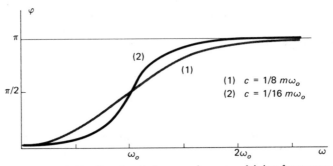

FIGURE 2.11    Graphs of phase angle versus driving frequency.

$\varphi$ as a function of $\omega$. We see that the phase difference is small for small $\omega$, so the response is in phase with the driving force. At the resonance frequency, $\varphi$ has increased to $\pi/2$ and thus the response is 90° out of phase with the driving force at resonance. Finally, for very large values of $\omega$, the value of $\varphi$ approaches $\pi$, hence the motion of the system is just 180° out of phase with the driving force.

*Electrical–Mechanical Analogs*

When an electric current flows in a circuit comprised of inductive, capacitative, and resistive elements, there is a precise analogy with a moving mechanical system of masses and springs with frictional forces of the type studied previously. Thus if a current $i = dq/dt$ ($q$ being the charge) flows through an inductance $L$, the potential difference across the inductance is $L\ddot{q}$ and the stored energy is $\frac{1}{2}L\dot{q}^2$. Hence inductance and charge are analogous with mass and displacement, respectively, and potential difference is analogous with force.   Similarly, if a capacitance $C$ carries a charge $q$, the potential difference is $C^{-1}q$ and the stored energy is $\frac{1}{2}C^{-1}q^2$.   Consequently we see that the reciprocal of $C$ is analogous with the stiffness constant of a spring.   Finally, for an electric current $i$ flowing through a resistance $R$, the potential difference is $iR = \dot{q}R$, and the rate of energy dissipation is $i^2R = \dot{q}^2R$ in analogy with the quantity $c\dot{x}^2$ for a mechanical system, Equation (2.51). Table 2.1 summarizes the situation.

TABLE 2.1

| | *Mechanical* | | *Electrical* |
|---|---|---|---|
| $x$ | Displacement | $q$ | Charge |
| $\dot{x}$ | Velocity | $\dot{q} = i$ | Current |
| $m$ | Mass | $L$ | Inductance |
| $k$ | Stiffness | $C^{-1}$ | Reciprocal of capacitance |
| $c$ | Damping resistance | $R$ | Resistance |
| $F$ | Force | $V$ | Potential difference |

## EXAMPLES

1. Determine the resonance frequency and the quality factor for the damped oscillator of Example 1, p. 66.   We have

$$\omega_r = (\omega_0^2 - 2\gamma^2)^{1/2}$$

$$= \left(\omega_0^2 - \frac{2\omega_0^2}{16}\right)^{1/2}$$

$$= \omega_0 \sqrt{\frac{7}{8}} = \sqrt{\frac{k}{m}} \sqrt{\frac{7}{8}}$$

for the resonance frequency in angular measure.   The quality factor is given by

$$Q = \frac{\omega_r}{2\gamma} = \frac{\omega_0(\frac{7}{8})^{1/2}}{2(\omega_0/4)} = 2\sqrt{\frac{7}{8}} = 1.87$$

2. If the applied frequency is $\omega_0/2$ for the above oscillator, find the phase angle $\varphi$.   From Equation (2.57), we have

$$\tan\varphi = \frac{2(\omega_0/4)(\omega_0/2)}{\omega_0^2 - (\omega_0/2)^2} = \frac{\frac{1}{4}}{\frac{3}{4}} = \frac{1}{3}$$

Hence

$$\varphi = \tan^{-1}\left(\tfrac{1}{3}\right) = 18.5°$$

### 2.16.  Motion Under a Nonsinusoidal Periodic Driving Force

In order to determine the motion of a harmonic oscillator subject to a periodic, but nonsinusoidal, driving force, it is necessary to employ a more involved method than that of the previous section.  In this more general case it is convenient to use the *principle of superposition*.  This principle states that if the applied force $F(t)$ that acts on a harmonic oscillator can be expanded into a sum

$$F(t) = \sum_n F_n(t)$$

such that the differential equations

$$m\ddot{x}_n + c\dot{x}_n + kx_n = F_n(t)$$

are individually satisfied by the functions

$$x_n = x_n(t)$$

then the differential equation

$$m\ddot{x} + c\dot{x} + kx = F(t) = \sum_n F_n(t)$$

is satisfied by the function

$$x = \sum_n x_n(t)$$

That the above theorem is valid follows immediately from the linearity of the differential equation of motion.

In particular, when the driving force $F(t)$ is periodic of angular frequency $\omega$, it can be resolved into a Fourier series.[11]   According to the theory of Fourier series, we can express $F(t)$ as a sum of sine and cosine terms, or alternatively, it can be written as a sum of complex exponentials, namely,

$$F(t) = \sum_n F_n e^{in\omega t} \qquad (n = 0, \pm 1, \pm 2, \ldots)$$

The coefficients are given by

$$F_n = \frac{\omega}{2\pi} \int F(t)e^{-in\omega t}\, dt$$

The limits of integration are $t = -\pi/\omega$ to $t = +\pi/\omega$.

As in the previous section, the actual motion is given by the sum of two parts, namely, a transient term, which we shall neglect, and a steady-state solution

$$x(t) = A_0 + A_1 e^{i\omega t} + A_2 e^{i2\omega t} + \cdots \qquad (2.67)$$

---

[11] See any standard textbook on Fourier methods.

The first term $A_0$ is a constant whose value depends on the form of $F(t)$. For a symmetrical driving force it has the value zero. The second term gives the response of the driven oscillator at the fundamental frequency $\omega$. The third term is the response at the second harmonic $2\omega$ of the applied force, and so on.

We can use the theory of the previous section to find the amplitudes $A_n$ in terms of the coefficients $F_n$. Thus, from Equation (2.58) we have

$$A_n = \frac{F_n/m}{\sqrt{(\omega_0{}^2 - n^2\omega^2)^2 + 4\gamma^2 n^2\omega^2}} \tag{2.68}$$

From the above analysis, we see that the final steady-state motion is periodic and that the particular harmonic $n\omega$ that is nearest to the resonant frequency $\omega_r$ has the greatest amplitude. In particular, if the damping constant $\gamma$ is very small and if the resonant frequency happens to coincide with one of the harmonics of the driving force so that, for some value of $n$, we have

$$\omega_r = n\omega$$

then the amplitude $A_n$ at this harmonic will be greatly dominant. Consequently the resulting motion of the oscillator may be very nearly sinusoidal even if a nonsinusoidal driving force is applied.

## DRILL EXERCISES

**2.1** A particle of mass $m$ is initially at rest. A constant force $F_0$ is suddenly applied at time $t = 0$. After a time $t_0$ the force suddenly doubles to the value $2F_0$ and remains constant thereafter. Find the speed of the particle and the total displacement at time $2t_0$.

**2.2** Find the velocity $v$ and the position $x$ as functions of $t$ for a particle of mass $m$ which starts from rest at time $t = 0$ and subject to the following forces:

    (a) $F = F_0$
    (b) $F = F_0 + bt$
    (c) $F = F_0 \cos \omega t$
    (d) $F = kt^2$

**2.3** Find the velocity $v$ as a function of the displacement $x$ for a particle of mass $m$ which starts from rest at $x = 0$ and subject to the following forces:

    (a) $F = F_0 + kx$
    (b) $F = F_0 e^{-kx}$
    (c) $F = F_0 + kv$

**2.4** The force acting on a particle varies with the distance $x$ according to the power law

$$F(x) = -kx^n$$

(a) Find the potential energy function.

(b) If $v = v_0$ at time $t = 0$ and $x = 0$, find $v$ as a function of $x$.

(c) Determine the turning points of the motion.

## PROBLEMS

2.5   A particle of mass $m$ is initially at rest.   A constant force $F_0$ acts on the particle for a time $t_0$.   The force then increases linearly with time such that after an additional interval $t_0$ the force is equal to $2F_0$.   Show that the total distance the particle goes in the total time $2t_0$ is $(13/6)F_0 t_0^2/m$.

2.6   A block is projected up an inclined plane with initial speed $v_0$.   If the inclination of the plane is $\theta$, and the coefficient of sliding friction between the plane and the block is $\mu$, find the total time required for the block to return to the point of projection.   For what value of $\mu$ will the block just come to rest as it returns to the initial point?

2.7   A block slides on a horizontal surface which has been lubricated with heavy oil such that the block suffers a viscous resistance that varies with speed $v$ according to the equation

$$F(v) = -cv^n$$

If the initial speed is $v_0$ at time $t = 0$, find $v$ and the displacement $x$ as functions of the time $t$.   Also find $v$ as a function of $x$.   In particular, show that for $n = \frac{1}{2}$, the block will not travel further than $2mv_0^{3/2}/3c$.

2.8   A particle of mass $m$ is released from rest a distance $b$ from a fixed origin of force that attracts the particle according to the inverse square law

$$F(x) = -kx^{-2}$$

Show that the time required for the particle to reach the origin is

$$\pi \left( \frac{mb^3}{8k} \right)^{1/2}$$

2.9   Find the relationship between the distance of fall and the speed for a falling body released from rest and subject to air resistance that is proportional to (a) the velocity and (b) the square of the velocity.

2.10   A projectile is fired vertically upward with initial speed $v_0$.   Assuming that the air resistance is proportional to the square of the speed, show that the speed that the projectile has when it hits the ground on its return is

$$\frac{v_0 v_t}{(v_0^2 + v_t^2)^{1/2}}$$

in which

$$v_t = \text{terminal speed} = \left( \frac{mg}{c} \right)^{1/2}$$

2.11 The velocity of a particle of mass $m$ varies with the displacement $x$ according to the equation

$$v = \frac{b}{x}$$

Find the force acting on the particle as a function of $x$.

2.12 Given that the force acting on a particle is the product of a function of the distance and a function of the velocity: $F(x,v) = f(x)g(v)$. Show that the differential equation of motion can be solved by integration. If the force is a product of a function of distance and a function of time, can the equation of motion be solved by simple integration? Can it be solved if the force is a product of a function of time and a function of velocity?

2.13 The force acting on a particle of mass $m$ is given by

$$F = kvx$$

in which $k$ is a constant. The particle passes through the origin with speed $v_0$ at time $t = 0$. Find $x$ as a function of $t$.

2.14 A particle executing simple harmonic motion of amplitude $A$ passes through the equilibrium position with speed $v_0$. What is the period of oscillation?

2.15 Two particles of mass $m_1$ and $m_2$, respectively, undergo simple harmonic motion of amplitude $A_1$ and $A_2$. If the total energy of particle 1 is twice that of particle 2, what is the ratio of their periods: $T_1/T_2$?

2.16 A particle undergoing simple harmonic motion has a speed $v_1$ when the displacement is $x_1$ and a speed $v_2$ when the displacement is $x_2$. Find the period and the amplitude of the motion in terms of the quantities given.

2.17 Two springs having stiffness $k_1$ and $k_2$, respectively, are used in a vertical position to support a single object of mass $m$. Show that the angular frequency of oscillation is $[(k_1 + k_2)/m]^{1/2}$ if the springs are tied in parallel, and $[k_1k_2/(k_1 + k_2)m]^{1/2}$ if the springs are tied in series.

2.18 A spring of stiffness $k$ supports a box of mass $M$ in which is placed a block of mass $m$. If the system is pulled downward a distance $d$ from the equilibrium position and then released, find the force of reaction between the block and the bottom of the box as a function of time. For what value of $d$ will the block just begin to leave the bottom of the box at the top of the vertical oscillations? Neglect any air resistance.

2.19 Show that the ratio of two successive maxima in the displacement of a damped harmonic oscillator is constant. [*Note:* The maxima do not occur at the points of contact of the displacement curve with the curve $Ae^{-\gamma t}$.]

2.20 Given that the amplitude of a damped harmonic oscillator drops to $1/e$ of its initial value after $n$ complete cycles. Show that the ratio of period of oscillation to the period of the same oscillator with no damping is given by

$$\frac{T}{T_0} = \left(1 + \frac{1}{4\pi^2 n^2}\right)^{1/2} \simeq 1 + \frac{1}{8\pi^2 n^2}$$

2.21   The terminal speed of a freely falling ball is $v_t$.   When the ball is supported by a light elastic spring the spring stretches by an amount $x_0$. Show that the natural frequency of oscillation of the system is given by

$$\omega_1 = \sqrt{\frac{g}{x_0} - \frac{g^2}{4v_t{}^2}}$$

in which a linear law of air resistance is assumed.

2.22   Show that the energy of the above system drops to $1/e$ of its initial value in a time $v_t/g$.

2.23   Show that the driving frequency $\omega$, for which the amplitude of a driven harmonic oscillator is one-half the amplitude at the resonant frequency, is approximately $\omega_0 \pm \gamma \sqrt{3}$.

2.24   Find the driving frequency for which the speed of the forced harmonic oscillator is greatest. [*Hint:* Maximize the quantity $v_{max} = \omega A(\omega)$.]

2.25   Show that the quality factor $Q$ of a driven harmonic oscillator is equal to the factor by which the response at zero driving frequency must be multiplied to give the response at the resonance frequency.

2.26   Solve the differential equation of motion of the harmonic oscillator subject to a damped harmonic driving force of the form

$$F_{ext} = F_0 e^{-at} \cos(\omega t)$$

2.27   Show that the Fourier series for a periodic "square wave" is

$$f(t) = \frac{4}{\pi}\left[ \sin(\omega t) + \frac{1}{3} \sin(3\omega t) + \frac{1}{5} \sin(5\omega t) + \cdots \right]$$

where

$$\begin{aligned} f(t) &= +1 \quad &\text{for} \quad & 0 < \omega t < \pi, \quad 2\pi < \omega t < 3\pi, \quad \text{and so on} \\ f(t) &= -1 \quad &\text{for} \quad & \pi < \omega t < 2\pi: \quad 3\pi < \omega t < 4\pi, \quad \text{and so on} \end{aligned}$$

2.28   Use the above result to find the steady-state motion of a damped harmonic oscillator that is driven by a periodic square-wave force of amplitude $F_0$.   In particular, find the relative amplitudes of the first three terms $A_1$, $A_3$, and $A_5$ of the response function $x(t)$ in the case that the third harmonic, $3\omega$, of the driving frequency coincides with the resonance frequency of the oscillator.   Let the quality factor $Q = 100$.

# 3. General Motion of a Particle in Three Dimensions

We turn our attention now to the general case of motion of a particle in space.

### 3.1. Linear Momentum

We have already seen that the vectorial form of the equation of motion of a particle is

$$\mathbf{F} = \frac{d\mathbf{p}}{dt}$$

or, equivalently

$$\mathbf{F} = \frac{d}{dt}(m\mathbf{v}) \tag{3.1}$$

This is essentially an abbreviation for three component equations in which the force components may involve the coordinates, their time derivatives, and the time. Unfortunately, no general method exists for finding analytical solutions in all possible cases. However, there are many physically important special types of force functions for which the differential equations of motion can be attacked by relatively simple methods. Some of these will be studied in the sections to follow.

In those cases where $\mathbf{F}$ is known as an explicit function of time, the momentum $\mathbf{p}$ can be found by finding the impulse, that is, by integrating with respect to time, as in the one-dimensional case, namely

$$\int \mathbf{F}(t)\,dt = \mathbf{p}(t) = m\mathbf{v}(t) \tag{3.2}$$

Similarly, a second integration will yield the position

$$\int \mathbf{v}(t)\,dt = \mathbf{r}(t) \tag{3.3}$$

Although the above method is perfectly valid, it is not a typical situation in particle dynamics that the force is known in advance as a function of time. Of course, in the special case of zero force, the momentum and velocity are constant and the preceding equations hold. We shall have occasion to discuss the concept of constant momentum under zero force in a more general form later when we take up the study of systems of particles in Chapter 6.

### 3.2. Angular Momentum

Consider the general equation of motion of a particle $\mathbf{F} = d\mathbf{p}/dt$. Let us multiply both sides by the operator $\mathbf{r} \times$ to obtain

$$\mathbf{r} \times \mathbf{F} = \mathbf{r} \times \frac{d\mathbf{p}}{dt}$$

The left-hand side of the above equation is, by definition, the moment of the force about the origin of the coordinate system. The right-hand side turns out to be the time derivative of the quantity $\mathbf{r} \times \mathbf{p}$. To prove this statement we differentiate

$$\frac{d}{dt}(\mathbf{r} \times \mathbf{p}) = \mathbf{v} \times \mathbf{p} + \mathbf{r} \times \frac{d\mathbf{p}}{dt}$$

But $\mathbf{v} \times \mathbf{p} = \mathbf{v} \times m\mathbf{v} = m\mathbf{v} \times \mathbf{v} = 0$. Thus, we can write

$$\mathbf{r} \times \mathbf{F} = \frac{d}{dt}(\mathbf{r} \times \mathbf{p}) \tag{3.4}$$

The quantity $\mathbf{r} \times \mathbf{p}$ is called the *angular momentum* of the particle about the origin. Our result, stated in words, is that *the time rate of change of the angular momentum of a particle is equal to the moment of force acting on the particle.*

The important concept of angular momentum will be found to be particularly useful in the study of planetary-type motion which we shall take up in Chapter 5, and in the study of systems of particles and rigid bodies, Chapters 6–9.

### 3.3.   The Work Principle

In the general equation of motion let us take the dot product of both sides with the velocity **v**

$$\mathbf{F}\cdot\mathbf{v} = \frac{d\mathbf{p}}{dt}\cdot\mathbf{v} = \frac{d(m\mathbf{v})}{dt}\cdot\mathbf{v}$$

Now from the rule for differentiation of a dot product, we have $d(\mathbf{v}\cdot\mathbf{v})/dt = 2\mathbf{v}\cdot d\mathbf{v}/dt$. Hence, if we assume that the mass $m$ is constant, we see that the above equation is equivalent to

$$\mathbf{F}\cdot\mathbf{v} = \frac{d}{dt}\left(\frac{1}{2}\,m\mathbf{v}\cdot\mathbf{v}\right) = \frac{dT}{dt} \tag{3.5}$$

in which we have introduced the kinetic energy $T = \frac{1}{2}mv^2$. Further, since $\mathbf{v}\,dt = d\mathbf{r}$, we can integrate to obtain

$$\int\mathbf{F}\cdot d\mathbf{r} = \int dT \tag{3.6}$$

Now the left-hand side of the above equation is a *line integral*. It represents the work done on the particle by the force **F** as the particle moves along the path of motion. The right-hand side is just the net change in the particle's kinetic energy. Hence the equation merely states that *the work done on the particle is equal to the increment in the kinetic energy.*

### 3.4.   Conservative Forces and Force Fields

Generally, the value of a line integral, the work in this case, depends on the path of integration, see Figure 3.1   In other words, the work done usually depends on the particular route the particle takes in going from one point to

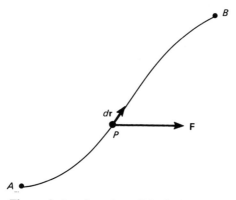

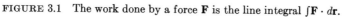

FIGURE 3.1   The work done by a force **F** is the line integral $\int\mathbf{F}\cdot d\mathbf{r}$.

another.  This means that if we were given the problem of calculating the value of the work integral we would normally need to know the path of motion of the particle beforehand.  However, the usual kinds of problems that are of interest in particle dynamics are those in which the path of motion is not known in advance, rather, the path is one of the things to be calculated.  It would appear then that the work principle expressed by Equation (3.6) might not prove very useful for our purposes.  However, it turns out that the work principle is indeed very useful in the study of the motion of a particle under the action of a particular kind of force known as *conservative force*.  Fortunately, many of the physically important forces are of this type.

When the force **F** is a function of the positional coordinates only, it is said to define a static *force field*.  Among the possible kinds of fields, there is an important class for which the work integral $\int \mathbf{F} \cdot d\mathbf{r}$ is independent of the path of integration.  Such force fields are conservative.  Mathematically, a conservative field is one in which the expression $\mathbf{F} \cdot d\mathbf{r}$ is an *exact differential*.  When a particle moves in a conservative field, the work integral and hence the kinetic energy increment can be known in advance.  This knowledge can be of use in predicting the motion of the particle.

### 3.5.   The Potential Energy Function
### In Three-Dimensional Motion

If a particle moves under the action of a conservative force **F**, the statement that the work increment $\mathbf{F} \cdot d\mathbf{r}$ is an exact differential means that it must be expressible as the differential of a scalar function of the position **r**, namely

$$\mathbf{F} \cdot d\mathbf{r} = -\, dV(\mathbf{r}) \tag{3.7}$$

This is analogous to the one-dimensional case where $F\, dx = -\, dV$, Section 2.7.  The function $V$ is the potential energy.  The work principle, Equation (3.6), then is simply expressed as $dT = -\, dV$, or

$$d(T + V) = 0 \tag{3.8}$$

This implies that the quantity $T + V$ remains constant as the particle moves.  We call it the total energy $E$, and write

$$\tfrac{1}{2}mv^2 + V(\mathbf{r}) = E \tag{3.9}$$

In the case of a nonconservative force the work increment is not an exact differential and therefore cannot be equated to a quantity $-\, dV$.  A common example of a nonconservative force is friction.  When nonconservative forces are present we can express the total force as a sum $\mathbf{F} + \mathbf{F}'$ where **F** is conservative and $\mathbf{F}'$ is nonconservative.  The work principle is then given by $dT = \mathbf{F} \cdot d\mathbf{r} + \mathbf{F}' \cdot d\mathbf{r} = -\, dV + \mathbf{F}' \cdot d\mathbf{r}$ or

$$d(T + V) = \mathbf{F}' \cdot d\mathbf{r}$$

We see that the quantity $T + V$ is not constant, but increases or decreases as the particle moves depending on the sign of $\mathbf{F}' \cdot d\mathbf{r}$. In the case of dissipative forces the direction of $\mathbf{F}'$ is opposite to that of $d\mathbf{r}$, hence $\mathbf{F}' \cdot d\mathbf{r}$ is negative and the total energy $T + V$ diminishes as the particle moves.

## 3.6. Gradient and the Del Operator in Mechanics

If rectangular coordinates are employed, the statement

$$\mathbf{F} \cdot d\mathbf{r} = - dV$$

is expressed as

$$F_x \, dx + F_y \, dy + F_z \, dz = - \frac{\partial V}{\partial x} \, dx - \frac{\partial V}{\partial y} \, dy - \frac{\partial V}{\partial z} \, dz$$

This clearly implies that

$$F_x = - \frac{\partial V}{\partial x} \qquad F_y = - \frac{\partial V}{\partial y} \qquad F_z = - \frac{\partial V}{\partial z} \qquad (3.10)$$

Stated in words, if the force field is conservative, then the components of the force are given by the negative partial derivatives of a potential energy function.

We can now express $\mathbf{F}$ vectorially as

$$\mathbf{F} = -\mathbf{i} \frac{\partial V}{\partial x} - \mathbf{j} \frac{\partial V}{\partial y} - \mathbf{k} \frac{\partial V}{\partial z} \qquad (3.11)$$

This equation can be written in a convenient abbreviated form

$$\mathbf{F} = -\nabla V \qquad (3.12)$$

Here we have introduced the vector differentiation operator

$$\nabla = \mathbf{i} \frac{\partial}{\partial x} + \mathbf{j} \frac{\partial}{\partial y} + \mathbf{k} \frac{\partial}{\partial z} \qquad (3.13)$$

It is called the del operator. The expression $\nabla V$ is also called the "gradient of V" and is sometimes written grad $V$. Mathematically, the gradient of a function is a vector that represents the spatial derivative of the function in direction and magnitude. Physically, the negative gradient of the potential energy function gives the direction and magnitude of the force that acts on a particle located in a field created by other particles. The meaning of the negative sign is that the particle is urged to move in the direction of *decreasing* potential energy rather than in the opposite direction. An illustration of the gradient is shown in Figure 3.2. Here the potential function is plotted out

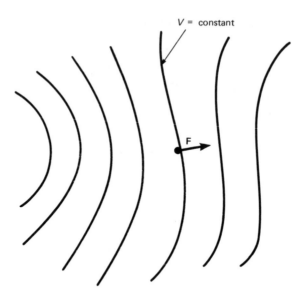

FIGURE 3.2   A force field represented by contour lines of potential energy.

in the form of contour lines representing the curves of constant potential energy.   The force at any point is always normal to the equipotential curve or surface passing through the point in question.

### 3.7.   Conditions for the Existence of a Potential Function

In Chapter 2 we found that one-dimensional motion of a particle is always conservative if the force is a function of position only.   The question naturally arises as to whether or not the corresponding statement is true for the general case of two- and three-dimensional motion.   That is, if the force acting on a particle is a function of the position coordinates only, is there always a function $V$ which satisfies Equations (3.10) above?   The answer to this question is *no;* only if the force components satisfy certain criteria does a potential function exist.

Let us assume that a potential function *does* exist, that is, that Equations (3.10) hold. Then we have

$$\frac{\partial F_x}{\partial y} = -\frac{\partial^2 V}{\partial y\,\partial x} \qquad \frac{\partial F_y}{\partial x} = -\frac{\partial^2 V}{\partial x\,\partial y}$$

This order of differentiation can be reversed, the two expressions are equal. Hence we can write

$$\frac{\partial F_x}{\partial y} = \frac{\partial F_y}{\partial x} \qquad \frac{\partial F_x}{\partial z} = \frac{\partial F_z}{\partial x} \qquad \frac{\partial F_y}{\partial z} = \frac{\partial F_z}{\partial y} \qquad\qquad (3.14)$$

These are the *necessary* conditions, then, on $F_x$, $F_y$, and $F_z$ for a potential function to exist; they express the condition that $\mathbf{F} \cdot d\mathbf{r} = F_x \, dx + F_y \, dy + F_z \, dz$ is an exact differential. It is also possible to show that they are sufficient conditions, that is, if Equations (3.14) hold at all points, then the force components are indeed derivable from a potential function $V(x,y,z)$, and the sum of the kinetic energy and the potential energy is a constant.[1]

The criteria for a force field to be conservative are conveniently expressed in terms of the del operator. In this application we introduce the cross product of the del operator:

$$\nabla \times \mathbf{F} = \mathbf{i} \left( \frac{\partial F_z}{\partial y} - \frac{\partial F_y}{\partial z} \right) + \mathbf{j} \left( \frac{\partial F_x}{\partial z} - \frac{\partial F_z}{\partial x} \right) + \mathbf{k} \left( \frac{\partial F_y}{\partial x} - \frac{\partial F_x}{\partial y} \right) \quad (3.15)$$

The cross product as defined above is called the "curl of $\mathbf{F}$." According to Equations (3.14), we see that the components of the curl each vanish if the force $\mathbf{F}$ is conservative. Thus the condition for a force to be conservative can be written in the compact form

$$\nabla \times \mathbf{F} = 0 \quad (3.16)$$

Mathematically, the above equation represents the condition that the expression $\mathbf{F} \cdot d\mathbf{r}$ is an exact differential, or in other words, that the integral $\int \mathbf{F} \cdot d\mathbf{r}$ is independent of the path of integration. Physically, the vanishing of the curl of $\mathbf{F}$ means that the work done by $\mathbf{F}$ on a moving particle is independent of the path of the particle in going from one given point to another.

There is a third expression involving the del operator, namely the dot product $\nabla \cdot \mathbf{F}$. This is called the "divergence of $\mathbf{F}$." In the case of a force field, the divergence gives a measure of the density of the sources of the field at a given point. The divergence is of particular importance in the theory of electricity and magnetism.

Expressions for the gradient, curl, and divergence in cylindrical and spherical coordinates are given in Appendix IV.

## EXAMPLES

1. Find the force field of the potential function $V = x^2 + xy + xz$. Applying the del operator, we have

$$\mathbf{F} = -\nabla V = -\mathbf{i}(2x + y + z) - \mathbf{j}x - \mathbf{k}x$$

---

[1] See any advanced calculus textbook, for example, A. E. Taylor, *Advanced Calculus*, Ginn, Boston, 1955. An interesting discussion of the conservancy criteria when this field contains singularities has been given by Feng in *Amer. J. Phys.* **37**, 616 (1969).

2. Is the force field $\mathbf{F} = \mathbf{i}xy + \mathbf{j}xz + \mathbf{k}yz$ conservative?   The curl of $\mathbf{F}$ is

$$\nabla \times \mathbf{F} = \begin{vmatrix} \mathbf{i} & \mathbf{j} & \mathbf{k} \\ \partial/\partial x & \partial/\partial y & \partial/\partial z \\ xy & xz & yz \end{vmatrix} = \mathbf{i}(z - x) + \mathbf{j}0 + \mathbf{k}(z - x)$$

The final expression is not zero, hence the field is not conservative.


3. For what values of the constants $a$, $b$, and $c$ is the force $\mathbf{F} =$ $\mathbf{i}(ax + by^2) + \mathbf{j}cxy$ conservative?   Taking the curl, we have

$$\nabla \times \mathbf{F} = \begin{vmatrix} \mathbf{i} & \mathbf{j} & \mathbf{k} \\ \partial/\partial x & \partial/\partial y & \partial/\partial z \\ ax + by^2 & cxy & 0 \end{vmatrix} = \mathbf{k}(c - 2b)y$$

This shows that the force is conservative, provided $c = 2b$.   The value of $a$ is immaterial.


4. Show that the inverse-square law of force in three dimensions $\mathbf{F} =$ $(-k/r^2)\mathbf{e}_r$ is conservative by the use of the curl.   Use spherical coordinates. The curl is given in Appendix IV as

$$\nabla \times \mathbf{F} = \frac{1}{r^2 \sin \theta} \begin{vmatrix} \mathbf{e}_r & \mathbf{e}_\theta r & \mathbf{e}_\phi r \sin \theta \\ \dfrac{\partial}{\partial r} & \dfrac{\partial}{\partial \theta} & \dfrac{\partial}{\partial \phi} \\ F_r & rF_\theta & rF_\phi \sin \theta \end{vmatrix}$$

We have $F_r = -k/r^2$, $F_\theta = 0$, $F_\phi = 0$.   The curl then reduces to

$$\nabla \times \mathbf{F} = \frac{\mathbf{e}_\theta}{r \sin \theta} \frac{\partial}{\partial \phi}\left(\frac{-k}{r^2}\right) - \frac{\mathbf{e}_\phi}{r} \frac{\partial}{\partial \theta}\left(\frac{-k}{r^2}\right) = 0$$

which, of course, vanishes since both partial derivatives are zero.   Thus the force in question is conservative.


### 3.8.   Forces of the Separable Type

It is often the case that a coordinate system can be chosen such that the components of a force field involve the respective coordinates alone, that is

$$\mathbf{F} = \mathbf{i}F_x(x) + \mathbf{j}F_y(y) + \mathbf{k}F_z(z) \tag{3.17}$$

Forces of this type are said to be *separable*. It is readily verified that the curl of such a force is identically zero and hence, that the field is conservative regardless of the particular forms of the force components as long as each is a function of only the one coordinate involved. The integration of the differential equations of motion is then very simple, because each component equation is of the type $m\ddot{x} = F(x)$. In this case the equations can be solved by the methods described under rectilinear motion in the previous chapter.

In the event that the force components involve the time and the time derivatives of the respective coordinates, then it is no longer true that the force is necessarily conservative. Nevertheless, if the force is separable, then the component equations of motion are of the form $m\ddot{x} = F(x,\dot{x},t)$ and may be solved by the methods used in the previous chapter. Some examples of separable forces, both conservative and nonconservative, will be discussed in the sections to follow.

### 3.9.   Motion of a Projectile in a Uniform Gravitational Field

One of the famous classical problems of particle dynamics is the motion of a projectile. We shall study this problem in some detail because it illustrates most of the general principles that have been cited in the foregoing sections.

#### No Air Resistance

First, for simplicity, we consider the case of a projectile moving with no air resistance. In this idealized situation there is only one force acting, namely the force of gravity. Choosing the $z$ axis to be vertical, we have the differential equation of motion

$$m\frac{d^2\mathbf{r}}{dt^2} = -mg\mathbf{k}$$

If we further idealize the problem and assume that the acceleration of gravity $g$ is constant, then the force function is clearly of the separable type and is also conservative since it is a special case of that expressed by Equation (3.17). We shall particularize the problem further by choosing the initial speed to be $v_0$ and the initial position to be at the origin at time $t = 0$. The energy equation (3.9) then reads

$$\tfrac{1}{2}m(\dot{x}^2 + \dot{y}^2 + \dot{z}^2) + mgz = \tfrac{1}{2}mv_0^2$$

or, equivalently

$$v^2 = v_0^2 - 2gz \tag{3.18}$$

thus giving the speed as a function of height.   This is all the information we
can obtain directly from the energy equation.

In order to proceed further, we must go back to the differential equation
of motion.   This can be written

$$\frac{d}{dt}\left(\frac{d\mathbf{r}}{dt}\right) = -g\mathbf{k}$$

which is of the form discussed in Section 1.20.   It can be integrated directly.
A single integration gives the velocity as

$$\frac{d\mathbf{r}}{dt} = -gt\mathbf{k} + \mathbf{v}_0$$

in which the constant of integration $\mathbf{v}_0$ is the initial velocity.   Another integration yields the position vector

$$\mathbf{r} = -\tfrac{1}{2}gt^2\mathbf{k} + \mathbf{v}_0 t \qquad\qquad (3.19)$$

The constant of integration $\mathbf{r}_0$, in this case, is zero since the initial position of
the projectile is taken to be the origin.   In components, the above equation is

$$x = \dot{x}_0 t$$
$$y = \dot{y}_0 t$$
$$z = \dot{z}_0 t - \tfrac{1}{2}gt^2$$

Here $\dot{x}_0$, $\dot{y}_0$, and $\dot{z}_0$ are the components of the initial velocity $\mathbf{v}_0$.   We have thus
solved the problem of determining the position of the projectile as a function
of time.

Concerning the path or trajectory of the projectile, we notice that if
the time $t$ is eliminated from the $x$ and $y$ equations, the result is

$$y = bx$$

in which the constant $b$ is given by

$$b = \frac{\dot{y}_0}{\dot{x}_0}$$

Thus the path lies entirely in a plane.   In particular, if $\dot{y}_0 = 0$, then the path
lies in the $xz$ plane.   Next, if we eliminate $t$ between the $x$ and $z$ equations,
we find the equation of the path to be of the form

$$z = \alpha x - \beta x^2$$

where $\alpha = \dot{z}_0/\dot{x}_0$ and $\beta = g/2\dot{x}_0^2$. Hence the path is a parabola lying in the
plane $y = bx$.   This is shown in Figure 3.3.

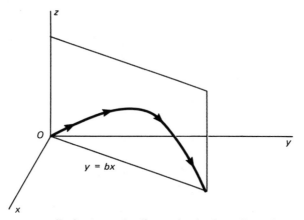

FIGURE 3.3   Path of a projectile moving in three dimensions.

### Linear Air Resistance

We now consider the motion of a projectile for the more realistic situation in which there is a retarding force due to air resistance.   In this case the motion is not conservative.   The total energy continually diminishes as a result of frictional loss.

For simplicity, let us assume that the law of air resistance is linear so that the resisting force varies directly with the velocity $\mathbf{v}$.   It will be convenient to write the constant of proportionality as $m\gamma$ where $m$ is the mass of the projectile.   Thus we have two forces acting on the projectile, namely the air resistance $-m\gamma\mathbf{v}$, and the force of gravity which is equal to $-mg\mathbf{k}$, as before.   The differential equation of motion is then

$$m \frac{d^2\mathbf{r}}{dt^2} = -m\gamma\mathbf{v} - mg\mathbf{k}$$

or, upon cancelling the $m$'s, we have

$$\frac{d^2\mathbf{r}}{dt^2} = -\gamma\mathbf{v} - g\mathbf{k}$$

The integration of the above equation is conveniently accomplished by expressing it in component form as follows:

$$\ddot{x} = -\gamma\dot{x}$$
$$\ddot{y} = -\gamma\dot{y}$$
$$\ddot{z} = -\gamma\dot{z} - g$$

We now see that the equations are separated.   Hence each can be solved individually by the methods of the previous chapter.   Using our results from Section 2.9, we can write down the solutions immediately.   They are

$$\dot{x} = \dot{x}_0 e^{-\gamma t}$$
$$\dot{y} = \dot{y}_0 e^{-\gamma t} \tag{3.20}$$
$$\dot{z} = \dot{z}_0 e^{-\gamma t} - \frac{g}{\gamma}(1 - e^{-\gamma t})$$

for the velocity components, and

$$x = \frac{\dot{x}_0}{\gamma}(1 - e^{-\gamma t})$$
$$y = \frac{\dot{y}_0}{\gamma}(1 - e^{-\gamma t}) \tag{3.21}$$
$$z = \left(\frac{\dot{z}_0}{\gamma} + \frac{g}{\gamma^2}\right)(1 - e^{-\gamma t}) - \frac{g}{\gamma}t$$

for the positional coordinates. Here, as before, the initial velocity components are $\dot{x}_0$, $\dot{y}_0$, and $\dot{z}_0$, and the initial position of the projectile is taken as the origin.

As in the case of zero air resistance, the motion remains entirely in the plane $y = bx$ with $b = \dot{y}_0/\dot{x}_0$. The path in this plane is not a parabola, however, but is a curve that lies below the corresponding parabolic trajectory. This is illustrated in Figure 3.4. Inspection of the $x$ and $y$ equations shows that, for large $t$, the values of $x$ and $y$ approach the limiting values

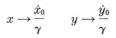

$$x \to \frac{\dot{x}_0}{\gamma} \qquad y \to \frac{\dot{y}_0}{\gamma}$$

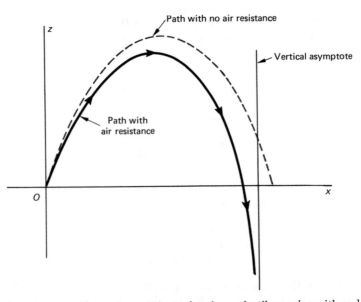

FIGURE 3.4   Comparison of the paths of a projectile moving with and without air resistance.

This means that the complete trajectory has a vertical asymptote as shown in the figure.

The final solution of the motion of a projectile with linear air resistance, expressed by Equations (3.21), can be written vectorially in the following way:

$$\mathbf{r} = \left( \frac{\mathbf{v}_0}{\gamma} + \frac{\mathbf{k}g}{\gamma^2} \right)(1 - e^{-\gamma t}) - \mathbf{k}\frac{gt}{\gamma}$$

That it is a solution of the vector differential equation of motion is easily verified by differentiation.

It is instructive to consider the case in which the air resistance is very small, that is, when the value of the quantity $\gamma t$ in the exponential factors is much smaller than unity. For this purpose we use the exponential series

$$e^u = 1 + u + \frac{u^2}{2!} + \frac{u^3}{3!} + \cdots$$

in which we let $u = -\gamma t$. The result, after cancellation and collection of terms, is expressible in the form

$$\mathbf{r} = \mathbf{v}_0 t - \tfrac{1}{2}gt^2\mathbf{k} - \Delta\mathbf{r}$$

where

$$\Delta\mathbf{r} = \gamma\left[ \mathbf{v}_0\left( \frac{t^2}{2!} - \frac{\gamma t^3}{3!} + \cdots \right) - \mathbf{k}g\left( \frac{t^3}{3!} - \frac{\gamma t^4}{4!} + \cdots \right) \right]$$

The quantity $\Delta\mathbf{r}$ can be regarded as a correction to the zero-resistance path that gives the true path.

In the actual motion of a projectile through the atmosphere, the law of resistance is by no means linear, but is a very complicated function of the velocity. An accurate calculation of the trajectory can be done by means of numerical integration methods aided by the use of high-speed computers.

### 3.10.   The Harmonic Oscillator in Two and Three Dimensions

In this section we consider the motion of a particle that is subject to a linear restoring force which is always directed toward a fixed point, the origin of our coordinate system. Such a force can be represented by the expression

$$\mathbf{F} = -k\mathbf{r}$$

Accordingly, the differential equation of motion is simply expressed as

$$m\frac{d^2\mathbf{r}}{dt^2} = -k\mathbf{r} \tag{3.22}$$

The situation can be represented approximately by a particle attached to a set of elastic springs as shown in Figure 3.5. This is the three-dimensional generalization of the linear oscillator studied earlier in Section 2.12.

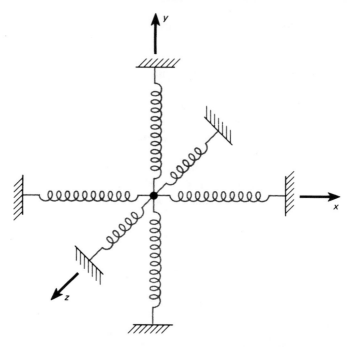

FIGURE 3.5   Model of a three-dimensional harmonic oscillator.

### The Two-Dimensional Oscillator

For the case of motion in a single plane, the above differential equation is equivalent to the two component equations

$$m\ddot{x} = -kx$$

$$m\ddot{y} = -ky$$

These are separated, and we can immediately write down the solutions in the form

$$x = A \cos (\omega t + \alpha) \qquad y = B \cos (\omega t + \beta) \tag{3.23}$$

in which

$$\omega = \left(\frac{k}{m}\right)^{1/2}$$

The constants of integration $A$, $B$, $\alpha$, and $\beta$ are determined from the initial conditions in any given case.

In order to find the equation of the path, we eliminate the time $t$ between the two equations. To do this, let us write the second equation in the form

$$y = B \cos (\omega t + \alpha + \Delta)$$

where

$$\Delta = \beta - \alpha$$

Then

$$y = B[\cos(\omega t + \alpha)\cos\Delta - \sin(\omega t + \alpha)\sin\Delta]$$

From the first of Equations (4.23), we then have

$$\frac{y}{B} = \frac{x}{A}\cos\Delta - \left(1 - \frac{x^2}{A^2}\right)^{1/2}\sin\Delta$$

or, upon squaring and transposing terms, we obtain

$$\frac{x^2}{A^2} - xy\frac{2\cos\Delta}{AB} + \frac{y^2}{B^2} = \sin^2\Delta \tag{3.24}$$

which is a quadratic equation in $x$ and $y$. Now the general quadratic

$$ax^2 + bxy + cy^2 + dx + ey = f$$

represents an ellipse, a parabola, or a hyperbola, depending on whether the discriminant

$$b^2 - 4ac$$

is negative, zero, or positive, respectively. In our case the discriminant is equal to $-(2\sin\Delta/AB)^2$ which is negative, so the path is an ellipse as shown in Figure 3.6.

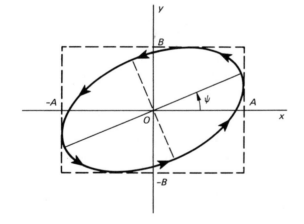

FIGURE 3.6   Elliptical path of motion of a two-dimensional harmonic oscillator.

In particular, if the phase difference $\Delta$ is equal to $\pi/2$, then the equation of the path reduces to the equation

$$\frac{x^2}{A^2} + \frac{y^2}{B^2} = 1$$

which is the equation of an ellipse whose axes coincide with the coordinate axes. On the other hand, if the phase difference is 0 or $\pi$, then the equation of the path reduces to that of a straight line, namely

$$y = \pm \frac{B}{A} x$$

The positive sign is taken if $\Delta = 0$, and the negative sign if $\Delta = \pi$. In the general case, it is possible to show that the axis of the elliptical path is inclined to the $x$ axis by the angle $\psi$ where

$$\tan 2\psi = \frac{2AB \cos \Delta}{A^2 - B^2} \tag{3.25}$$

The derivation is left as an exercise.

### The Three-Dimensional Harmonic Oscillator

In the case of three-dimensional motion the differential equation of motion is equivalent to the three equations

$$m\ddot{x} = -kx \qquad m\ddot{y} = -ky \qquad m\ddot{z} = -kz$$

which are separated. Hence the solutions may be written in the form of Equations (3.23) or, alternately, we may write

$$
\begin{aligned}
x &= A_1 \sin \omega t + B_1 \cos \omega t \\
y &= A_2 \sin \omega t + B_2 \cos \omega t \\
z &= A_3 \sin \omega t + B_3 \cos \omega t
\end{aligned}
\tag{3.26}
$$

The six constants of integration are determined from the initial position and velocity of the particle. Now Equations (3.26) can be expressed vectorially as

$$\mathbf{r} = \mathbf{A} \sin \omega t + \mathbf{B} \cos \omega t$$

in which the components of $\mathbf{A}$ are $A_1$, $A_2$, and $A_3$, and similarly for $\mathbf{B}$. It is clear that the motion takes place entirely in a single plane which is common to the two constant vectors $\mathbf{A}$ and $\mathbf{B}$, and that the path of the particle in that plane is an ellipse, as in the two-dimensional case. Hence the analysis concerning the shape of the elliptical path under the two-dimensional case also applies to the three-dimensional case.

### Nonisotropic Oscillator

The above discussion considered the motion of the so-called three-dimensional *isotropic oscillator*, wherein the restoring force is independent of the direction of the displacement. If the restoring force depends on the direction of the displacement, we have the case of the *nonisotropic oscillator*.

For a suitable choice of axes, the differential equations for the nonisotropic case can be written

$$m\ddot{x} = -k_1 x$$
$$m\ddot{y} = -k_2 y \tag{3.27}$$
$$m\ddot{z} = -k_3 z$$

Here we have a case of *three* different frequencies of oscillation: $\omega_1 = \sqrt{k_1/m}$, $\omega_2 = \sqrt{k_2/m}$, $\omega_3 = \sqrt{k_3/m}$, and the motion is given by the solutions

$$x = A \cos(\omega_1 t + \alpha)$$
$$y = B \cos(\omega_2 t + \beta) \tag{3.28}$$
$$z = C \cos(\omega_3 t + \gamma)$$

Again, the six constants of integration in the above equations are determined from the initial conditions. The resulting oscillation of the particle lies entirely within a rectangular box (whose sides are $2A$, $2B$, and $2C$) centered on the origin. In the event that $\omega_1$, $\omega_2$, and $\omega_3$ are commensurate, that is, if

$$\frac{\omega_1}{n_1} = \frac{\omega_2}{n_2} = \frac{\omega_3}{n_3} \tag{3.29}$$

where $n_1$, $n_2$, and $n_3$ are integers, the path, called a *Lissajous* figure, will be closed, because after a time $2\pi n_1/\omega_1 = 2\pi n_2/\omega_2 = 2\pi n_3/\omega_3$ the particle will return to its initial position and the motion will be repeated. [In Equation (3.29) it is assumed that any common integral factor is canceled out.] On the other hand, if the $\omega$'s are *not* commensurate, the path is not closed. In this case the path may be said to fill completely the rectangular box mentioned above, at least in the sense that if we wait long enough, the particle will come arbitrarily close to any given point.

The net restoring force exerted on a given atom in a solid crystalline substance is approximately linear in the displacement in many cases. The resulting frequencies of oscillation usually lie in the infrared region of the spectrum: $10^{12}$ to $10^{14}$ vibrations per second.

## 3.11.  Motion of Charged Particles in Electric and Magnetic Fields

When an electrically charged particle is in the vicinity of other electric charges, it experiences a force. This force $\mathbf{F}$ is said to be due to the electric field $\mathbf{E}$ which arises from these other charges. We write

$$\mathbf{F} = q\mathbf{E}$$

where $q$ is the electric charge carried by the particle in question.[2] The equation of motion of the particle is then

$$m \frac{d^2\mathbf{r}}{dt^2} = q\mathbf{E} \tag{3.30}$$

or, in component form,

$$m\ddot{x} = qE_x$$
$$m\ddot{y} = qE_y$$
$$m\ddot{z} = qE_z$$

The field components are, in general, functions of the position coordinates $x$, $y$, and $z$. In the case of time-varying fields (that is, if the charges producing $\mathbf{E}$ are moving) the components, of course, also involve $t$.

Let us consider a simple case, namely that of a uniform constant electric field. We can choose one of the axes, say the $z$ axis, to be in the direction of the field. Then $E_x = E_y = 0$, and $E = E_z$. The differential equations of motion of a particle of charge $q$ moving in this field are then

$$\ddot{x} = 0 \qquad \ddot{y} = 0 \qquad \ddot{z} = \frac{qE}{m} = \text{constant}$$

These are of exactly the same form as those for a projectile in a uniform gravitational field. The path is therefore a parabola.

It is shown in textbooks dealing with electromagnetic theory[3] that

$$\nabla \times \mathbf{E} = 0$$

if $\mathbf{E}$ is due to static charges. This means that motion in such a field is conservative, and that there exists a potential function $\Phi$ such that $\mathbf{E} = -\nabla\Phi$. The potential energy of a particle of charge $q$ in such a field is then $q\Phi$, and the total energy is constant and is equal to $\frac{1}{2}mv^2 + q\Phi$.

In the presence of a static magnetic field $\mathbf{B}$ (called the magnetic induction) the force acting on a moving particle is conveniently expressed by means of the cross product, namely,

$$\mathbf{F} = q(\mathbf{v} \times \mathbf{B}) \tag{3.31}$$

---

[2] In mks units $F$ is in newtons, $q$ in coulombs, and $E$ in volts per meter. In cgs units $F$ is in dynes, $q$ in electrostatic units, and $E$ in statvolts per centimeter.

[3] For example, J. C. Slater and N. H. Frank, *Electromagnetism*, McGraw-Hill, New York, 1947.

where $\mathbf{v}$ is the velocity, and $q$ is the charge.[4] The differential equation of motion of a particle moving in a purely magnetic field is then

$$m\frac{d^2\mathbf{r}}{dt^2} = q(\mathbf{v} \times \mathbf{B}) \tag{3.32}$$

The above equation states that the acceleration of the particle is always at right angles to the direction of motion. This means that the tangential component of the acceleration ($\dot{v}$) is zero, and so the particle moves with constant speed. This is true even if $\mathbf{B}$ is a varying function of the position $\mathbf{r}$ as long as it does not vary with time.

## EXAMPLE

Let us examine the motion of a charged particle in a uniform constant magnetic field. Suppose we choose the $z$ axis to be in the direction of the field; that is, we shall write

$$\mathbf{B} = \mathbf{k}B$$

The differential equation of motion now reads

$$m\frac{d^2\mathbf{r}}{dt^2} = q(\mathbf{v} \times \mathbf{k}B) = qB\begin{vmatrix} \mathbf{i} & \mathbf{j} & \mathbf{k} \\ \dot{x} & \dot{y} & \dot{z} \\ 0 & 0 & 1 \end{vmatrix}$$

$$m(\mathbf{i}\ddot{x} + \mathbf{j}\ddot{y} + \mathbf{k}\ddot{z}) = qB(\mathbf{i}\dot{y} - \mathbf{j}\dot{x})$$

Equating components, we have

$$\begin{aligned} m\ddot{x} &= qB\dot{y} \\ m\ddot{y} &= -qB\dot{x} \\ \ddot{z} &= 0 \end{aligned} \tag{3.33}$$

Here, for the first time, we meet a set of differential equations of motion which are *not* of the separated type. The solution is relatively simple, however, for we can integrate at once with respect to $t$ to obtain

$$\begin{aligned} m\dot{x} &= qBy + c_1 \\ m\dot{y} &= -qBx + c_2 \\ \dot{z} &= \text{constant} = \dot{z}_0 \end{aligned}$$

or

$$\dot{x} = \omega y + C_1 \qquad \dot{y} = -\omega x + C_2 \qquad \dot{z} = \dot{z}_0 \tag{3.34}$$

---

[4] Equation (3.31) is valid for mks units: $F$ is in newtons, $q$ in coulombs, $v$ in meters per second, and $B$ in webers per square meter. In cgs units we must write $F = (q/c)$ $(\mathbf{v} \times \mathbf{B})$, where $F$ is in dynes, $q$ in electrostatic units, $c$ is the speed of light—$3 \times 10^{10}$ cm per sec—and $B$ is in gauss. (See Slater and Frank, footnote 3.)

where we have used the abbreviation $\omega = qB/m$. The $c$'s are constants of integration, and $C_1 = c_1/m$, $C_2 = c_2/m$. Upon inserting the expression for $\dot{y}$ from the second part of Equation (3.34) into the first part of Equation (3.33), we obtain the following separated equation for $x$:

$$\ddot{x} + \omega^2 x = \omega^2 a \tag{3.35}$$

where $a = C_2/\omega$. The solution is clearly

$$x = a + A \cos(\omega t + \theta_0) \tag{3.36}$$

where $A$ and $\theta_0$ are constants of integration. Now, if we differentiate with respect to $t$, we have

$$\dot{x} = -A\omega \sin(\omega t + \theta_0) \tag{3.37}$$

The above expression for $\dot{x}$ may be substituted for the left side of the first of Equations (3.34) and the resulting equation solved for $y$. The result is

$$y = b - A \sin(\omega t + \theta_0) \tag{3.38}$$

where $b = -C_1/\omega$. To find the form of the path of motion, we eliminate $t$ between Equation (3.36) and Equation (3.38) to get

$$(x - a)^2 + (y - b)^2 = A^2 \tag{3.39}$$

Thus the projection of the path of motion on the $xy$ plane is a circle of radius

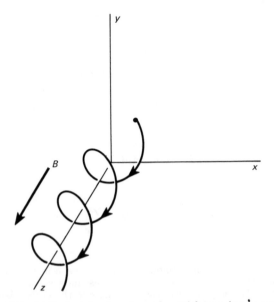

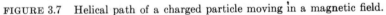

FIGURE 3.7    Helical path of a charged particle moving in a magnetic field.

$A$ centered at the point $(a,b)$. Since, from the third of Equations (3.34), the speed in the $z$ direction is constant, we conclude that the path is a *spiral*. The axis of the spiral path is in the direction of the magnetic field, as shown in Figure 3.7. From Equation (3.37) we have

$$\dot{y} = -A\omega \cos (\omega t + \theta_0) \tag{3.40}$$

Upon eliminating $t$ between Equation (3.37) and Equation (3.40), we find

$$\dot{x}^2 + \dot{y}^2 = A^2\omega^2 = A^2 \left(\frac{qB}{m}\right)^2 \tag{3.41}$$

Letting $v_1 = (\dot{x}^2 + \dot{y}^2)^{1/2}$, we see that the radius $A$ of the spiral is given by

$$A = \frac{v_1}{\omega} = v_1 \frac{m}{qB} \tag{3.42}$$

If there is no component of the velocity in the $z$ direction, the path is a circle of radius $A$. It is evident that $A$ is directly proportional to the speed $v_1$, and that the angular frequency $\omega$ of motion in the circular path is *independent* of the speed. $\omega$ is known as the cylotron frequency. The cyclotron, invented by Ernest Lawrence, depends for its operation on the fact that $\omega$ is independent of the speed of the charged particle.

### 3.12.  Constrained Motion of a Particle

When a moving particle is restricted geometrically in the sense that it must stay on a certain definite surface or curve, the motion is said to be *constrained*. A piece of ice sliding around in a bowl, or a bead sliding on a wire, are examples of constrained motion. The constraint may be complete, as with the bead, or it may be one sided, as in the former example. Constraints may be fixed, or they may be moving. In this chapter we shall study only fixed constraints.

*The Energy Equation for Smooth Constraints*

The total force acting on a particle moving under constraint can be expressed as the vector sum of the external force $\mathbf{F}$ and the force of constraint $\mathbf{R}$. The latter force is the reaction of the constraining agent upon the particle. The equation of motion may therefore be written

$$m\frac{d\mathbf{v}}{dt} = \mathbf{F} + \mathbf{R} \tag{3.43}$$

If we take the dot product with the velocity $\mathbf{v}$ we have

$$m \frac{d\mathbf{v}}{dt} \cdot \mathbf{v} = \mathbf{F} \cdot \mathbf{v} + \mathbf{R} \cdot \mathbf{v} \tag{3.44}$$

Now in the case of a *smooth* constraint—for example, a frictionless surface—the reaction $\mathbf{R}$ is normal to the surface or curve while the velocity $\mathbf{v}$ is tangent to the surface. Hence $\mathbf{R}$ is perpendicular to $\mathbf{v}$ and the dot product $\mathbf{R} \cdot \mathbf{v}$ vanishes. Equation (3.44) then reduces to

$$\frac{d}{dt} \left( \frac{1}{2} m\mathbf{v} \cdot \mathbf{v} \right) = \mathbf{F} \cdot \mathbf{v}$$

Consequently, if $\mathbf{F}$ is conservative, we can integrate as in Section 3.5, and we find the same energy relation as Equation (3.9), namely,

$$\tfrac{1}{2}mv^2 + V(x,y,z) = \text{constant} = E$$

Thus the particle, although remaining on the surface or curve, moves in such a way that the total energy is constant. We might, of course, have expected this to be the case for frictionless constraints.

### EXAMPLE

A particle is placed on top of a smooth sphere of radius $a$. If the particle is slightly disturbed, at what point will it leave the sphere?

The forces acting on the particle are the downward force of gravity and the reaction $\mathbf{R}$ of the spherical surface. The equation of motion is

$$m \frac{d\mathbf{v}}{dt} = m\mathbf{g} + \mathbf{R}$$

Let us choose coordinate axes as shown in Figure 3.8. The potential energy is then $mgz$, and the energy equation reads

$$\tfrac{1}{2}mv^2 + mgz = E$$

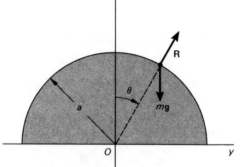

FIGURE 3.8   Forces acting on a particle sliding on a smooth sphere.

From the initial conditions ($v = 0$ for $z = a$) we have $E = mga$, so, as the particle slides down, its speed is given by the equation

$$v^2 = 2g(a - z)$$

Now, if we take radial components of the equation of motion, we can write the force equation as

$$-\frac{mv^2}{a} = -mg \cos \theta + R = -mg\frac{z}{a} + R$$

Hence

$$R = mg\frac{z}{a} - \frac{mv^2}{a} = mg\frac{z}{a} - \frac{m}{a} 2g(a - z)$$

$$= \frac{mg}{a} (3z - 2a)$$

Thus $R$ vanishes when $z = \frac{2}{3}a$, at which point the particle will leave the sphere. This may be argued from the fact that the sign of $R$ changes from positive to negative there.

*Motion on a Curve*

For the case in which a particle is constrained to move on a certain curve, the energy equation together with the equations of the curve in parametric form

$$x = x(s) \qquad y = y(s) \qquad z = z(s)$$

suffice to determine the motion. (The parameter $s$ is the distance measured along the curve from some arbitrary reference point.) The motion may be found by consideration of the fact that the potential energy can be expressed as a function of $s$ alone, while the kinetic energy is just $\frac{1}{2}m\dot{s}^2$. Thus the energy equation may be written

$$\tfrac{1}{2}m\dot{s}^2 + V(s) = E$$

from which $s$ (hence $x$, $y$, and $z$) can be obtained by integration. Alternately, by differentiating the above equation with respect to $t$ and canceling the common factor $\dot{s}$, we obtain the following differential equation of motion for the particle:

$$m\ddot{s} + \frac{dV}{ds} = 0 \tag{3.45}$$

This equation is equivalent to the equation

$$m\ddot{s} - F_s = 0$$

where $F_s$ is the component of the external force $\mathbf{F}$ in the direction of $s$. This means that $F_s = -dV/ds$.

### 3.13.  The Simple Pendulum

The above considerations are well illustrated by the simple pendulum—a heavy particle attached to the end of a light inextensible rod or cord, the motion being in a vertical plane.   The simple pendulum is also dynamically equivalent to a bead sliding on a smooth wire in the form of a vertical circular loop.   As shown in Figure 3.9, let $\theta$ be the angle between the vertical and

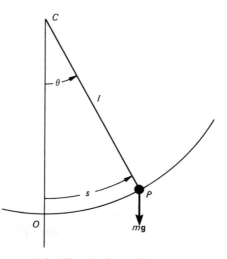

FIGURE 3.9   The simple pendulum.

the line $CP$ where $C$ is the center of the circular path and $P$ is the instantaneous position of the particle.   The distance $s$ is measured from the equilibrium position $O$.   From the figure, we see that the component $F_s$ of the force of gravity $m\mathbf{g}$ in the direction of $s$ is equal to $-mg \sin \theta$.   If $l$ is the length of the pendulum, then $\theta = s/l$.   The differential equation of motion then reads

$$m\ddot{s} + mg \sin \left(\frac{s}{l}\right) = 0$$

or, in terms of $\theta$, we may write

$$\ddot{\theta} + \frac{g}{l} \sin \theta = 0$$

It should be noted that the potential energy $V$ can be expressed as $mgz$ where $z$ is the vertical distance of the particle from $O$, namely,

$$V = mgz = mgl(1 - \cos \theta)$$
$$= mgl - mgl \cos \left(\frac{s}{l}\right)$$

Hence $-dV/ds = -mg \sin (s/l) = -mg \sin \theta = F_s$.

In order to find an approximate solution of the differential equation of motion, let us assume that $\theta$ remains small.   In this case

$$\sin \theta \simeq \theta$$

so we have

$$\ddot{\theta} + \frac{g}{l}\,\theta = 0$$

This is the differential equation of the harmonic oscillator.   The solution, as we have seen in Section 2.12, is

$$\theta = \theta_0 \cos (\omega_0 t + \varphi_0)$$

where $\omega_0 = \sqrt{g/l}$.   $\theta_0$ is the amplitude of oscillation, and $\varphi_0$ is a phase factor. Thus, to the extent that $\theta$ is a valid approximation for $\sin \theta$, the motion is simple harmonic, and the period of oscillation $T_0$ is given by

$$T_0 = \frac{2\pi}{\omega_0} = 2\pi \sqrt{\frac{l}{g}} \tag{3.46}$$

the well-known elementary formula.

## 3.14.   More Accurate Solution of the Simple Pendulum Problem and the Nonlinear Oscillator

The differential equation of motion of the simple pendulum

$$\ddot{\theta} + \frac{g}{l} \sin \theta = 0$$

is a special case of the general differential equation for motion under a non-linear restoring force, that is, a force which varies in some manner other than in direct proportion to the displacement.   The equation of the general one-dimensional problem with no damping may be written

$$\ddot{\xi} + f(\xi) = 0 \tag{3.47}$$

where $\xi$ is the variable denoting the displacement from the equilibrium position, so that

$$f(0) = 0$$

Nonlinear differential equations usually require some method of approximation for their solution.   Suppose that the function $f(\xi)$ is expanded as a power series in $\xi$, namely,

$$f(\xi) = a_1\xi + a_2\xi^2 + a_3\xi^3 + \cdots$$

The differential equation of motion is then

$$\frac{d^2\xi}{dt^2} + a_1\xi + a_2\xi^2 + a_3\xi^3 + \cdots = 0 \tag{3.48}$$

This is the expanded form of the general equation of motion of the nonlinear oscillator without damping.   The term $a_1\xi$ in the above equation is the *linear* term.   If this term is predominant, that is, if $a_1$ is much larger than the other coefficients, then the motion will be approximately simple harmonic with angular frequency $a_1^{1/2}$.   A more accurate solution must take into account the remaining nonlinear terms.

To illustrate, let us return to the problem of the simple pendulum.   If we use the series expansion

$$\sin \theta = \theta - \frac{\theta^3}{3!} + \frac{\theta^5}{5!} - \cdots$$

and retain only the first two terms, we obtain

$$\ddot{\theta} + \frac{g}{l}\,\theta - \frac{g}{6l}\,\theta^3 = 0 \tag{3.49}$$

as a second approximation to the differential equation of motion.   We know that the motion is periodic.   Suppose we *try* a solution in the form of a simple sinusoidal function

$$\theta = A \cos \omega t$$

Inserting this into the differential equation, we obtain

$$-A\omega^2 \cos \omega t + \frac{g}{l}\,A\cos\omega t - \frac{g}{6l}\,A^3 \cos^3\omega t = 0$$

or, upon using the trigonometric identity

$$\cos^3 u = \tfrac{3}{4}\cos u + \tfrac{1}{4}\cos 3u$$

we have, after collecting terms,

$$\left(-A\omega^2 + \frac{g}{l}\,A - \frac{gA^3}{8l}\right)\cos\omega t - \frac{gA^3}{24l}\cos 3\omega t = 0$$

Excluding the trivial case $A = 0$, we see that the above equation cannot hold for all values of $t$.   Hence our trial function $A\cos\omega t$ cannot be a solution. From the fact that the term in $\cos 3\omega t$ appears in the above equation, however, we might suspect that a trial solution of the form

$$\theta = A\cos\omega t + B\cos 3\omega t \tag{3.50}$$

will represent a better approximation than $A\cos\omega t$.   This turns out to be the case.   If we insert the above solution into Equation (3.49), we find, after a procedure similar to that above, the following equation:

$$\left(-A\omega^2 + \frac{g}{l}\,A - \frac{gA^3}{8l}\right)\cos\omega t + \left(-9B\omega^2 + \frac{g}{l}\,B - \frac{gA^3}{24l}\right)\cos 3\omega t$$
$$+ \text{ (terms in higher powers of } B \text{ and higher multiples of } \omega t) = 0$$

Again the equation will not hold for all values of $t$, but our approximate solution will be reasonably accurate if the coefficients of the first two cosine terms can be made to vanish separately:

$$-A\omega^2 + \frac{g}{l}A - \frac{gA^3}{8l} = 0 \qquad -9B\omega^2 + \frac{g}{l}B - \frac{gA^3}{24l} = 0$$

From the first equation

$$\omega^2 = \frac{g}{l}\left(1 - \frac{A^2}{8}\right) \tag{3.51}$$

With this value of $\omega^2$, we find from the second equation

$$B = -A^3\frac{1}{3(64 + 27A^2)} \simeq -\frac{A^3}{192}$$

Now, from our trial solution Equation (3.50) we see that the amplitude $\theta_0$ of the oscillation of the pendulum is given by

$$\theta_0 = A + B$$
$$= A - \frac{A^3}{192}$$

or, if $A$ is small,

$$\theta_0 \simeq A$$

The meaning of Equation (3.51) is now clear.  The frequency of oscillation depends on the amplitude $\theta_0$.  In fact, we can write

$$\omega \simeq \sqrt{\frac{g}{l}}\left(1 - \frac{1}{8}\theta_0^2\right)^{1/2}$$

or, for the period, we have

$$T = \frac{2\pi}{\omega} \simeq 2\pi\sqrt{\frac{l}{g}}\left(1 - \frac{1}{8}\theta_0^2\right)^{-1/2}$$
$$\simeq 2\pi\sqrt{\frac{l}{g}}\left(1 + \frac{1}{16}\theta_0^2 + \cdots\right)$$
$$\simeq T_0\left(1 + \frac{1}{16}\theta_0^2 + \cdots\right) \tag{3.52}$$

where $T_0$ is the period for zero amplitude.

The above analysis, although it is admittedly very crude, brings out two essential features of free oscillation under a nonlinear restoring force; that is, the period of oscillation is a function of the amplitude of vibration, and the oscillation is not strictly sinusoidal but can be considered as the superposition of a mixture of harmonics.  It can be shown that the vibration of a nonlinear system driven by a purely sinusoidal driving force will also be distorted; that is, it will contain harmonics.  The loudspeaker of a radio receiver or a "hi-fi" system, for example, may introduce distortion (harmonics) over and above that introduced by the electronic amplifying system.

### 3.15.   Exact Solution of the Motion of the Simple Pendulum by Means of Elliptic Integrals

From the expression for the potential energy of the simple pendulum we can write the energy equation as follows:

$$\tfrac{1}{2}m(l\dot\theta)^2 + mgl(1 - \cos\theta) = E \qquad (3.53)$$

If the pendulum is pulled aside at an angle $\theta_0$ (the amplitude) and released $(\dot\theta_0 = 0)$, then $E = mgl(1 - \cos\theta_0)$.   The above equation then reduces to

$$\dot\theta^2 = \frac{2g}{l}(\cos\theta - \cos\theta_0) \qquad (3.54)$$

after transposing terms and dividing by $ml^2$.   By use of the identity $\cos\theta = 1 - 2\sin^2(\theta/2)$, we can further write

$$\dot\theta^2 = \frac{4g}{l}\left(\sin^2\frac{\theta_0}{2} - \sin^2\frac{\theta}{2}\right) \qquad (3.55)$$

It is expedient to express the motion in terms of the variable $\varphi$ defined by the equation

$$\sin\varphi = \frac{\sin(\theta/2)}{\sin(\theta_0/2)} = \frac{1}{k}\sin\frac{\theta}{2} \qquad (3.56)$$

Upon differentiating with respect to $t$, we have

$$(\cos\varphi)\dot\varphi = \frac{1}{k}\cos\left(\frac{\theta}{2}\right)\frac{\dot\theta}{2} \qquad (3.57)$$

From Equations (3.56) and (3.57) we can readily transform Equation (3.55) into the corresponding equation in $\varphi$, namely,

$$\dot\varphi^2 = \frac{g}{l}(1 - k^2\sin^2\varphi) \qquad (3.58)$$

The relationship between $\varphi$ and $t$ is then found by separating variables and integrating:

$$t = \sqrt{\frac{l}{g}}\int_0^\varphi \frac{d\varphi}{\sqrt{1 - k^2\sin^2\varphi}} = \sqrt{\frac{l}{g}}\,F(k,\varphi) \qquad (3.59)$$

The function $F(k,\varphi) = \int_0^\varphi (1 - k^2\sin^2\varphi)^{-1/2}\,d\varphi$ is known as the *incomplete elliptic integral of the first kind*.   The period of the pendulum is obtained by noting that $\theta$ increases from 0 to $\theta_0$ in one quarter of a cycle.   Thus we see that $\varphi$ goes from 0 to $\pi/2$ in the same time interval.   Therefore, we may write for the period $T$

$$T = 4\sqrt{\frac{l}{g}}\int_0^{\pi/2} \frac{d\varphi}{\sqrt{1 - k^2\sin^2\varphi}} = 4\sqrt{\frac{l}{g}}\,K(k) \qquad (3.60)$$

The function $K(k) = \int_0^{\pi/2} (1 - k^2 \sin^2 \varphi)^{-1/2} d\varphi = F(k,\pi/2)$ is called the *complete elliptic integral of the first kind*.   Values of the elliptic integrals are tabulated.[5]   An approximate expression may be obtained, however, by expanding the integrand in Equation (3.60) by the binomial theorem and integrating term by term.   The result is

$$T = 4 \sqrt{\frac{l}{g}} \int_0^{\pi/2} \left(1 + \frac{k^2}{2} \sin^2 \varphi + \cdots \right) d\varphi = 2\pi \sqrt{\frac{l}{g}} \left(1 + \frac{k^2}{4} + \cdots \right) \quad (3.61)$$

Now, for small values of the amplitude $\theta_0$, we have

$$k^2 = \sin^2 \frac{\theta_0}{2} \simeq \frac{\theta_0^2}{4}$$

Thus we may write approximately

$$T \simeq 2\pi \sqrt{\frac{l}{g}} \left(1 + \frac{\theta_0^2}{16} + \cdots \right) \quad (3.62)$$

which agrees with the value of $T$ found in the previous section.

## EXAMPLE

Find the period of a simple pendulum swinging with an amplitude of $20°$. Use tables of elliptic functions, and also compare with the values calculated by the above approximations.

For an amplitude of $20°$, $k = \sin 10° = 0.17365$, and $\theta_0/2 = 0.17453$ radians.   The results are as follows:

From tables and Equation (3.60) $T = 4 \sqrt{l/g} \, K(10°) = \sqrt{l/g} \, (6.3312)$
From Equation (3.61) $T = 2\pi \sqrt{l/g} \, (1 + \frac{1}{4} \sin^2 10°) = \sqrt{l/g} \, (6.3306)$
From Equation (3.62) $T = 2\pi \sqrt{l/g} \, (1 + \theta_0^2/16) = \sqrt{l/g} \, (6.3310)$
Elementary formula $T_0 = 2\pi \sqrt{l/g} = \sqrt{l/g} \, (6.2832)$

## 3.16.   The Isochronous Problem

It is interesting to investigate the question of whether or not there is a curve of constraint for which a particle will oscillate under gravity *isochronously*, that is, with a period that is independent of the amplitude.

Let $\theta$ be the angle between the horizontal and the tangent to the constraining curve (Figure 3.10).   Then the component of the gravitational force in the direction of motion is $-mg \sin \theta$.   The differential equation of

---

[5] See, for example, L. M. Milne-Thomson, *Jacobian Elliptic Function Tables*, Dover, New York, 1950; or B. O. Peirce, *A Short Table of Integrals*, Ginn, Boston, 1929.   See also Appendix III.

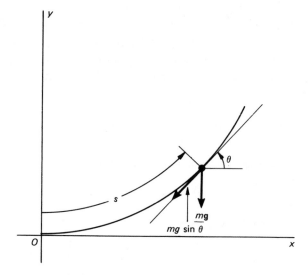

FIGURE 3.10   Forces involved in the isochronous case.

motion along the path of constraint (assumed smooth) is then

$$m\ddot{s} = -mg \sin \theta$$

But if the above equation represents simple harmonic notion along the curve, we must have

$$m\ddot{s} = -ks$$

Therefore, a constraining curve which satisfies the equation

$$s = c \sin \theta$$

will produce simple harmonic motion.

Now we can find $x$ and $y$ in terms of $\theta$ from the above equation, as follows:

$$\frac{dx}{d\theta} = \frac{dx}{ds}\frac{ds}{d\theta} = (\cos \theta)(c \cos \theta)$$

Hence

$$x = \int c \cos^2 \theta \, d\theta = \frac{c}{4} (2\theta + \sin 2\theta) \tag{3.63}$$

Similarly,

$$\frac{dy}{d\theta} = \frac{dy}{ds}\frac{ds}{d\theta} = (\sin \theta)(c \cos \theta)$$

So

$$y = \int c \sin \theta \cos \theta \, d\theta = -\frac{c}{4} \cos 2\theta \tag{3.64}$$

Equations (3.63) and (3.64) are the parametric equations of a *cycloid*. Thus a constraining curve in the form of a cycloid will produce motion such that $s$ varies harmonically with time, and the period of oscillation will be independent of the amplitude. As a corollary, we see that a particle starting from rest on a smooth cycloidal curve takes the same time to reach the bottom regardless of the point at which it begins.

The Dutch physicist and mathematician Christiaan Huygens discovered the above facts in connection with attempts to improve the accuracy of pendulum clocks. He also discovered the theory of evolutes and found that the evolute of a cycloid is also a cycloid. Hence, by providing cycloidal "cheeks" for a pendulum, the motion of the bob must follow a cycloidal path and the period is thus independent of the amplitude. Though ingenious, the invention never found extensive practical use.

### 3.17.   The Spherical Pendulum

A classic problem in constrained motion is that of a particle which is required to move on a smooth spherical surface, such as a small mass sliding around inside a smooth spherical bowl. The case is perhaps more aptly illustrated by a heavy bob attached to a light inextensible rod or cord which is free to swing in any direction about a fixed point, see Figure 3.11. This is the so-called spherical pendulum.

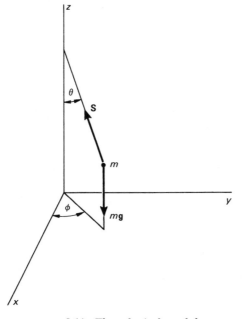

FIGURE 3.11   The spherical pendulum.

*Approximate Solution in Rectangular Coordinates*

There are two forces acting on the particle, namely the downward force of gravity, and the tension **S** in the constraining rod or cord. The differential equation of motion then reads

$$m\ddot{\mathbf{r}} = m\mathbf{g} + \mathbf{S}$$

If we choose the $z$ axis to be vertical, the rectangular components of the equation of motion are as follows:

$$m\ddot{x} = S_x$$
$$m\ddot{y} = S_y$$
$$m\ddot{z} = S_z - mg$$

An approximate solution is readily obtained for the case in which the displacement from the equilibrium position is very small. The magnitude of the tension is then very nearly constant and equal to $mg$, and we have $|x| \ll l$, $|y| \ll l$, $z \simeq 0$. The $x$ and $y$ components of **S** are then given by the approximate relations

$$S_x \simeq -mg\,\frac{x}{l}$$

$$S_y \simeq -mg\,\frac{y}{l}$$

which are easily verified from the geometry of the figure. The $x$-$y$ differential equations of motion then reduce to

$$\ddot{x} + \frac{g}{l}\,x = 0$$

$$\ddot{y} + \frac{g}{l}\,y = 0$$

These are similar to the equations of the two-dimensional harmonic oscillator treated earlier in Section 3.10. The solutions are, as we have seen,

$$x = A\,\cos(\omega t + \alpha)$$
$$y = B\,\cos(\omega t + \beta)$$

in which

$$\omega = \left(\frac{g}{l}\right)^{1/2}$$

as in the simple plane pendulum.

To the extent that our approximations are valid, the motion is such that the projection on the $xy$ plane is an ellipse. There are, of course, special cases in which the projection is a straight line, or a circle, depending on the initial conditions.

*Solution in Spherical Coordinates*

For a more accurate treatment of the spherical pendulum than that given above, we shall employ spherical coordinates as defined in Figure 3.11.    The tension **S** has only a radial component, but the weight $m\mathbf{g}$ has both a radial component $mg \cos \theta$ and a transverse component $-mg \sin \theta$.   Hence the differential equation of motion can be resolved into spherical components as follows:

$$ma_r = F_r = mg \cos \theta - S$$
$$ma_\theta = F_\theta = -mg \sin \theta$$
$$ma_\varphi = F_\varphi = 0$$

The three acceleration components $a_r$, $a_\theta$, and $a_\varphi$ are given in Chapter 1, Section 1.25.   Since the constraint is that

$$r = l = \text{constant}$$

we can ignore the radial component of the acceleration.   The other two components reduce to

$$a_\theta = l\ddot{\theta} - l\dot{\varphi}^2 \sin \theta \cos \theta$$
$$a_\varphi = l\ddot{\varphi} \sin \theta + 2l\dot{\varphi}\dot{\theta} \cos \theta$$

Thus, after transposing terms and performing the obvious cancellations, the differential equations in $\theta$ and $\varphi$ become

$$\ddot{\theta} - \dot{\varphi}^2 \sin \theta \cos \theta + \frac{g}{l} \sin \theta = 0 \tag{3.65}$$

$$\frac{1}{\sin \theta} \frac{d}{dt} (\dot{\varphi} \sin^2 \theta) = 0 \tag{3.66}$$

The second equation implies that the quantity in parentheses is constant. Let us call it $h$.   It is, in fact, the angular momentum (per unit mass) about the vertical axis. The reason that it is constant stems from the absence of any moment of force about that axis.    Then we can write

$$\dot{\varphi} = \frac{h}{\sin^2 \theta} \tag{3.67}$$

Upon inserting the above value of $\dot{\varphi}$ into Equation (3.65), we obtain the following *separated* equation in $\theta$:

$$\ddot{\theta} + \frac{g}{l} \sin \theta - h^2 \frac{\cos \theta}{\sin^3 \theta} = 0 \tag{3.68}$$

It is instructive to consider some special cases at this point.   First, if the angle $\varphi$ is constant, then $\dot{\varphi} = 0$ and so $h = 0$.   Consequently, Equation (3.68) reduces to

$$\ddot{\theta} + \frac{g}{l} \sin \theta = 0$$

which, of course, is just the differential equation of the simple pendulum. The motion takes place in the plane $\varphi = \varphi_0 = \text{constant}$.

The second special case is that of the *conical pendulum*; $\theta = \theta_0 =$ constant. In this case $\dot\theta = 0$ and $\ddot\theta = 0$, so Equation (3.68) reduces to

$$\frac{g}{l} \sin \theta_0 - h^2 \frac{\cos \theta_0}{\sin^3 \theta_0} = 0$$

or

$$h^2 = \frac{g}{l} \sin^4 \theta_0 \sec \theta_0 \qquad (3.69)$$

From the value of $h$ given by the above equation, we find from Equation (3.67) that

$$\dot\varphi_0{}^2 = \frac{g}{l} \sec \theta_0 \qquad (3.70)$$

as the condition for conical motion of the pendulum.

The above equation can also be obtained by considering the forces acting on the particle in its circular motion as shown in Figure 3.12. The acceleration is constant in magnitude, namely $\rho\dot\varphi_0{}^2 = (l \sin \theta_0)\dot\varphi_0{}^2$, and it is directed toward the center of the circular path. Hence, upon taking horizontal and vertical components, we have

$$S \sin \theta_0 = (ml \sin \theta_0)\dot\varphi_0{}^2$$
$$S \cos \theta_0 = mg$$

which reduce to Equation (3.70) upon elimination of $S$.

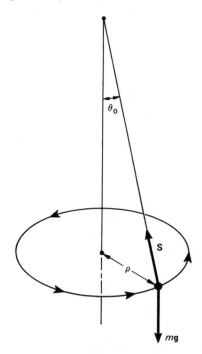

FIGURE 3.12  The conical case of the spherical pendulum.

Let us now consider the case in which the motion is *almost* conical; that is, the value of $\theta$ remains close to the value $\theta_0$. If we insert the expression for $h^2$ given in Equation (3.69) into the separated differential equation for $\theta$, Equation (3.68), the result is

$$\ddot{\theta} + \frac{g}{l}\left(\sin\theta - \frac{\sin^4\theta_0 \cos\theta}{\cos\theta_0 \sin^3\theta}\right) = 0 \qquad (3.71)$$

It is convenient at this point to introduce the new variable $\xi$ defined as

$$\xi = \theta - \theta_0$$

The expression in parentheses in Equation (3.71) may be expanded as a power series in $\xi$ according to the standard formula

$$f(\xi) = f(0) + f'(0)\xi + f''(0)\frac{\xi^2}{2!} + \cdots$$

We find, after performing the indicated operations, that $f(0) = 0$ and $f'(0) = 3\cos\theta_0 + \sec\theta_0$. Since we are concerned with the case of small values of $\xi$, we shall neglect higher powers of $\xi$ than the first, and so we can write Equation (3.71) as

$$\ddot{\xi} + \frac{g}{l}b\xi = 0 \qquad (3.72)$$

where $b = 3\cos\theta_0 + \sec\theta_0$. The motion in $\xi$ or $\theta$ is therefore given by

$$\xi = \theta - \theta_0 = \xi_0 \cos\left(\sqrt{\frac{gb}{l}}\,t + \epsilon\right) \qquad (3.73)$$

Thus $\theta$ oscillates harmonically about the value $\theta_0$ with a period

$$T_1 = 2\pi\sqrt{\frac{l}{gb}} = 2\pi\sqrt{\frac{l}{g(3\cos\theta_0 + \sec\theta_0)}} \qquad (3.74)$$

Now the value of $\dot{\varphi}$, from Equation (3.67), does not vary greatly from the value given by the purely conical motion $\dot{\varphi}_0$, so $\varphi$ increases steadily during the oscillation of $\theta$ about $\theta_0$. The path of the particle is shown in Figure 3.13. During one complete oscillation of $\theta$ the value of the azimuth angle $\varphi$ increases by the amount

$$\varphi_1 \cong \dot{\varphi}_0 T_1$$

From the values of $\dot{\varphi}_0$ and $T_1$ given above, we readily find

$$\varphi_1 = 2\pi(3\cos^2\theta_0 + 1)^{-1/2}$$

Let $\rho$ denote the radius of the circle for $\theta = \theta_0$, as in Figure 3.12. Then $\cos^2\theta_0 = 1 - \rho^2/l^2$, and consequently $\varphi_1 = 2\pi(4 - 3\rho^2/l^2)^{-1/2}$. Thus $\varphi_1$ is slightly greater than $\pi$. By expanding in powers of $\rho$, we find that the excess $\Delta\varphi$ is given by

$$\Delta\varphi \cong \frac{3\pi}{8}\frac{\rho^2}{l^2} + \cdots$$

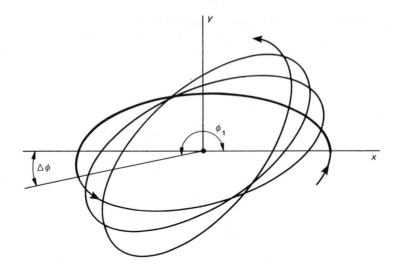

FIGURE 3.13  Projection on the $xy$ plane of the path of motion of the spherical pendulum.

Earlier in this section we proved that the projection of the path of the pendulum bob on the $xy$ plane is approximately an ellipse if the angle $\theta$ is small. We can now interpret the above result to mean that the major axis of the ellipse is not steady, but precesses in the direction of increasing $\varphi$. The axis of the ellipse turns through the angle $\Delta\varphi$ during each complete oscillation in $\theta$. This is illustrated in Figure 3.13.

### Energy Considerations. Limits of the Vertical Motion

In order to relate the amplitude of the vertical oscillation of the spherical pendulum to the parameters of the problem, it is advantageous to use the energy equation. In our notation, the potential energy is given by $V = -mgl \cos\theta$.

To find the kinetic energy, we use the fact that the components of the velocity in spherical coordinates are $(\dot{r}, r\dot{\theta}, r\dot{\varphi}\sin\theta)$. Thus, since $r = l =$ constant, we have $T = \frac{1}{2}mv^2 = \frac{1}{2}m(l^2\dot{\theta}^2 + l^2\dot{\varphi}^2 \sin^2\theta)$. The energy equation then reads

$$E = \frac{ml^2}{2}(\dot{\theta}^2 + \dot{\varphi}^2 \sin^2\theta) - mgl \cos\theta$$

Let us solve the above equation for $\dot{\theta}^2$. To do this, we use the relation, derived earlier, that $\dot{\varphi} = h/\sin^2\theta$. Let us also set $\cos\theta = u$. The result is

$$\dot{\theta}^2 = \frac{2E}{ml^2} + \frac{2g}{l}u - \frac{h^2}{1-u^2} = f(u) \tag{3.75}$$

The roots of the equation $f(u) = 0$ determine the limits, or turning points of the oscillation in $\theta$, since $\dot\theta$ vanishes at those roots.    The motion is confined to those values of $\theta$ such that $f(u)$ is nonnegative.    Thus the vertical oscillations lie between two horizontal circles, see Figure 3.14.    If, in particular, the two real roots which lie between $+1$ and $-1$ are equal, then the motion is confined to a single horizontal circle, that is, we have the case of a conical pendulum.

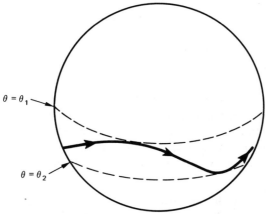

**FIGURE 3.14**    Illustrating the vertical motion of the spherical pendulum.

## DRILL EXERCISES

**3.1**    Determine which of the following forces are conservative by finding the curl:

(a) $\mathbf{F} = \mathbf{i}x + \mathbf{j}y$
(b) $\mathbf{F} = \mathbf{i}y + \mathbf{j}x$
(c) $\mathbf{F} = \mathbf{i}y - \mathbf{j}x$
(d) $\mathbf{F} = \mathbf{i}xy + \mathbf{j}yz + \mathbf{k}zx$
(e) $\mathbf{F} = \mathbf{i}yz + \mathbf{j}zx + \mathbf{k}xy$

**3.2**    Find the value of the constant $c$ such that the following forces are conservative:

(a) $\mathbf{F} = \mathbf{i}xy + \mathbf{j}cx^2$

(b) $\mathbf{F} = \mathbf{i}\dfrac{z}{y} + \mathbf{j}c\dfrac{xz}{y^2} + \mathbf{k}\dfrac{x}{y}$

**3.3**    Find the force for each of the following potential energy functions:

(a) $V = xyz$
(b) $V = k(x^\alpha + y^\beta + z^\gamma)$
(c) $V = kx^\alpha y^\beta z^\gamma$
(d) $V = ke^{(\alpha x + \beta y + \gamma z)}$

3.4  Find the potential function for those forces in Exercise 3.1 that are conservative.

3.5  A particle of mass $m$ moves in the force field given by the potential function of Exercise 3.3(a). If the particle passes through the origin with speed $v_0$, what will its speed be if and when it passes through the point (2,2,2)?

## PROBLEMS

3.6  Consider the two force functions
    (a) $\mathbf{F} = \mathbf{i}x + \mathbf{j}y$
    (b) $\mathbf{F} = \mathbf{i}y - \mathbf{j}x$

Verify that (a) is conservative and that (b) is nonconservative by showing that the integral $\int \mathbf{F} \cdot d\mathbf{r}$ is independent of the path of integration for (a), but not for (b), by taking two paths in which the starting point is the origin (0,0), and the end point is (1,1). For one path take the line $x = y$. For the other path, take the $x$ axis out to the point (0,1) and then the line $x = 1$ up to the point (1,1).

3.7  Show that the variation of gravity with height can be accounted for approximately by the following potential energy function:

$$V = mgz\left(1 - \frac{z}{R}\right)$$

in which $R$ is the radius of the earth. Find the force given by the above potential function. From this, find the component differential equations of motion of a projectile under such a force.

3.8  A projectile is fired from the origin with initial speed $v_0$ at an angle of elevation $\theta$ with the horizontal. If air resistance is neglected, show that if the ground is level, the projectile hits the ground a distance

$$\frac{v_0^2 \sin 2\theta}{g}$$

from the origin. This is called the *horizontal range*. Also show that the decrease in the horizontal range is approximately

$$\frac{4v_0^3\gamma \sin\theta \sin 2\theta}{3g}$$

in the case of a linear air resistance.

3.9  Particles of mud are thrown from the rim of a rolling wheel. If the forward speed of the wheel is $v_0$, and the radius of the wheel is $b$, show that the greatest height above the ground that the mud can go is

$$b + \frac{v_0^2}{2g} + \frac{gb^2}{2v_0^2}$$

At what point on the rolling wheel does this mud leave?

3.10   A gun is located at the bottom of a hill of constant slope $\varphi$.  Show that the range of the gun measured up the slope of the hill is

$$\frac{2v_0{}^2 \cos \theta \sin (\theta - \varphi)}{g \cos^2 \varphi}$$

where $\theta$ is the angle of elevation of the gun, and that the maximum value of the slope range is

$$\frac{v_0{}^2}{g(1 + \sin \varphi)}$$

3.11   Write down the component form of the differential equations of motion of a projectile if the air resistance is proportional to the square of the speed.   Are the equations separated?   Show that the $x$ component of the velocity is given by

$$\dot{x} = \dot{x}_0 e^{-\gamma s}$$

where $s$ is the distance the projectile has traveled along the path of motion.

3.12   The initial conditions for a two-dimensional isotropic oscillator are $x(0) = A$, $\dot{x}(0) = 0$, $y(0) = B$, $\dot{y}(0) = \omega C$ where $\omega$ is the frequency. Show that the motion takes place entirely within a rectangle of dimensions $2A$ and $2(B^2 + C^2)^{1/2}$.   Find the inclination $\psi$ of the axis of the elliptical path in terms of $A$, $B$, and $C$.

3.13   A particle of unit mass moves in the three-dimensional *nonisotropic* harmonic oscillator potential

$$V = x^2 + 4y^2 + 9z^2$$

If the particle passes through the origin with unit speed in the (1,1,1) direction at time $t = 0$, determine $x$, $y$, and $z$ as functions of time.

3.14   An atom is situated in a simple cubic crystal lattice.   If the potential energy of interaction between any two atoms is of the form $cr^{-\alpha}$ where $c$ and $\alpha$ are constants and $r$ is the distance between the two atoms, show that the total energy of interaction of a given atom with its six nearest neighbors is approximately that of the three-dimensional harmonic oscillator potential

$$V = A + B(x^2 + y^2 + z^2)$$

where $A$ and $B$ are constants.   [*Note:* Assume that the six neighboring atoms are fixed and are located at the points $(\pm d,0,0)$, $(0,\pm d,0)$, $(0,0,\pm d)$, and that the displacement $(x,y,z)$ of the given atom from the equilibrium position $(0,0,0)$ is small compared to $d$.]   Then $V = \sum cr_i{}^{-\alpha}$ where $r_1 = [(d - x)^2 + y^2 + z^2]^{1/2}$ with similar expressions for $r_2, r_3, \ldots, r_6$.   Use the binomial theorem to obtain the required result.

3.15   An electron moves in a force field composed of a uniform electric field **E** and a uniform magnetic field **B** which is at right angles to **E**.   Let

$\mathbf{E} = \mathbf{j}E$ and $\mathbf{B} = \mathbf{k}B$. Take the initial position of the electron at the origin with initial velocity $\mathbf{v}_0 = \mathbf{i}v_0$ in the $x$ direction. Find the resulting motion of the particle. Show that the path of motion is a cycloid:

$$x = a \sin \omega t + bt$$
$$y = c(1 - \cos \omega t)$$
$$z = 0$$

Cycloidal motion of electrons is utilized in the *magnetron*——an electronic tube used to produce high-frequency radio waves.

3.16    A particle is placed on the side a smooth sphere of radius $b$ at a distance $b/2$ above the central plane. As the particle slides down the side of the sphere, at what point will it leave?

3.17    A bead slides on a smooth wire bent into the form of a circular loop of radius $b$. If the plane of the loop is vertical, and if the bead starts from rest at a point which is level with the center of the loop, find the speed of the bead at the bottom and the reaction of the wire on the bead at that point.

3.18    In the above problem, determine the time for the bead to slide to the bottom.

3.19    In a laboratory experiment a simple pendulum is used to determine the value of $g$. If the amplitude of oscillation of the pendulum is 30°, find the error incurred in the use of the elementary formula

$$T = 2\pi \sqrt{\frac{l}{g}}$$

3.20    A spherical pendulum of length 1 m is undergoing small oscillations about a conical angle $\theta_0$. If $\theta_0 = 30°$, find the period of the conical motion, the period of the oscillation of $\theta$ about $\theta_0$, and the angle of precession $\Delta\varphi$.

3.21    Prove that the two real roots that lie between $+1$ and $-1$ of the equation $f(u) = 0$, Equation (3.75), are equal in the case of the conical pendulum.

3.22    The string of a spherical pendulum of length $l$ is held initially at an angle of 90° with the vertical. The bob is started with a horizontal velocity $v_0$ perpendicular to the string. If

$$v_0^2 = \tfrac{1}{2}gl$$

find the lowest level to which the pendulum bob descends during its motion. [*Hint:* From the initial conditions, $u = 0$ is one root of the equation $f(u) = 0$.]

# 4. Noninertial Reference Systems

It is frequently very convenient and sometimes necessary, in describing the motion of a particle, to use a coordinate system which is not inertial. A coordinate system fixed to the earth, for example, is the most convenient one to use in expressing the motion of a projectile, although the earth is accelerating and rotating.

## 4.1. Translation of the Coordinate System

The simplest type of motion of the coordinate system is that of pure translation. In Figure 4.1 $OXYZ$ are the primary coordinate axes (assumed fixed), and $Oxyz$ are the moving axes. In the case of pure translation, the respective axes $OX$ and $Ox$, and so on, remain parallel. The position vector of a particle $P$ is denoted by $\mathbf{R}$ in the primary or fixed system, and by $\mathbf{r}$ in the moving system. The displacement $OO$ of the moving origin is denoted by $\mathbf{R}_0$. Thus

$$\mathbf{R} = \mathbf{r} + \mathbf{R}_0 \tag{4.1}$$

Taking the first and second derivatives with respect to the time $t$, we find the velocity and acceleration vectors to be given by

$$\mathbf{V} = \mathbf{v} + \mathbf{V}_0 \tag{4.2}$$
$$\mathbf{A} = \mathbf{a} + \mathbf{A}_0 \tag{4.3}$$

in which $\mathbf{V}_0$ and $\mathbf{A}_0$ are the velocity and acceleration, respectively, of the moving

117

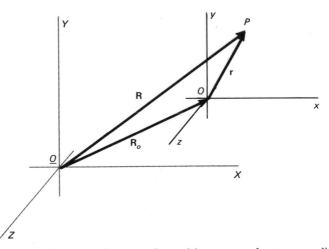

FIGURE 4.1   Relationship between the position vectors for two coordinate
systems undergoing pure translation relative to one another.

origin, and **v** and **a** are the velocity and acceleration, respectively, of $P$ in the
moving system.

In particular, if the moving system is not accelerating, so that $\mathbf{A}_0 = 0$,
then

$$\mathbf{A} = \mathbf{a}$$

That is, the acceleration is the same with respect to either system.   Consequently, if the primary system is inertial, then the moving system is also
inertial.   This statement is true only for the case in which there is no rotation
of the moving system.   The subject of rotation will be studied in Section 4.3
below.


### 4.2.  Inertial Forces

If the primary system is inertial so that Newton's second law

$$\mathbf{F} = m\mathbf{A}$$

is valid, then, from Equation (4.3), the equation of motion in the moving
system is

$$\mathbf{F} - m\mathbf{A}_0 = m\mathbf{a} \tag{4.4}$$

Thus an acceleration $\mathbf{A}_0$ of the reference system can be accounted-for by the
addition of a term $-m\mathbf{A}_0$ to the force **F**.   We shall call this term the *inertial
term*.   If we wish, we can write

$$\text{"F"} = m\mathbf{a} \tag{4.5}$$

for the equation of motion in the moving system if we include the inertial
term as a part of the force "F."   This term is not due to interactions with

other bodies, as are ordinary forces, but stems from the choice of a reference system.   An inertial reference system, as discussed in Chapter 2, is, by definition, one in which there are no inertial terms in the equation of motion.

Sometimes the inertial terms in the equations of motion are called *inertial forces* or *fictitious forces*.   Whether or not one wishes to call them "forces" is purely a matter of terminology.   In any case, these terms are present if an accelerated coordinate system is used to describe the motion of a particle.

## EXAMPLE

A block of wood rests on a rough horizontal table.   If the table is accelerated in a horizontal direction, under what conditions will the block slip?   Let $\mu$ be the coefficient of friction between the block and the table top.   Then the force of friction **F** has a maximum value of $\mu mg$, where $m$ is the mass of the block.   The condition for slipping is that the inertial force $-m\mathbf{A_0}$ exceeds the frictional force where $\mathbf{A_0}$ is the acceleration of the table.   Hence the condition for slipping is

$$|-m\mathbf{A_0}| > \mu mg$$

or

$$A_0 > \mu g$$

### 4.3.   General Motion of the Coordinate System

We now consider the case in which the reference system undergoes both translation and rotation relative to an inertial system.   The position vector of the particle in the inertial system is denoted by **R**, as above, and in the moving system by **r**, Figure 4.2.

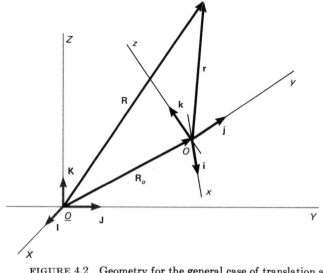

FIGURE 4.2   Geometry for the general case of translation and rotation of the coordinate system.

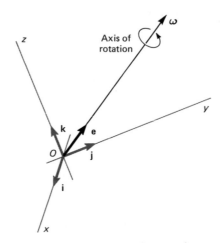

**FIGURE 4.3**   The angular velocity vector of a rotating coordinate system.

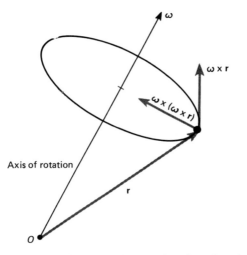

**FIGURE 4.4**   Illustrating the centripetal acceleration.

Let the direction of the axis of rotation of the moving system be specified by the unit vector **e**, as shown in Figure 4.3, and let $\omega$ be the angular speed about this axis.   Then the angular velocity of the moving system is

$$\boldsymbol{\omega} = \omega \mathbf{e}$$

We have previously shown in Section 1.25 that the velocity imparted to a particle due to rotation about an axis can be expressed by the cross product

$$\mathbf{v}_{\text{rot}} = \boldsymbol{\omega} \times \mathbf{r}. \tag{4.6}$$

Consequently Equation (4.2) can be generalized to include rotation by writing

$$\frac{d\mathbf{R}}{dt} = \dot{\mathbf{r}} + \boldsymbol{\omega} \times \mathbf{r} + \mathbf{V}_0 \tag{4.7}$$

In the above equation *the dot above a vector denotes the time derivative of that vector in the rotating coordinate system.* We shall employ this convention in the remainder of this chapter.

To summarize, the velocity of a moving particle measured in a primary inertial coordinate system can be expressed as the sum of three vectors (1) the velocity $\dot{\mathbf{r}}$ of the particle in the moving system, (2) the rotational velocity $\boldsymbol{\omega} \times \mathbf{r}$ that the particle has as a result of being in the rotating system, and (3) the velocity $\mathbf{V}_0$ of the origin of the moving system.

Now a little reflection will show that for the case of any vector quantity $\mathbf{q}$, the time derivative in the primary system is given by adding the term $\boldsymbol{\omega} \times \mathbf{q}$ to the time derivative in the rotating system, namely

$$\left(\frac{d\mathbf{q}}{dt}\right)_{\text{fixed}} = \dot{\mathbf{q}} + \boldsymbol{\omega} \times \mathbf{q} \tag{4.8}$$

This is a very important result and the reader should take time to convince himself that it is correct. A very direct but slightly laborious proof consists of expressing $\mathbf{q}$ in terms of its components in the rotating system. See Problem 4.13 at the end of the chapter.

As a particular example, if we let $\mathbf{q} = \boldsymbol{\omega}$, then we find

$$\left(\frac{d\boldsymbol{\omega}}{dt}\right)_{\text{fixed}} = \dot{\boldsymbol{\omega}}$$

since $\boldsymbol{\omega} \times \boldsymbol{\omega} = 0$. That is, angular acceleration is the same in both systems.

Given Equation (4.8) it is easy to derive the relationships for higher time derivatives. Thus for the second derivative we find

$$\left(\frac{d^2\mathbf{q}}{dt^2}\right)_{\text{fixed}} = \ddot{\mathbf{q}} + (\boldsymbol{\omega} \times \dot{\mathbf{q}}) + (\dot{\boldsymbol{\omega}} \times \mathbf{q}) + \boldsymbol{\omega} \times (\dot{\mathbf{q}} + \boldsymbol{\omega} \times \mathbf{q})$$

$$= \ddot{\mathbf{q}} + 2\boldsymbol{\omega} \times \dot{\mathbf{q}} + \dot{\boldsymbol{\omega}} \times \mathbf{q} + \boldsymbol{\omega} \times (\boldsymbol{\omega} \times \mathbf{q}) \tag{4.9}$$

Let us apply Equation (4.9) to find the relationship between the acceleration vectors. Let $\mathbf{q}$ be equal to the quantity $d\mathbf{R}/dt - \mathbf{V}_0 = \dot{\mathbf{r}} + \boldsymbol{\omega} \times \mathbf{r}$. We then find the following result:

$$\frac{d^2\mathbf{R}}{dt^2} = \ddot{\mathbf{r}} + 2\boldsymbol{\omega} \times \dot{\mathbf{r}} + \dot{\boldsymbol{\omega}} \times \mathbf{r} + \boldsymbol{\omega} \times (\boldsymbol{\omega} \times \mathbf{r}) + \mathbf{A}_0 \tag{4.10}$$

The steps are left as an exercise. The first term on the right-hand side is just the acceleration of the particle in the moving system. The next three terms are rotational terms for the acceleration of the particle as seen in the fixed system. The last term is the acceleration of the moving origin.

The term $2\boldsymbol{\omega} \times \dot{\mathbf{r}}$ is known as the *Coriolis acceleration*. The term $\dot{\boldsymbol{\omega}} \times \mathbf{r}$ is the *transverse acceleration*. The last term $\boldsymbol{\omega} \times (\boldsymbol{\omega} \times \mathbf{r})$ is the *centripetal acceleration*. It is always directed toward the axis of rotation and is perpendicular to that axis, as shown in Figure 4.4

## EXAMPLES

1. A wheel of radius $b$ rolls along the ground with constant forward speed $v$. Find the acceleration, relative to the ground, of any point on the rim. Let us choose a coordinate system fixed to the rotating wheel, and let the moving origin be at the center with the $x$ axis passing through the point in question, as shown in Figure 4.5. Then we have

$$\mathbf{r} = \mathbf{i}b \qquad \ddot{\mathbf{r}} = 0 \qquad \dot{\mathbf{r}} = 0$$

The angular velocity vector is given by

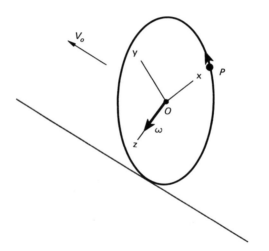

FIGURE 4.5   Rotating coordinate system fixed to a rolling wheel.

$$\boldsymbol{\omega} = \mathbf{k}\omega = \mathbf{k}\frac{v}{b}$$

for the choice of coordinates shown. Hence all terms in the expression for acceleration vanish except the centripetal term. It is given by

$$\begin{aligned}
\mathbf{A} = \boldsymbol{\omega} \times (\boldsymbol{\omega} \times \mathbf{r}) &= \mathbf{k}\omega \times (\mathbf{k}\omega \times \mathbf{i}b) \\
&= \frac{v^2}{b}\mathbf{k} \times (\mathbf{k} \times \mathbf{i}) \\
&= \frac{v^2}{b}\mathbf{k} \times \mathbf{j} \\
&= \frac{v^2}{b}(-\mathbf{i})
\end{aligned}$$

Thus **A** is of magnitude $v^2/b$ and is always directed toward the center of the rolling wheel.

   2. A bicycle travels with constant speed around a track of radius $\rho$. What is the acceleration of the highest point on one of its wheels?   Let $v$ denote the speed of the bicycle and $b$ the radius of the wheel.   We choose a coordinate system with origin at the center of the wheel and with the $x$ axis horizontal pointing toward the center of curvature $C$ of the track.   Rather than have the moving coordinate system rotate with the wheel, we choose a system in which the $z$ axis remains vertical as shown in Figure 4.6.   Thus the $Oxyz$ system rotates with angular velocity

$$\boldsymbol{\omega} = \mathbf{k}\,\frac{v}{\rho}$$

and the acceleration of the moving origin is

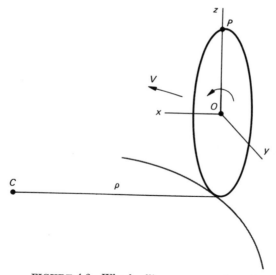

FIGURE 4.6   Wheel rolling on a curved track.

$$\mathbf{A}_0 = \mathbf{i}\,\frac{v^2}{\rho}$$

Since each point on the wheel is moving in a circle of radius $b$ with respect to the moving origin, the acceleration in the $Oxyz$ system of any point on the wheel is directed toward $O$ and has magnitude $v^2/b$. Thus, in the moving system we have

$$\ddot{\mathbf{r}} = -\mathbf{k}\,\frac{v^2}{b}$$

for the point at the top of the wheel. Also, the velocity of this point in the moving system is given by

$$\dot{\mathbf{r}} = -\mathbf{j}v$$

so the Coriolis acceleration is

$$2\boldsymbol{\omega} \times \dot{\mathbf{r}} = 2\left(\frac{v}{\rho}\mathbf{k}\right) \times (-\mathbf{j}v) = 2\frac{v^2}{\rho}\mathbf{i}$$

Since the angular velocity $\boldsymbol{\omega}$ is constant, the transverse acceleration is zero. The centripetal acceleration is also zero, because

$$\boldsymbol{\omega} \times (\boldsymbol{\omega} \times \mathbf{r}) = \frac{v^2}{\rho^2}\mathbf{k} \times (\mathbf{k} \times b\mathbf{k}) = 0$$

Thus the total acceleration of the highest point on the wheel is

$$\mathbf{A} = 3\frac{v^2}{\rho}\mathbf{i} - \frac{v^2}{b}\mathbf{k}$$

### 4.4. Dynamics of a Particle in a Rotating Coordinate System

Since the primary coordinate system is assumed to be an inertial system, then the fundamental equation of motion is

$$\mathbf{F} = m\frac{d^2\mathbf{R}}{dt^2}$$

In view of Equation (4.10), we can now write the equation of motion in terms of the moving coordinates as follows:

$$\mathbf{F} - m\mathbf{A}_0 - 2m\boldsymbol{\omega} \times \dot{\mathbf{r}} - m\dot{\boldsymbol{\omega}} \times \mathbf{r} - m\boldsymbol{\omega} \times (\boldsymbol{\omega} \times \mathbf{r}) = m\ddot{\mathbf{r}} \qquad (4.11)$$

The terms have been transposed in order to display them in the form of inertial forces to be added to the physical force $\mathbf{F}$. The inertial terms have been given names.

The *Coriolis force:*

$$\mathbf{F}_{\text{Cor}} = -2m\boldsymbol{\omega} \times \dot{\mathbf{r}}$$

The *transverse force:*

$$\mathbf{F}_{\text{trans}} = -m\dot{\boldsymbol{\omega}} \times \mathbf{r}$$

The *centrifugal force:*

$$\mathbf{F}_{\text{cent}} = -m\boldsymbol{\omega} \times (\boldsymbol{\omega} \times \mathbf{r})$$

The remaining force $-m\mathbf{A}_0$ is the inertial term due to translation of the coordinate system and has been discussed in Section 4.2 above.

Again, as in the previous discussion of the inertial term $-m\mathbf{A}_0$, we can write the equation of motion in the moving system as

$$\text{``F''} = m\ddot{\mathbf{r}}$$

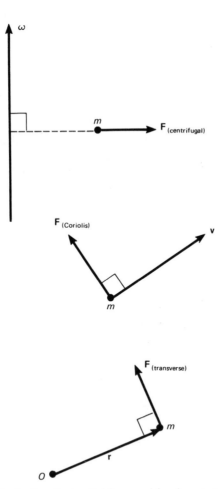

**FIGURE 4.7**  Illustrating the inertial forces arising from rotation of the coordinate system.  The forces are drawn separately for clarity.

in which the total "force" is given by

$$\text{"F"} = \mathbf{F} + \mathbf{F}_{\text{Cor}} + \mathbf{F}_{\text{trans}} + \mathbf{F}_{\text{cent}} - m\mathbf{A}_0 \tag{4.12}$$

The four inertial terms on the right-hand side all depend on the particular coordinate system in which the motion is described. They arise from the inertial properties of matter rather than from the presence of other bodies.

The Coriolis force is particularly interesting. It is present only if a particle is *moving in a rotating coordinate system.* Its direction is always perpendicular to the velocity vector of the particle in the moving system. The Coriolis force thus seems to deflect a moving particle at right angles to its direction of motion. This force is important, for example, in computing the trajectory of a projectile. Coriolis effects are also responsible for the circulation of air around high- or low-pressure areas on the earth's surface. Thus in the case of a high-pressure area the air tends to flow outward and to the right in the northern hemisphere, so that the circulation is clockwise. In the southern hemisphere the reverse is true.

The transverse force is present only if there is an angular acceleration of the rotating coordinate system. This force is always perpendicular to the radius vector $\mathbf{r}$, hence the name transverse.

Finally, the centrifugal force is the familiar force arising from rotation about an axis. This force is always directed outward away from the axis of rotation and is perpendicular to that axis. If $\theta$ is the angle between the radius vector $\mathbf{r}$ and the rotation vector $\boldsymbol{\omega}$, then the magnitude of the centrifugal force is clearly $mr\omega^2 \sin\theta$ or $m\rho\omega^2$ where $\rho$ is the perpendicular distance from the moving particle to the axis of rotation. The various forces are illustrated in Figure 4.7.

## EXAMPLES

1. A bug crawls outward with constant speed $v$ along the spoke of a wheel which is rotating with constant angular velocity $\boldsymbol{\omega}$ about a vertical axis. Find all the forces acting on the bug. First, let us choose a coordinate system fixed on the wheel, and let the $x$ axis point along the spoke in question. Then we have

$$\mathbf{r} = \mathbf{i}x = \mathbf{i}vt$$
$$\dot{\mathbf{r}} = \mathbf{i}\dot{x} = \mathbf{i}v$$
$$\ddot{\mathbf{r}} = 0$$

for the equations of motion of the bug as described in the rotating system. If we choose the $z$ axis to be vertical, then

$$\boldsymbol{\omega} = \mathbf{k}\omega$$

The various forces are then given by the following:

Coriolis force

$$-2m\boldsymbol{\omega} \times \dot{\mathbf{r}} = -2m\omega v(\mathbf{k} \times \mathbf{i}) = -2m\omega v\mathbf{j}$$

Transverse force

$$-m\dot{\boldsymbol{\omega}} \times \mathbf{r} = 0 \qquad (\omega = \text{constant})$$

Centrifugal force

$$\begin{aligned}
-m\boldsymbol{\omega} \times (\boldsymbol{\omega} \times \mathbf{r}) &= -m\omega^2[\mathbf{k} \times (\mathbf{k} \times i x)] \\
&= -m\omega^2(\mathbf{k} \times \mathbf{j}x) \\
&= m\omega^2 x\mathbf{i}
\end{aligned}$$

Thus, Equation (4.11) reads

$$\mathbf{F} - 2m\omega v\mathbf{j} + m\omega^2 x\mathbf{i} = 0$$

Here $\mathbf{F}$ is the real force exerted on the bug by the spoke. The forces are shown in Figure 4.8.

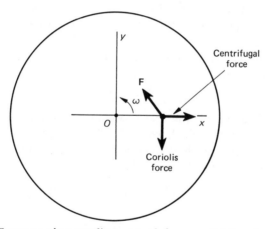

FIGURE 4.8   Forces on a bug crawling outward along a radial line of a rotating turntable.

2. In the above problem, find how far the bug can crawl before it starts to slip, given the coefficient of friction $\mu$ between the bug and the spoke. Since the force of friction $\mathbf{F}$ has a maximum value of $\mu mg$, slipping will start when

$$|\mathbf{F}| = \mu mg$$

or

$$[(2m\omega v)^2 + (m\omega^2 x)^2]^{1/2} = \mu mg$$

Upon solving for $x$, we find

$$x = \frac{(\mu^2 g^2 - 4\omega^2 v^2)^{1/2}}{\omega^2}$$

for the distance the bug can crawl before slipping.

### 4.5.  Effects of the Earth's Rotation

Let us apply the theory developed in the foregoing sections to a coordinate system which is moving with the earth.   Since the angular speed of the earth's rotation is $2\pi$ radians per day, or about $7.3 \times 10^{-5}$ radian per sec, we might expect the effects of such rotation to be relatively small.   Nevertheless, it is the spin of the earth that produces the equatorial bulge; the equatorial radius is some 13 miles greater than the polar radius.

*Static Effects.   The Plumb Line*

We consider first the case of a particle which is at rest on the surface of the earth.   For definiteness, we shall take the particle to be the bob at the end of a plumb line.   Let us choose the origin of our coordinate system to be at the position of the bob, so that $\mathbf{r} = 0$.   Now the angular velocity vector $\boldsymbol{\omega}$ is in the direction of the earth's axis and is very nearly constant; that is, the angular acceleration $\dot{\omega}$ is zero.   For the static case, then, all terms in the equation of motion Equation (4.11), vanish except the applied force $\mathbf{F}$ and the inertial term $-m\mathbf{A}_0$.   The result is

$$\mathbf{F} - m\mathbf{A}_0 = 0$$

The force $\mathbf{F}$ is given by the vector sum of two forces: the true gravitational attraction of the earth (which we shall call $m\mathbf{G}$) and the vertical tension of the plumb line (which we shall denote by $-m\mathbf{g}$).   The forces are shown in Figures 4.9 and 4.10.   We have then

$$m\mathbf{G} - m\mathbf{g} - m\mathbf{A}_0 = 0 \tag{4.13}$$

or

$$\mathbf{g} = \mathbf{G} - \mathbf{A}_0$$

Now the vector $m\mathbf{G}$ is in the direction of the center of the earth.   The acceleration $\mathbf{A}_0$ is just the centripetal acceleration of our moving origin.   Its magnitude is $\rho\omega^2$ or $(r_e \cos \lambda)\omega^2$, where $r_e$ is the radius of the earth, and $\lambda$ is the *geocentric latitude*.   The term $-m\mathbf{A}_0$ (the centrifugal force) is of magnitude $(mr_e \cos \lambda)\omega^2$.   It is directed away from and is perpendicular to the earth's axis, as indicated in Figure 4.9.   Thus the plumb line does not point to the earth's center, but deviates by a small angle $\epsilon$.   From Equation (4.13) the vector $m\mathbf{g}$ may be represented diagrammatically as the third side of a triangle, the other two sides of which are $m\mathbf{G}$ and $-m\mathbf{A}_0$ (Figure 4.10).   Applying the law of sines, we have

$$\frac{\sin \epsilon}{mr_e\omega^2 \cos \lambda} = \frac{\sin \lambda}{mg}$$

or, since $\epsilon$ is small,

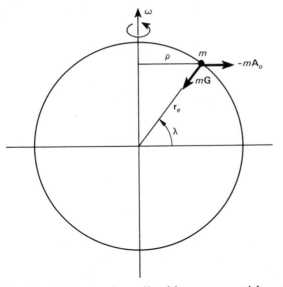

FIGURE 4.9    Gravitational and centrifugal forces on a particle on the surface
of the earth.

$$\sin \epsilon \simeq \epsilon = \frac{r_e \omega^2}{g} \sin \lambda \cos \lambda = \frac{r_e \omega^2}{2g} \sin 2\lambda \qquad (4.14)$$

Thus $\epsilon$ vanishes at the equator ($\lambda = 0$) and at the poles ($\lambda = \pm 90°$), as we would expect.   The maximum deviation of the plumb line from the "true" vertical is at $\lambda = 45°$ where

$$\epsilon_{max} = \frac{r_e \omega^2}{2g} \simeq 1.7 \times 10^{-3} \, \text{radian} \simeq \frac{1}{10} \, \text{degree}$$

The shape of the earth is such that the plumb line is normal to the surface of the earth at any point.   The resulting cross section is approximately elliptical (Figure 4.11).   In the above analysis it is assumed that the gravitational force $m\mathbf{G}$ is constant and is directed toward the center of the earth.

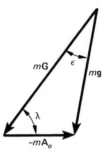

FIGURE 4.10    Vector diagram defining the quantity $m\mathbf{g}$.

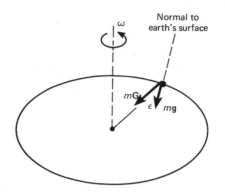

FIGURE 4.11   Exaggerated diagram showing the flattening of the earth due to rotation.

This assumption is not strictly valid, because the earth is not a true sphere. Local variations owing to mountains, mineral deposits, and so on, also affect the direction of the plumb line to a slight extent.

*Dynamic Effects.   Motion of a Projectile*

The equation of motion Equation (4.11) can be written

$$m\ddot{\mathbf{r}} = \mathbf{F} + (m\mathbf{G} - m\mathbf{A}_0) - 2m\boldsymbol{\omega} \times \dot{\mathbf{r}} - m\boldsymbol{\omega} \times (\boldsymbol{\omega} \times \mathbf{r})$$

where $\mathbf{F}$ represents any applied forces other than gravity. But, from the static case considered above, the combination $m\mathbf{G} - m\mathbf{A}_0$ is called $m\mathbf{g}$, hence we can write the equation of motion as

$$m\ddot{\mathbf{r}} = \mathbf{F} + m\mathbf{g} - 2m\boldsymbol{\omega} \times \dot{\mathbf{r}} - m\boldsymbol{\omega} \times (\boldsymbol{\omega} \times \mathbf{r})$$

Let us consider the motion of a projectile. If we neglect air resistance, then $\mathbf{F} = 0$. Furthermore, the term $-m\boldsymbol{\omega} \times (\boldsymbol{\omega} \times \mathbf{r})$ is very small compared to the other terms, so we shall neglect it. The equation of motion then reduces to

$$m\ddot{\mathbf{r}} = m\mathbf{g} - 2m\boldsymbol{\omega} \times \dot{\mathbf{r}} \tag{4.15}$$

in which the last term is the Coriolis force.

    To solve the above equation we shall choose the directions of the coordinate axes $Oxyz$ such that the $z$ axis is vertical (in the direction of the plumb line), the $x$ axis is to the east, and the $y$ axis points north (Figure 4.12). With this choice of axes, we have

$$\mathbf{g} = -\mathbf{k}g$$

and

$$\boldsymbol{\omega} = \omega_x\mathbf{i} + \omega_y\mathbf{j} + \omega_z\mathbf{k}$$
$$= (\omega \cos \lambda)\mathbf{j} + (\omega \sin \lambda)\mathbf{k}$$

Therefore

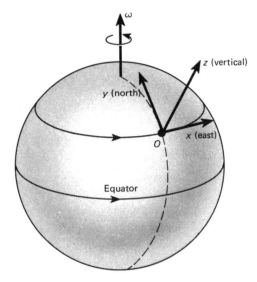

FIGURE 4.12   Coordinate axes for analyzing projectile motion.

$$\boldsymbol{\omega} \times \dot{\mathbf{r}} = \begin{vmatrix} \mathbf{i} & \mathbf{j} & \mathbf{k} \\ \omega_x & \omega_y & \omega_z \\ \dot{x} & \dot{y} & \dot{z} \end{vmatrix}$$

$$= \mathbf{i}(\omega\dot{z}\cos\lambda - \omega\dot{y}\sin\lambda) + \mathbf{j}(\omega\dot{x}\sin\lambda) + \mathbf{k}(-\omega\dot{x}\cos\lambda) \quad (4.16)$$

Upon using the above expressions for $\boldsymbol{\omega} \times \dot{\mathbf{r}}$ in Equation (4.15) and canceling the $m$'s and equating components, we find

$$\ddot{x} = -2\omega(\dot{z}\cos\lambda - \dot{y}\sin\lambda) \quad (4.17)$$
$$\ddot{y} = -2\omega(\dot{x}\sin\lambda) \quad (4.18)$$
$$\ddot{z} = -g + 2\omega\dot{x}\cos\lambda \quad (4.19)$$

for the component differential equations of motion. These equations are not of the separated type, but we can integrate once with respect to $t$ to obtain

$$\dot{x} = -2\omega(z\cos\lambda - y\sin\lambda) + \dot{x}_0 \quad (4.20)$$
$$\dot{y} = -2\omega x\sin\lambda + \dot{y}_0 \quad (4.21)$$
$$\dot{z} = -gt + 2\omega x\cos\lambda + \dot{z}_0 \quad (4.22)$$

The constants of integration $\dot{x}_0$, $\dot{y}_0$, and $\dot{z}_0$ are the initial components of the velocity. The values of $\dot{y}$ and $\dot{z}$ from the last two equations above may be substituted into Equation (4.17). The result is

$$\ddot{x} = 2\omega gt\cos\lambda - 2\omega(\dot{z}_0\cos\lambda - \dot{y}_0\sin\lambda) \quad (4.23)$$

where terms involving $\omega^2$ have been neglected. We now integrate again to get

$$\dot{x} = \omega gt^2\cos\lambda - 2\omega t(\dot{z}_0\cos\lambda - \dot{y}_0\sin\lambda) + \dot{x}_0$$

and therefore

$$x = \tfrac{1}{3}\omega gt^3\cos\lambda - \omega t^2(\dot{z}_0\cos\lambda - \dot{y}_0\sin\lambda) + \dot{x}_0 t \quad (4.24)$$

The above value of $x$ may be inserted into Equations (4.21) and (4.22). The resulting equations, when integrated, yield

$$y = \dot{y}_0 t - \omega \dot{x}_0 t^2 \sin \lambda \qquad (4.25)$$
$$z = -\tfrac{1}{2}gt^2 + \dot{z}_0 t + \omega \dot{x}_0 t^2 \cos \lambda \qquad (4.26)$$

where, again, terms of order $\omega^2$ have been ignored, and the projectile is assumed to be at the origin at time $t = 0$.

Let us consider some special cases. First, if a particle is dropped from rest ($\dot{x}_0 = \dot{y}_0 = \dot{z}_0 = 0$), we have

$$x = \tfrac{1}{3}\omega g t^3 \cos \lambda$$
$$y = 0$$
$$z = -\tfrac{1}{2}gt^2$$

Thus the particle drifts to the *east*. If it falls through a vertical distance $h$, then $t^2 \simeq 2h/g$, and so the eastward drift is

$$\frac{1}{3} \omega \cos \lambda \left( \frac{8h^3}{g} \right)^{1/2}$$

Since the earth turns to the east, common sense would seem to say that the particle should drift westward. Can the reader think of an explanation?

As a second special case, consider a projectile fired with a very high velocity in a nearly horizontal direction, and let us take this direction to be east. Then $\dot{x}_0 = v_0$, and $\dot{y}_0 = \dot{z}_0 = 0$. From Equation (4.25) we have

$$y = -\omega v_0 t^2 \sin \lambda$$

which means that the projectile drifts to the right. If $H$ is the horizontal range, then $H \simeq v_0 t_1$, where $t_1$ is the time of flight. The drift of the projectile to the right (in traversing the eastward distance $H$) is then approximately

$$\frac{\omega H^2}{v_0} \sin \lambda$$

It can be shown that this is the amount of drift, regardless of the direction in which the projectile is initially aimed, provided the trajectory is flat.

### 4.6.   The Foucault Pendulum

In this section we shall study the effect of the earth's rotation on the motion of a spherical pendulum. As in the approximate treatment of the spherical pendulum given in Section 3.17, we shall use rectangular coordinates. As shown in Figure 4.13, the force acting on the pendulum bob is the vector sum of the vertical term $m\mathbf{g}$ and the tension $\mathbf{S}$ in the cord. The differential equation of motion is then

$$m\ddot{\mathbf{r}} = m\mathbf{g} + \mathbf{S} - 2m\boldsymbol{\omega} \times \dot{\mathbf{r}} \qquad (4.27)$$

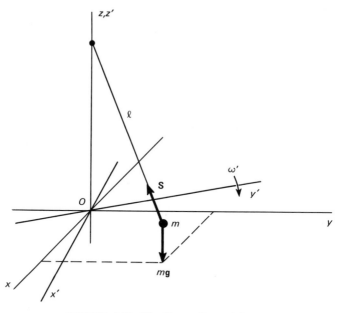

FIGURE 4.13   The Foucault pendulum.

where the term $-m\boldsymbol{\omega} \times (\boldsymbol{\omega} \times \mathbf{r})$ has been neglected. The components of $\boldsymbol{\omega} \times \dot{\mathbf{r}}$ are given by Equation (4.16) above, and the $x$-$y$ components of $\mathbf{S}$ are, as in Section 3.17,

$$S_x = \frac{-x}{l}\,S \qquad S_y = \frac{-y}{l}\,S$$

Equation (4.27) then resolves into

$$m\ddot{x} = \frac{-x}{l}\,S - 2m\omega(\dot{z}\cos\lambda - \dot{y}\sin\lambda) \tag{4.28}$$

$$m\ddot{y} = \frac{-y}{l}\,S - 2m\omega\dot{x}\sin\lambda \tag{4.29}$$

$$m\ddot{z} = S_z - mg + 2m\omega\dot{x}\cos\lambda \tag{4.30}$$

We are interested in the case where the displacement from the vertical is small, so that the tension $S$ is very nearly constant and equal to $mg$. Also, in this case, we can neglect $\dot{z}$ compared to $\dot{y}$ in Equation (4.28). The $x$-$y$ motion is then given by the following differential equations:

$$\ddot{x} = -\frac{g}{l}\,x + 2\omega'\dot{y}$$

$$\ddot{y} = -\frac{g}{l}\,y - 2\omega'\dot{x}$$

where $\omega' = \omega\sin\lambda$.

A convenient method of solving the above pair of differential equations is to multiply the second by $i$ and add the two together. The result is the single equation

$$\ddot{u} + 2i\omega'\dot{u} + \frac{g}{l}u = 0$$

in the complex variable $u = x + iy$.

We have already met a differential equation of this type in the study of harmonic motion, Section 2.14, Equation (2.42). The general solution may be written

$$u = A_1 e^{q_1 t} + A_2 e^{q_2 t}$$

in which $A_1$ and $A_2$ are complex constants of integration, and $q_1, q_2$ are the roots of the auxiliary equation

$$q^2 + 2i\omega'q + \frac{g}{l}u = 0$$

These roots are found to be expressible as

$$q_1, q_2 = -i(\omega' \pm \omega_0)$$

in which $\omega_0{}^2 = g/l$. Here we have neglected $\omega'^2$ compared to $\omega_0{}^2$. We can now write the solution in the form

$$u = (A_1 e^{i\omega_0 t} + A_2 e^{-i\omega_0 t})e^{-i\omega' t}$$

In order to interpret the above result, let us temporarily set $\omega' = 0$. Remembering that $x$ and $y$ are the real and imaginary parts of $u$, we can easily show that the quantity in parentheses represents an elliptical path composed of two perpendicular harmonic motions of frequency $\omega_0$. If we now include the term $e^{-i\omega' t}$, we see that the result is merely to rotate the complex vector $u$ through an angle $-\omega' t$. This is the effect of the earth's rotation, and is thus seen to cause the elliptical path of the spherical pendulum to precess at an angular rate $\omega' = \omega \sin \lambda$. This precession is, of course, superimposed on the natural precession discussed in Section 3.17. The natural precession is ordinarily much larger than the rotational precession under discussion. However, if the pendulum is carefully started by drawing it aside with a thread and letting it start from rest by burning the thread, the natural precession is rendered negligibly small compared to the rotational effect.[1]

The rotational precession is clockwise in the Northern Hemisphere and counterclockwise in the Southern. The period is $2\pi/\omega' = (24/\lambda)$ hr. Thus at a latitude of $45°$ the period is about 34 hr. The result was first demonstrated by the French physicist Jean Foucault in Paris in the year 1851. The

---

[1] For a quantitative treatment of the relative amounts of the two precessions, see J. L. Synge and B. A. Griffith, *Principles of Mechanics*, McGraw-Hill, New York, 1959.

Foucault pendulum has come to be a traditional display in the major planetariums throughout the world.

## DRILL EXERCISES

**4.1** A noninertial coordinate system $Oxyz$ is accelerating with unit acceleration in the direction of the $x$ axis and is also rotating with constant unit angular velocity about that axis. Determine the absolute acceleration of a particle moving with unit speed along the $y$ axis in terms of its distance $y$ from the origin.

**4.2** A plumb line is carried along in a moving train. If $m$ is the mass of the plumb bob, find the tension in the cord and the deflection from the local vertical if (a) the train is moving with constant acceleration $a_0$ in a given direction, and (b) the train is rounding a curve of radius $\rho$ with constant speed $v_0$. Neglect any effects due to the earth's rotation.

**4.3** Find the magnitude and direction of the Coriolis force on a racing car of mass 10 metric tons traveling due south at a speed of 400 km/hr at a latitude of 45°N.

**4.4** A particle is dropped from a height of 200 m at a latitude of 40°N. Find the deflection due to the Coriolis effect.

## PROBLEMS

**4.5** An automobile is traveling with constant forward acceleration $a_0$. At a given instant the forward speed is $v_0$. Find which point on the tire has the greatest absolute acceleration, relative to the ground, and find the direction and magnitude of this acceleration.

**4.6** In the motion of the bicycle wheel, Example 2, p. 122, what is the acceleration of the lowest point on the wheel?

**4.7** Work Example 2, p. 122, by using a coordinate system with the origin at the center of the turning radius, the $x$ axis passing through the center of the wheel, and the $z$ axis vertical.

**4.8** An insect crawls with speed $v$ in a circular path of radius $b$ on a phonograph turntable that revolves with constant angular velocity $\omega$. Describe the motion in a coordinate system fixed to the turntable. Find the acceleration $\mathbf{A}$ of the insect relative to the outside, and find the force of friction $\mathbf{F}$ exerted on the insect. In particular, find $\mathbf{A}$ and $\mathbf{F}$ for the two cases

$$v = b\omega \qquad \text{and} \qquad v = -b\omega$$

Note that in the latter case, the insect is stationary relative to the outside.

**4.9** Derive an expression for the third derivative of the position vector $d^3\mathbf{R}/dt^3$ in terms of the components in a rotating coordinate system.

**4.10**   A projectile is shot vertically with initial speed $v_0$.   Neglecting air resistance, and assuming that $g$ is constant, find where the projectile lands when it hits the ground.

**4.11**   A spherical pendulum of length $l$ undergoes small oscillations about the conical angle $\theta_0$.   For what value of $\theta_0$ will the precession due to the earth's rotation just cancel the natural precession discussed in Chapter 4?   Assume that $\theta_0$ is small.   Find the approximate value for $l = 10$ m and $\lambda = 45°$N.

**4.12**   The differential equation of motion of a charged particle in an electric field **E** and a magnetic field **B** is

$$m\ddot{\mathbf{r}} = q\mathbf{E} + q\mathbf{v} \times \mathbf{B}$$

in an inertial coordinate system.   Show that if the motion is referred to a coordinate system rotating with angular velocity $(q/2m)\mathbf{B}$, the equation of motion becomes

$$m\ddot{\mathbf{r}} = q\mathbf{E}$$

where it is assumed that $B$ is small enough so that terms of order $B^2$ can be neglected.   This result is known as *Larmor's theorem*.

**4.13**   Derive Equation (4.8) by using rectangular coordinates.

# 5. Central Forces and Celestial Mechanics

A force whose line of action passes through a single point or center and whose magnitude depends only on the distance from that center is called a *central force*. Central forces are of fundamental importance in physics, for they include such forces as gravity, electrostatic forces, and others. The forces of interaction between the fundamental particles of nature are mostly central in the sense that, for two particles, either particle acts as a center of force for the other. The main purpose of the present chapter is to study the motion of a particle in a central force field with particular emphasis on gravitational fields.

### 5.1. The Law of Gravity

Newton announced his law of universal gravitation in 1666. It is no exaggeration to state that this marked the beginning of.modern astronomy, for the law of gravity accounts for the motions of the planets of the solar system, their satellites, binary or double stars, and even stellar systems. The law may be stated:

Every particle in the universe attracts every other particle with a force that varies directly as the product of the masses of the two particles and inversely as the square of their distance apart. The direction of the force is along the straight line joining the two particles.

We can express the law vectorially by the equation

$$\mathbf{F}_{ij} = G \frac{m_i m_j}{r_{ij}^2} \left( \frac{\mathbf{r}_{ij}}{r_{ij}} \right) \tag{5.1}$$

where $\mathbf{F}_{ij}$ is the force on particle $i$, of mass $m_i$ exerted by particle $j$, of mass $m_j$. The vector $\mathbf{r}_{ij}$ is the directed line segment running from particle $i$ to particle $j$, as shown in Figure 5.1. The law of action and reaction requires that $\mathbf{F}_{ij} = -\mathbf{F}_{ji}$. The constant of proportionality $G$ is known as the universal constant of gravitation. Its value is determined in the laboratory by carefully measuring the force between two spherical bodies of known mass. The

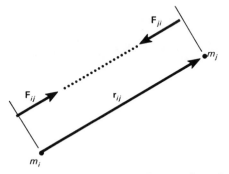

FIGURE 5.1   Action and reaction in Newton's law of gravity.

currently accepted value of $G$, as obtained at the U.S. National Bureau of Standards, is

$$G = (6.673 \pm 0.003) \times 10^{-8} \frac{\text{dyne cm}^2}{\text{g}^2}$$

All of our present knowledge of the masses of astronomical bodies, including the earth, is based on the value of $G$.

## 5.2.   Gravitational Force Between a Uniform Sphere and a Particle

In Chapter 2, where we discussed the motion of a falling body, it was asserted that the gravitational force of the earth on a particle above the earth's surface is inversely proportional to the square of the particle's distance from the center of the earth; that is, the earth attracts as if all of its mass were concentrated at a single point. We shall now prove that this is true for any uniform spherical body, or any spherically symmetric distribution of matter.

Consider first a thin uniform shell of mass $M$ and radius $R$. Let $r$ be the distance from the center $O$ to a test particle $P$ of mass $m$ (Figure 5.2). It is assumed that $r > R$. We shall divide the shell into circular rings of width $R \, \Delta\theta$ where, as shown in the figure, the angle $POQ$ is denoted by $\theta$, $Q$ being a point on the ring. The circumference of our representative ring element is therefore $2\pi R \sin \theta$, and its mass $\Delta M$ is given by

$$\Delta M \simeq \rho 2\pi R^2 \sin\theta \, \Delta\theta$$

where $\rho$ is the mass per unit area of the shell.

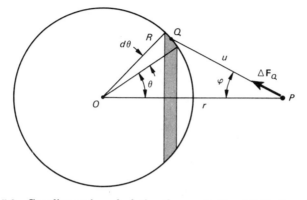

FIGURE 5.2   Coordinates for calculating the gravitational field of a spherical shell.

Now the gravitational force exerted on $P$ by a small subelement $Q$ of the ring (which we shall regard as a particle) is in the direction $PQ$. Let us resolve this force $\Delta\mathbf{F}_q$ into two components, one component along $PO$, of magnitude $\Delta F_q \cos\varphi$, the other perpendicular to $PO$, of magnitude $\Delta F_q \sin\varphi$. Here $\varphi$ is the angle $OPQ$, as shown in the figure. From symmetry we can easily see that the vector sum of all of the perpendicular components exerted on $P$ by the whole ring vanishes. The force $\Delta\mathbf{F}$ exerted by the entire ring is therefore in the direction $PO$, and its magnitude $\Delta F$ is obtained by summing the components $\Delta F_q \cos\varphi$. The result is clearly

$$\Delta F = G\frac{m\Delta M}{u^2}\cos\varphi = G\frac{m2\pi\rho R^2 \sin\theta \cos\varphi}{u^2}\Delta\theta$$

where $u$ is the distance $PQ$ (the distance from the particle $P$ to the ring) as shown. The magnitude of the force exerted on $P$ by the whole shell is then obtained by taking the limit of $\Delta\theta$ and integrating:

$$F = Gm2\pi\rho R^2 \int_0^\pi \frac{\sin\theta \cos\varphi \, d\theta}{u^2}$$

The integral is most easily evaluated by expressing the integrand in terms of $u$. From the triangle $OPQ$ we have, from the law of cosines,

$$r^2 + R^2 - 2rR\cos\theta = u^2$$

Differentiating, we have, since both $R$ and $r$ are constant,

$$rR\sin\theta \, d\theta = u \, du$$

Also, in the same triangle $OPQ$, we can write

$$\cos\varphi = \frac{u^2 + r^2 - R^2}{2ru}$$

Upon performing the substitutions given by the above two equations, we obtain

$$F = Gm2\pi\rho R^2 \int_{\theta=0}^{\theta=\pi} \frac{u^2 + r^2 - R^2}{2Rr^2u^2}\, du$$

$$= \frac{GmM}{4Rr^2} \int_{r-R}^{r+R} \left(1 + \frac{r^2 - R^2}{u^2}\right) du$$

$$= \frac{GmM}{r^2}$$

where $M = 4\pi\rho R^2$ is the mass of the shell. We can then write vectorially

$$\mathbf{F} = -G\frac{Mm}{r^2}\, \mathbf{e}_r \tag{5.2}$$

where $\mathbf{e}_r$ is the unit radial vector from the origin $O$. The above result means that a uniform spherical shell of matter attracts an external particle as if the whole mass of the shell were concentrated at its center. This will be true for every concentric spherical portion of a solid uniform sphere. A uniform spherical body, therefore, attracts an external particle as if the entire mass of the sphere were located at the center. The same is true also for a nonuniform sphere as long as the distribution of mass is radially symmetric.

It can be shown that the gravitational force on a particle located *inside* a uniform spherical shell is zero. The proof is left as an exercise.

### 5.3.  Potential Energy in a Gravitational Field. Gravitational Potential

In Chapter 2, Section 2.11, we proved that the inverse-square law of force leads to an inverse first power law for the potential energy function. In this section we shall derive this same relationship in a more physical way.

Let us consider the work $W$ required to move a test particle of mass $m$ along some prescribed path in the gravitational field of another particle of mass $M$.

We shall place the particle of mass $M$ at the origin of our coordinate system, as shown in Figure 5.3(a). Since the force $\mathbf{F}$ on the test particle is given by $\mathbf{F} = -(GMm/r^2)\mathbf{e}_r$, then, to overcome this force, an external force $-\mathbf{F}$ must be applied. The work $dW$ done in moving the test particle through a distance $d\mathbf{r}$ is thus given by

$$dW = -\mathbf{F}\cdot d\mathbf{r} = \frac{GMm}{r^2}\, \mathbf{e}_r \cdot d\mathbf{r} \tag{5.3}$$

Now we can resolve $d\mathbf{r}$ into two components: $\mathbf{e}_r\, dr$ parallel to $\mathbf{e}_r$ (the radial component) and the other at right angles to $\mathbf{e}_r$ [Figure 5.3(b)]. Clearly,

$$\mathbf{e}_r\cdot d\mathbf{r} = dr$$

and so $W$ is given by

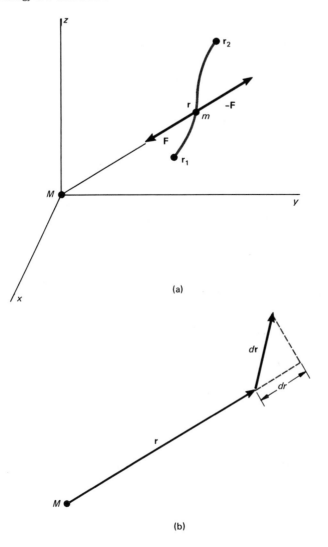

(a)

(b)

FIGURE 5.3  Diagram for finding the work required to move a test particle
from one point to another in a gravitational field.

$$W = GMm \int_{r_1}^{r_2} \frac{dr}{r^2} = -GMm \left( \frac{1}{r_2} - \frac{1}{r_1} \right) \tag{5.4}$$

where $r_1$ and $r_2$ are the radial distances of the particle at the beginning and
end, respectively, of the path.   Thus the work is independent of the particular path taken; it depends only on the end points.   This verifies a fact we
already knew, namely that the inverse-square law of force is conservative.

We can define the potential energy of a particle of mass at a given point
in the gravitational field of another particle as the work done in moving the

test particle from some (arbitrary) reference position to the point in question. It is convenient to take the reference position at infinity.[1]  Putting $r_1 = \infty$ and $r_2 = r$ in Equation (6.4), we have

$$V(r) = GMm \int_{\infty}^{r} \frac{dr}{r^2} = -\frac{GMm}{r} \tag{5.5}$$

It is sometimes convenient to define a quantity $\Phi$, called the *gravitational potential*, as the gravitational potential energy per unit mass:

$$\Phi = \frac{V}{m}$$

Thus the gravitational potential in the field of a particle of mass $M$ is given by

$$\Phi = -\frac{GM}{r} \tag{5.6}$$

If we have a number of particles $M_1$, $M_2$, . . . $M_i$, . . . located at the positions $\mathbf{r}_1$, $\mathbf{r}_2$, . . . $\mathbf{r}_i$ . . . , then the gravitational potential at the point $(x,y,z)$ is the sum of the gravitational potentials of all the particles, that is

$$\Phi(x,y,z) = \Sigma\Phi_i = -G\Sigma\frac{M_i}{u_i} \tag{5.7}$$

in which $u_i$ is the distance from the particle $i$, of mass $M_i$, to the field point $\mathbf{r}(x,y,z)$.  Thus

$$u_i = |\mathbf{r} - \mathbf{r}_i|$$

The ratio of the gravitational force on a given particle to the mass of that particle is called the *gravitational field intensity*.  It is denoted by $\mathsf{g}$.  Then

$$\mathsf{g} = \frac{\mathbf{F}}{m}$$

The relationship between field intensity and the potential is the same as that between the force $\mathbf{F}$ and the potential energy $V$, namely

$$\mathsf{g} = -\nabla\Phi \tag{5.8}$$
$$\mathbf{F} = -\nabla V$$

The gravitational field intensity can be calculated by first finding the potential function from Equation (5.7) and then calculating the gradient.  This

---

[1] It is important to note that it is not legitimate to *define* potential energy as the integral of $\mathbf{F} \cdot d\mathbf{r}$ unless we know in advance that $\mathbf{F}$ is conservative, that is, that a potential function exists.

method is usually simpler than the method of calculating the field directly from the inverse-square law. The reason is that the potential is a scalar sum whereas the field is given by a vector sum. The situation is quite analogous to the theory of electrostatic fields. In fact, one can apply any of the corresponding results from electrostatics to find gravitational fields and potentials with the proviso, of course, that there are no negative masses.

### Potential of a Uniform Spherical Shell

As an example, let us find the potential function for a uniform spherical shell. By using the same notation as that of Figure 5.2, we have

$$\Phi = -G \int \frac{dM}{u} = -G \int \frac{2\pi \rho R^2 \sin \theta \, d\theta}{u}$$

From the same relation between $u$ and $\theta$ that we used earlier, we find that the above equation may be simplified to read

$$\Phi = -G \frac{2\pi \rho R^2}{rR} \int_{r-R}^{r+R} du = -\frac{GM}{r} \tag{5.9}$$

where $M$ is the mass of the shell. This is the same potential function as that of a single particle of mass $M$ located at $O$. Hence the gravitational field outside the shell is the same as if the entire mass were concentrated at the center. It is left as a problem to show that, with an appropriate change of the integral and its limits, the potential inside the shell is constant and hence that the field there is zero.

### Potential and Field of a Thin Ring

We now wish to find the potential function and the gravitational field intensity in the plane of a thin circular ring. Let the ring be of radius $R$ and mass $M$. Then, for an exterior point lying in the plane of the ring, Figure 5.4, we have

$$\Phi = -G \int \frac{dM}{u} = -G \int_0^{2\pi} \frac{\mu R \, d\theta}{u}$$

in which $\mu$ is the linear density of the ring. In order to evaluate the integral, we shall express the integrand in terms of the angle $\psi$ shown. In the triangle $OPQ$ we have

$$R \sin \psi = r \sin \varphi$$

Differentiating,

$$R \cos \psi \, d\psi = r \cos \varphi \, d\varphi = r \cos \varphi(-d\theta - d\psi)$$

The last step follows from the fact that $\theta + \varphi + \psi = \pi$. Upon transposing terms and using the relation $u = R \cos \psi + r \cos \varphi$, we obtain

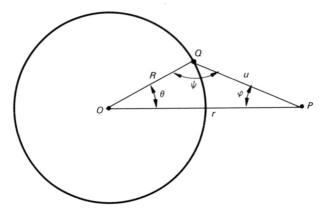

FIGURE 5.4   Coordinates for calculating the gravitational field of a ring.

$$u \, d\psi = -r \cos \varphi \, d\theta = -(r^2 - R^2 \sin^2 \psi)^{1/2} \, d\theta$$

Hence the integral above becomes

$$\Phi = -G\mu R4 \int_0^{\pi/2} (r^2 - R^2 \sin^2 \psi)^{-1/2} \, d\psi = -G \frac{4\mu R}{r} K\left(\frac{R}{r}\right) \quad (5.10)$$

where $K$ is the complete elliptic integral as defined in Section 3.15.   By expanding the integrand as a series and integrating term by term, we can also
write

$$\Phi = -G \frac{4\mu R}{r} \left(\frac{\pi}{2} + \frac{\pi R^2}{8r^2} + \cdots \right)$$

$$= -\frac{GM}{r} \left(1 + \frac{R^2}{4r^2} + \cdots \right)$$

The field intensity at a distance $r$ from the center of the ring is then in the
radial direction (since $\Phi$ is not a function of $\theta$), and is given by

$$\mathbf{G} = -\frac{\partial \Phi}{\partial r} \mathbf{e}_r = \left(-\frac{GM}{r^2} - \frac{3GMR^2}{4r^2} - \cdots \right) \mathbf{e}_r$$

Thus the field is *not* given by an inverse-square law.   If $r$ is very large compared to $R$, however, the first term predominates, and the field is approximately of the inverse-square type.   In fact, the same is true for a finite body
of any shape; that is, *for distances large compared to the linear dimensions of the
body, the field tends to become prodominantly inverse square.*

### 5.4.   Potential Energy in a General Central Field

We have previously shown that a central field of the inverse-square type is conservative.   Let us now consider the question as to whether or not *any* central field of force is conservative.   A general isotropic central field can be expressed in the following way:

$$\mathbf{F} = f(r)\mathbf{e}_r \tag{5.11}$$

in which $\mathbf{e}_r$ is the unit radial vector.   To apply the test for conservativeness, we calculate the curl of $\mathbf{F}$.   It is convenient here to employ spherical coordinates for which the curl is given in Appendix IV. We find

$$\nabla \times \mathbf{F} = \frac{1}{r^2 \sin \theta} \begin{vmatrix} \mathbf{e}_r & \mathbf{e}_\theta r & \mathbf{e}_\phi r \sin \theta \\ \dfrac{\partial}{\partial r} & \dfrac{\partial}{\partial \theta} & \dfrac{\partial}{\partial \phi} \\ F_r & rF_\theta & rF_\phi \sin \theta \end{vmatrix}$$

For our central force $F_r = f(r)$, $F_\theta = 0$, $F_\phi = 0$.   The curl then reduces to

$$\nabla \times \mathbf{F} = \frac{\mathbf{e}_\phi}{r \sin \theta}\frac{\partial f}{\partial \phi} - \frac{\mathbf{e}_\phi}{r}\frac{\partial f}{\partial \theta} = 0$$

The two partial derivatives both vanish since $f(r)$ does not depend on the angular coordinates $\phi$ and $\theta$.   Thus the curl vanishes and so the general central field defined by Equation (5.11) is conservative. We recall that the same test was applied to the inverse-square field in Section 3.7, Example 4.

We can now define a potential energy

$$V(r) = -\int_\infty^r \mathbf{F}\cdot d\mathbf{r} = -\int_\infty^r f(r)\,dr \tag{5.12}$$

This allows us to calculate the potential energy function, given the force function.   Conversely, if we know the potential energy function, we have

$$f(r) = -\frac{\partial V(r)}{\partial r} \tag{5.13}$$

giving the force function for a central field.

## 5.5.   Angular Momentum in Central Fields

We previously proved in Section 3.2 that the time rate of change of the quantity $\mathbf{r} \times \mathbf{p}$, the angular momentum, is equal to the moment of the force acting on a particle about a given origin.   Let us denote the angular momentum by the symbol $\mathbf{L}$. Then the angular momentum theorem states that

$$\frac{d\mathbf{L}}{dt} = \mathbf{r} \times \mathbf{F} \tag{5.14}$$

Let us apply the above general rule to the particular case of a particle moving in a central field.   Here the force $\mathbf{F}$ acts in the direction of the radius vector $\mathbf{r}$.   Hence the cross product $\mathbf{r} \times \mathbf{F}$ vanishes, that is, there is zero moment.   Consequently, for any central field

$$\frac{d\mathbf{L}}{dt} = 0$$

and therefore

$$\mathbf{L} = \text{constant}$$

*The angular momentum of a particle moving in a central field always remains constant.*

As a corollary, it follows that the path of motion of a particle in a central field remains in a *single plane*, because the constant angular momentum vector $\mathbf{L}$ is normal to both $\mathbf{r}$ and $\mathbf{v}$, and therefore is normal to the plane in which the particle moves.   Thus it is possible, without loss of generality, to employ plane polar coordinates in treating central motion.

*Magnitude of the Angular Momentum*

In order to determine the magnitude of the angular momentum, it is convenient to resolve the velocity vector $\mathbf{v}$ into radial and transverse components in polar coordinates.   Thus we can write

$$\mathbf{v} = \dot{r}\mathbf{e}_r + r\dot{\theta}\mathbf{e}_\theta$$

in which $\mathbf{e}_r$ is the unit radial vector and $\mathbf{e}_\theta$ is the unit transverse vector.   The magnitude of the angular momentum is then given by

$$L = |\mathbf{r} \times m\mathbf{v}| = |r\mathbf{e}_r \times m(\dot{r}\mathbf{e}_r + r\dot{\theta}\mathbf{e}_\theta)|$$

Since $\mathbf{e}_r \times \mathbf{e}_r = 0$ and $\mathbf{e}_r \times \mathbf{e}_\theta = 1$, we find

$$L = mr^2\dot{\theta} = \text{constant}$$

for a particle moving in a central field of force.

### 5.6.   The Law of Areas.   Kepler's Laws of Planetary Motion

The angular momentum of a particle is related to the rate at which the position vector sweeps out area.   To show this, consider Figure 5.5 which illustrates two successive position vectors $\mathbf{r}$ and $\mathbf{r} + \Delta\mathbf{r}$ representing the

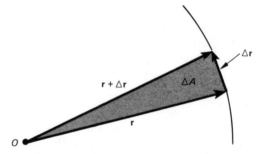

FIGURE 5.5   Area swept out by the radius vector.

motion of a particle in a time interval $\Delta t$.   The area $\Delta A$ of the shaded triangular segment lying between the two vectors is expressible as

$$\Delta A = \tfrac{1}{2} |\mathbf{r} \times \Delta \mathbf{r}|$$

Upon division by $\Delta t$ and taking the limit, we have

$$\frac{dA}{dt} = \frac{1}{2} |\mathbf{r} \times \mathbf{v}| \tag{5.15}$$

From the definition of $\mathbf{L}$, we can further write

$$\frac{dA}{dt} = \frac{1}{2m} |\mathbf{r} \times m\mathbf{v}| = \frac{L}{2m} \tag{5.16}$$

for the rate at which the radius vector sweeps out area.   Since the angular momentum $\mathbf{L}$ is constant in any central field, it follows that the *areal velocity* $dA/dt$ is also constant in a central field.

### Kepler's Laws

The fact that the planets move about the sun in such a way that the areal velocities are constant was discovered empirically by Johannes Kepler in 1609.   Kepler deduced this rule, and two others,[2] from a painstaking study of planetary positions recorded by Tycho Brahe.   Kepler's three laws are:

(1)  Each planet moves in an ellipse with the sun as a focus.
(2)  The radius vector sweeps out equal areas in equal times.
(3)  The square of the period of revolution about the sun is proportional to the cube of the major axis of the orbit.

Newton showed that Kepler's three laws are consequences of the law of gravity.   From the argument leading to Equation (5.16), we see that the second law comes about from the fact that the gravitational field of the sun is central. The other two laws, as we shall show later, are consequences of the fact that the force varies as the inverse square of the distance.

### 5.7.   Orbit of a Particle in a Central-Force Field

To study the motion of a particle in a central field, it is convenient to express the differential equation of motion

$$m\ddot{\mathbf{r}} = f(r)\,\mathbf{e}_r$$

in polar coordinates.   As shown in Chapter 1, the radial component of $\ddot{\mathbf{r}}$ is $\ddot{r} - r\dot{\theta}^2$, and the transverse component is $2\dot{r}\dot{\theta} + r\ddot{\theta}$.   The component differential equations of motion are then

$$m(\ddot{r} - r\dot{\theta}^2) = f(r) \tag{5.17}$$
$$m(2\dot{r}\dot{\theta} + r\ddot{\theta}) = 0 \tag{5.18}$$

---

[2] The third law was announced in 1619.

From the latter equation it follows that

$$\frac{d}{dt}(r^2\dot\theta) = 0$$

or

$$r^2\dot\theta = \text{constant} = h \tag{5.19}$$

From Equation (5.14) we see that

$$h = \frac{L}{m} \tag{5.20}$$

Thus $h$ is the angular momentum per unit mass. Its constancy is simply a restatement of a fact which we already know, namely, that the angular momentum of a particle is constant when it is moving under the action of a central force.

Given a certain radial force function $f(r)$, we could, in theory, solve the pair of differential equations [Equations (5.17) and (5.18)] to obtain $r$ and $\theta$ as functions of $t$. It is often the case that one is interested only in the path in space (the *orbit*) without regard to the time $t$. To find the equation of the orbit, we shall use the variable $u$ defined by

$$r = \frac{1}{u} \tag{5.21}$$

Then

$$\dot r = -\frac{1}{u^2}\dot u = -\frac{1}{u^2}\dot\theta\frac{du}{d\theta} = -h\frac{du}{d\theta} \tag{5.22}$$

The last step follows from the fact that

$$\dot\theta = hu^2 \tag{5.23}$$

according to Equations (5.19) and (5.21).

Differentiating a second time, we have

$$\ddot r = -h\frac{d}{dt}\frac{du}{d\theta} = -h\dot\theta\frac{d^2u}{d\theta^2} = -h^2u^2\frac{d^2u}{d\theta^2} \tag{5.24}$$

From these values of $r$, $\dot\theta$, and $\ddot r$, we readily find that Equation (5.17) transforms to

$$\frac{d^2u}{d\theta^2} + u = -\frac{1}{mh^2u^2}f(u^{-1}) \tag{5.25}$$

The above equation is the differential equation of the orbit of a particle moving under a central force. The solution gives $u$ (hence $r$) as a function of $\theta$. Conversely, if one is given the polar equation of the orbit, namely, $r = r(\theta) = u^{-1}$, then the force function can be found by differentiating to get $d^2u/d\theta^2$ and inserting this into the differential equation.


## EXAMPLES

1. A particle in a central field moves in the spiral orbit

$$r = c\theta^2$$

Determine the form of the force function. We have

$$u = \frac{1}{c\theta^2}$$

and

$$\frac{du}{d\theta} = \frac{-2}{c}\theta^{-3} \qquad \frac{d^2u}{d\theta^2} = \frac{6}{c}\theta^{-4} = 6cu^2$$

Then, from Equation (5.25),

$$6cu^2 + u = -\frac{1}{mh^2u^2}f(u^{-1})$$

Hence

$$f(u^{-1}) = -mh^2(6cu^4 + u^3)$$

and

$$f(r) = -mh^2\left(\frac{6c}{r^4} + \frac{1}{r^3}\right)$$

Thus the force is a combination of an inverse cube and inverse fourth power law.

   2. In the above problem, determine how the angle $\theta$ varies with time. Here we use the fact that $h = r^2\dot{\theta}$ is constant.   Thus

$$\dot{\theta} = hu^2 = h\,\frac{1}{c^2\theta^4}$$

or

$$\theta^4\,d\theta = \frac{h}{c^2}\,dt$$

and so, by integrating, we find

$$\frac{\theta^5}{5} = hc^{-2}t$$

where the constant of integration is taken to be zero.   Then

$$\theta = Ct^{1/5}$$

where

$$C = \text{constant} = (5hc^{-2})^{1/5}$$

## 5.8.   Energy Equation of the Orbit

   The square of the speed is given in polar coordinates by

$$v^2 = \dot{r}^2 + r^2\dot{\theta}^2$$

Since a central force is conservative, the total energy $T + V$ is constant and is given by

$$\tfrac{1}{2}m(\dot{r}^2 + r^2\dot{\theta}^2) + V(r) = E = \text{constant} \qquad (5.26)$$

We can also write the above equation in terms of the variable $u = 1/r$. From Equations (5.22) and (5.23) we obtain

$$\frac{1}{2}mh^2\left[\left(\frac{du}{d\theta}\right)^2 + u^2\right] + V(u^{-1}) = E \tag{5.27}$$

In the above equation the only variables occurring are $u$ and $\theta$. We shall call this equation, therefore, *the energy equation of the orbit.*

## EXAMPLE

In the example of the preceding section we had for the spiral orbit $r = c\theta^2$:

$$\frac{du}{d\theta} = \frac{-2}{c}\theta^{-3} = -2c^{1/2}u^{3/2}$$

so the energy equation of the orbit is

$$\tfrac{1}{2}mh^2(4cu^3 + u^2) + V = E$$

Thus

$$V(r) = E - \frac{1}{2}mh^2\left(\frac{4c}{r^3} + \frac{1}{r^2}\right)$$

This readily gives the force function of the example above, since $f(r) = -dV/dr$.

### 5.9.  Orbits in an Inverse-Square Field

The most important type of central field is that in which the force varies inversely as the square of the radial distance:

$$f(r) = -\frac{k}{r^2}$$

In the above equation, since we have included a minus sign, the constant of proportionality $k$ is positive for an attractive force, and vice versa. (As we have seen in Section 5.2, $k = GMm$ for a gravitational field.)   The equation of the orbit [Equation (5.25)] then becomes

$$\frac{d^2u}{d\theta^2} + u = \frac{k}{mh^2} \tag{5.28}$$

The general solution is clearly

$$u = A\cos(\theta - \theta_0) + \frac{k}{mh^2}$$

or

$$r = \frac{1}{A \cos (\theta - \theta_0) + k/mh^2} \tag{5.29}$$

The constants of integration $A$ and $\theta_0$ are determined from the initial conditions. The value of $\theta_0$ merely determines the orientation of the orbit, so we can, without loss of generality in discussing the form of the orbit, choose $\theta_0 = 0$.   Then

$$r = \frac{1}{A \cos \theta + k/mh^2} \tag{5.30}$$

This is the polar equation of the orbit.   It is the equation of a conic section (ellipse, parabola, or hyperbola) with the origin at a focus.   The equation can be written in the standard form

$$r = r_0 \frac{1 + e}{1 + e \cos \theta} \tag{5.31}$$

where

$$e = \frac{A m h^2}{k} \tag{5.32}$$

and

$$r_0 = \frac{m h^2}{k(1 + e)} \tag{5.33}$$

The constant $e$ is called the *eccentricity*.   The different cases, illustrated in Figure 5.6, are

$e < 1:ellipse$
$e = 0:circle\ (special\ case\ of\ an\ ellipse)$
$e = 1:parabola$
$e > 1:hyperbola$

From Equation (5.31), $r_0$ is the value of $r$ for $\theta = 0$.   The value of $r$ for $\theta = \pi$ is given by

$$r_1 = r_0 \frac{1 + e}{1 - e} \tag{5.34}$$

In reference to the elliptic orbits of the planets around the sun, the distance $r_0$ is called the *perihelion* distance (closest to the sun) and the distance $r_1$ is called the *aphelion* distance (farthest from the sun).   The corresponding distances for the orbit of the moon around the earth—and for the orbits of the earth's artificial satellites—are called the *perigee* and *apogee* distances, respectively.

The orbital eccentricities of the planets are quite small.   (See Table 5.1 below.)   For example, in the case of the earth's orbit $e = 0.017$, $r_0 = 91,000,000$ miles, and $r_1 = 95,000,000$ miles.   On the other hand, the comets generally have large orbital eccentricities (highly elongated orbits).   Halley's

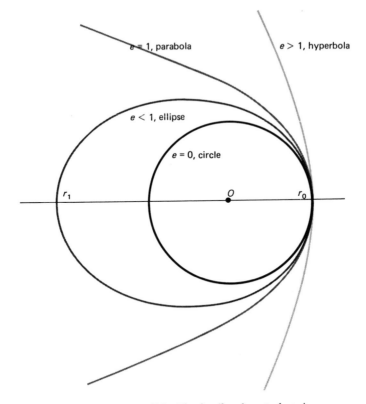

FIGURE 5.6  The family of central conics.

comet, for instance, has an orbital eccentricity of 0.967 with a perihelion distance of only 55,000,000 miles, while at aphelion it is beyond the orbit of Neptune.  Many comets (the nonrecurring type) have parabolic or hyperbolic orbits.

### Orbital Parameters from Initial Conditions

From Equation (5.33) we find the eccentricity can be expressed as

$$e = \frac{mh^2}{kr_0} - 1 \qquad (5.35)$$

Let $v_0$ be the speed of the particle at $\theta = 0$.  Then, from the definition of the constant $h$ we have

$$h = r^2\dot{\theta} = r_0^2\dot{\theta}_0 = r_0 v_0$$

The eccentricity is then given by

$$e = \frac{mr_0 v_0^2}{k} - 1 \qquad (5.36)$$

For a circular orbit $(e = 0)$ we have then $k = mr_0v_0^2$ or

$$\frac{k}{r_0^2} = \frac{mv_0^2}{r_0}$$

Now let us denote the quantity $k/mr_0$ by $v_c^2$, so that if $v_0 = v_c$, the orbit is a circle. The expression for the eccentricity, Equation (5.36), can then be written

$$e = (v_0/v_c)^2 - 1 \qquad (5.37)$$

and the equation of the orbit can be written as

$$r = r_0 \frac{(v_0/v_c)^2}{1 + [(v_0/v_c)^2 - 1]\cos\theta} \qquad (5.38)$$

The value of $r_1$ is given by $\theta = \pi$, thus

$$r_1 = r_0 \frac{(v_0/v_c)^2}{2 - (v_0/v_c)^2}$$

## EXAMPLE

A rocket satellite is going around the earth in a circular orbit of radius $r_0$. A sudden blast of the rocket motor increases the speed by 10 percent. Find the equation of the new orbit, and compute the apogee distance. Let $v_c$ be the speed in the circular orbit, and let $v_0$ be the new initial speed; that is

$$v_0 = 1.1v_c$$

Equation (5.38) of the new orbit then reads

$$r = r_0 \frac{1.21}{1 + 0.21\cos\theta}$$

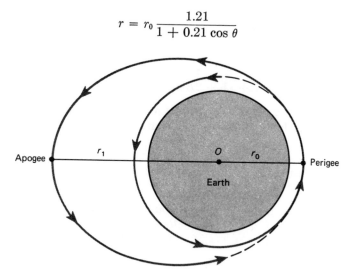

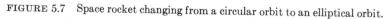

FIGURE 5.7   Space rocket changing from a circular orbit to an elliptical orbit.

and the apogee distance is

$$r_1 = r_0 \frac{1.21}{2 - 1.21} = 1.53 r_0$$

The orbits are shown in Figure 5.7.

### 5.10.   Orbital Energies in the Inverse-Square Field

Since the potential-energy function $V(r)$ for an inverse-square force field is given by

$$V(r) = -\frac{k}{r} = -ku$$

the energy equation of the orbit, Equation (5.27), then reads

$$\frac{1}{2} mh^2 \left[ \left( \frac{du}{d\theta} \right)^2 + u^2 \right] - ku = E$$

or, upon separating variables,

$$d\theta = \left( \frac{2E}{mh^2} + \frac{2ku}{mh^2} - u^2 \right)^{-1/2} du$$

Upon integrating, we find

$$\theta = \sin^{-1} \left[ \frac{mh^2 u - k}{(k^2 + 2Emh^2)^{1/2}} \right] + \theta_0$$

where $\theta_0$ is a constant of integration.   If we let $\theta_0 = -\pi/2$ and solve for $u$, we obtain

$$u = \frac{k}{mh^2} [1 + (1 + 2Emh^2 k^{-2})^{1/2} \cos \theta]$$

or

$$r = \frac{mh^2 k^{-1}}{1 + (1 + 2Emh^2 k^{-2})^{1/2} \cos \theta} \tag{5.39}$$

This is the polar equation of the orbit.   If we compare it with Equations (5.31) and (5.32), we see that the eccentricity is given by

$$e = (1 + 2Emh^2 k^{-2})^{1/2} \tag{5.40}$$

The above expression for the eccentricity allows us to classify the orbits according to the total energy $E$ as follows

$$E < 0 \quad e < 1 : \textit{closed orbits (ellipse or circle)}$$
$$E = 0 \quad e = 1 : \textit{parabolic orbit}$$
$$E > 0 \quad e > 1 : \textit{hyperbolic orbit}$$

Since $E = T + V$ and is constant, the closed orbits are those for which $T < |V|$, and the open orbits are those for which $T \geq |V|$.

### EXAMPLE

A comet is observed to have a speed $v_0$ when it is a distance $r_0$ from the sun, and its direction of motion makes angle $\varphi$ with the radius vector from the sun.   Find the eccentricity of the comet's orbit.

In the sun's gravitational field $k = GMm$, where $M$ is the mass of the sun, and $m$ is the mass of the body.   The total energy $E$ is then given by

$$E = \frac{1}{2}mv^2 - \frac{GMm}{r} = \frac{1}{2}mv_0^2 - \frac{GMm}{r_0} = \text{constant}$$

and the orbit will be elliptic, parabolic, or hyperbolic, according to whether $E$ is negative, zero, or positive.   Accordingly, if $v_0^2$ is less than, equal to, or greater than $2GM/r_0$, the orbit will be an ellipse, a parabola, or a hyperbola, respectively. Now

$$h = |\mathbf{r} \times \mathbf{v}| = r_0 v_0 \sin\varphi$$

The eccentricity $e$, from Equation (5.40), therefore has the value

$$e = \left[ 1 + \left( v_0^2 - \frac{2GM}{r_0} \right) \frac{r_0^2 v_0^2 \sin^2\varphi}{G^2 M^2} \right]^{1/2}$$

The product $GM$ may be expressed in terms of the earth's speed $v_e$ and orbital radius $r_e$ (assuming a circular orbit), namely,

$$GM = r_e v_e^2$$

The equation giving the eccentricity can then be written

$$e = \left[ 1 + \left( \frac{v_0^2}{v_e^2} - \frac{2r_e}{r_0} \right) \frac{r_0^2 v_0^2}{r_e^2 v_e^2} \sin^2\varphi \right]^{1/2}$$

*Limits of the Radial Motion*

From the radial equation of the orbit, Equation (5.39), we see that the values of $r$ for $\theta = 0$, $r_0$, and for $\theta = \pi$, $r_1$, are given by

$$r_0 = \frac{mh^2 k^{-1}}{1 + (1 + 2Emh^2 k^{-2})^{1/2}} \tag{5.41}$$

$$r_1 = \frac{mh^2 k^{-1}}{1 - (1 + 2Emh^2 k^{-2})^{1/2}} \tag{5.42}$$

Now in the case of an elliptical orbit, $E$ is negative, and the major axis $2a$ of the ellipse is given by

$$2a = r_0 + r_1$$

We then find that

$$2a = -\frac{k}{|E|}$$

Thus the value of $a$ is determined entirely from the total energy.

In the case of a circular orbit of radius $a$, we have

$$V = -\frac{k}{a} = \text{constant}$$

$$E = -\frac{k}{2a} = \text{constant}$$

Thus the kinetic energy is given by

$$T = \frac{1}{2}mv^2 = E - V = \frac{k}{2a}$$

It can be shown that the time average of the kinetic energy for elliptic motion in an inverse-square field is also $k/2a$, and that the time average of the potential energy is $-k/a$, where $a$ is the semimajor axis of the ellipse. The proof is left as an exercise.

### 5.11.  Periodic Time of Orbital Motion

In Section 5.6 we showed that the areal velocity $\dot{A}$ of a particle moving in any central field is constant. Consequently, from Equations (5.16) and (5.20), the time $t_{12}$ required for a particle to move from one point $P_1$ to any other point $P_2$ (Figure 5.8) is given by

$$t_{12} = \frac{A_{12}}{\dot{A}} = A_{12}\frac{2m}{L} = A_{12}\frac{2}{h}$$

where $A_{12}$ is the area swept out by the radius vector between $P_1$ and $P_2$.

Let us apply the above result to the case of an elliptic orbit of a particle in an inverse-square field. Since the area of an ellipse is $\pi ab$, where $a$ and $b$

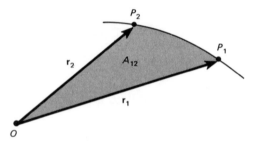

FIGURE 5.8   Area swept out by radius vector.

are the semimajor and the semiminor axes, respectively, then the time $\tau$ required for the particle to complete one orbital path is expressed by

$$\tau = \frac{2\pi ab}{h} \tag{5.43}$$

But for an ellipse

$$\frac{b}{a} = \sqrt{1 - e^2}$$

where $e$ is the eccentricity.   Thus we can write

$$\tau = \frac{2\pi a^2}{h} \sqrt{1 - e^2}$$

Furthermore, if we refer to Equations (5.33) and (5.34), we find that the major axis is given by

$$2a = r_0 + r_1 = \frac{mh^2}{k} \left( \frac{1}{1+e} + \frac{1}{1-e} \right) = \frac{2mh^2}{k(1-e^2)}$$

We can therefore express the period as

$$\tau = 2\pi \left( \frac{m}{k} \right)^{1/2} a^{3/2} \tag{5.44}$$

Thus, for a given inverse-square force field the period depends only on the size of the major axis of an elliptical orbit.

Since, for a planet of mass $m$ moving in the sun's gravitational field, $k = GMm$, we can write for the period of orbital motion of a planet

$$\tau = ca^{3/2} \tag{5.45}$$

where $c = 2\pi(GM)^{-1/2}$.   Clearly, $c$ is the same for all planets.   Equation (5.45) is a mathematical statement of Kepler's third law.   If $a$ is expressed in astronomical units (93,000,000 miles $= a_{earth} = 1$ astronomical unit) and

TABLE 5.1

| Planet | Semimajor Axis in Astronomical Units | Period in Years | Eccentricity |
|--------|-------------------------------------|-----------------|--------------|
| Mercury | 0.387 | 0.241 | 0.206 |
| Venus | 0.723 | 0.615 | 0.007 |
| Earth | 1.000 | 1.000 | 0.017 |
| Mars | 1.524 | 1.881 | 0.093 |
| Jupiter | 5.203 | 11.86 | 0.048 |
| Saturn | 9.539 | 29.46 | 0.056 |
| Uranus | 19.19 | 84.02 | 0.047 |
| Neptune | 30.06 | 164.8 | 0.009 |
| Pluto | 39.46 | 247.7 | 0.249 |

$r$ is in years, then the numerical value of $c$ is unity.  In Table 5.1 are listed the periods, semimajor axes in astronomical units, and the orbital eccentricities of the planets of the solar system.

### 5.12.  Motion in an Inverse-Square Repulsive Field.  Scattering of Atomic Particles

There is an important physical application involving motion of a particle in a central field in which the law of force is of the inverse-square repulsive type, namely the deflection of high-speed atomic particles (protons, alpha particles, and so on) by the positively charged nuclei of atoms.  The basic investigations underlying our present knowledge of atomic and nuclear structure are scattering experiments, the first of which were carried out by the British physicist Lord Rutherford in the early part of this century.

Consider a particle of charge $q$ and mass $m$ (the incident high-speed particle) passing near a heavy particle of charge $Q$ (the nucleus, assumed fixed).  The incident particle is repelled with a force given by Coulomb's law:

$$f(r) = \frac{Qq}{r^2}$$

where the position of $Q$ is taken to be the origin.  (We shall use cgs electro-static units for $Q$ and $q$.  Then $r$ is in centimeters, and the force is in dynes.)  The differential equation of the orbit then takes the form

$$\frac{d^2u}{d\theta^2} + u = -\frac{Qq}{mh^2}$$

and so the equation of the orbit is

$$u^{-1} = r = \frac{1}{A\cos(\theta - \theta_0) - Qq/mh^2}$$

We can also write the equation of the orbit in the form given by Equation (5.39), namely

$$r = \frac{mh^2 Q^{-1} q^{-1}}{-1 + (1 + 2Emh^2 Q^{-2} q^{-2})^{1/2} \cos(\theta - \theta_0)} \tag{5.46}$$

since $k = -Qq$.  The orbit is a hyperbola.  This may be seen from the physical fact that the energy $E$ is always greater than zero in a repulsive field of force.  (In our case $E = \frac{1}{2}mv^2 + Qq/r$.)  Hence the eccentricity $e$, the coefficient of $\cos(\theta - \theta_0)$, is greater than unity, which means that the orbit must be hyperbolic.

The incident particle approaches along one asymptote and recedes along the other, as shown in Figure 5.9.  We have chosen the direction of the polar axis such that the initial position of the particle is $\theta = 0$, $r = \infty$.  It is clear

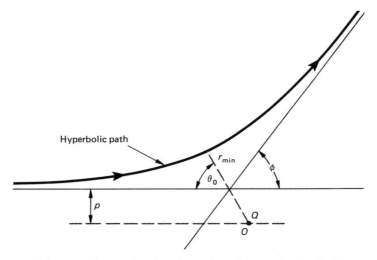

FIGURE 5.9   Hyperbolic path of a charged particle moving in the inverse-square repulsive field of another charged particle.

from either of the two equations of the orbit that $r$ assumes its minimum value when $\cos (\theta - \theta_0) = 1$, that is, when $\theta = \theta_0$.  Since $r = \infty$ when $\theta = 0$, then $r$ is also infinite when $\theta = 2\theta_0$.  Hence the angle between the two asymptotes of the hyperbolic path is $2\theta_0$, and the angle $\varphi$ through which the incident particle is deflected is given by

$$\varphi = \pi - 2\theta_0$$

Furthermore, in Equation (5.47) the denominator on the right vanishes at $\theta = 0$ and $\theta = 2\theta_0$.  Thus,

$$-1 + (1 + 2Emh^2Q^{-2}q^{-2})^{1/2} \cos \theta_0 = 0$$

from which we readily find

$$\tan \theta_0 = (2Em)^{1/2}hQ^{-1}q^{-1} = \cot \frac{\varphi}{2} \qquad (5.47)$$

The last step follows from the angle relationship given above.

In applying the above equation to scattering problems, it is convenient to express the constant $h$ in terms of another quantity $p$ called the *impact parameter*.  The impact parameter is the perpendicular distance from the origin (scattering center) to the initial line of motion of the particle, as shown in Figure 5.9.  We have then

$$h = |\mathbf{r} \times \mathbf{v}| = pv_0$$

where $v_0$ is the initial speed of the particle.  We know also that the energy $E$ is constant and is equal to the initial kinetic energy $\frac{1}{2}mv_0^2$, because the initial potential energy is zero $(r = \infty)$. Accordingly, we can write the scattering formula, Equation (5.47), in the form

$$\cot \frac{\varphi}{2} = \frac{pmv_0^2}{Qq} = \frac{2pE}{Qq} \tag{5.48}$$

## EXAMPLES

1. An alpha particle emitted by radium $(E = 5$ million electron volts $= 5 \times 10^6 \times 1.6 \times 10^{-12}$ erg) suffers a deflection of $90°$ upon passing near a gold nucleus. What is the value of the impact parameter? For alpha particles $q = 2e$, and for gold $Q = 79e$, where $e$ is the elementary charge. (The charge carried by a single electron is $-e$.) In our units $e = 4.8 \times 10^{-10}$ esu. Thus, from Equation (5.48),

$$p = \frac{Qq}{2E} \cot 45° = \frac{2 \times 79 \times (4.8)^2 + 10^{-20} \text{ cm}}{2 \times 5 \times 1.6 \times 10^{-6}}$$
$$= 2.1 \times 10^{-12} \text{ cm}$$

2. Calculate the distance of closest approach of the alpha particle in the above problem. The distance of closest approach is given by the equation of the orbit [Equation (5.46)] for $\theta = \theta_0$, thus

$$r_{\min} = \frac{mh^2 Q^{-1} q^{-1}}{-1 + (1 + 2Emh^2 Q^{-2} q^{-2})^{1/2}}$$

Upon using Equation (5.48), the above equation, after a little algebra, can be written

$$r_{\min} = \frac{p \cot (\varphi/2)}{-1 + [1 + \cot^2 (\varphi/2)]^{1/2}} = \frac{p \cos (\varphi/2)}{1 - \sin (\varphi/2)}$$

Thus, for $\varphi = 90$ degrees, we find $r_{\min} = 2.41 \ p = 5.1 \times 10^{-12}$ cm.

Notice that the expressions for $r_{\min}$ become indeterminate when $h = p = 0$. In this case the particle is aimed directly at the nucleus. It approaches the nucleus along a straight line, and, being continually repelled by the coulomb force, its speed is reduced to zero when it reaches a certain point, $r_{\min}$, from which point it returns along the same straight line. The angle of deflection is $180°$. The value of $r_{\min}$ in this case is found by using the fact that the energy $E$ is constant. At the turning point the potential energy is $Qq/r_{\min}$, and the kinetic energy is zero. Hence $E = \frac{1}{2}mv_0^2 = Qq/r_{\min}$, and

$$r_{\min} = \frac{Qq}{E}$$

For radium alpha particles and gold nuclei we find $r_{\min} \simeq 10^{-12}$ cm when the angle of deflection is $180°$. The fact that such deflections are actually observed shows that the order of magnitude of the radius of the nucleus is at least as small as $10^{-12}$ cm.

### 5.13.   Motion in a Nearly Circular Orbit.   Stability

A circular orbit is possible under any attractive central force, but not all central forces result in *stable* circular orbits.   We wish to investigate the following question: If a particle traveling in a circular orbit suffers a slight disturbance, will the ensuing orbit remain close to the original circular path?   In order to answer the query, we refer to the radial differential equation of motion. Since $\dot\theta = h/r^2$, we can write the radial equation as follows:

$$m\ddot r - \frac{mh^2}{r^3} = f(r) \tag{5.49}$$

Now for a circular orbit, $r$ is constant, and $\ddot r = 0$.   Thus, calling $a$ the radius of the circular orbit, we have

$$-\frac{mh^2}{a^3} = f(a) \tag{5.50}$$

for the force at $r = a$.

Now let us express the radial motion in terms of the variable $x$ defined by

$$x = r - a$$

The differential equation can then be written

$$m\ddot x - mh^2(x + a)^{-3} = f(x + a) \tag{5.51}$$

Expanding the two terms involving $x + a$ as power series in $x$, we obtain

$$m\ddot x - mh^2 a^{-3}\left(1 - 3\frac{x}{a} + \cdots\right) = f(a) + f'(a)x + \cdots$$

The above equation, by virtue of the relation shown in Equation (5.50), reduces to

$$m\ddot x + \left[\frac{-3}{a}f(a) - f'(a)\right]x = 0 \tag{5.52}$$

if we neglect terms involving $x^2$ and higher powers of $x$.   Now, if the coefficient of $x$ (the quantity in brackets) in the above equation is positive, then the equation is the same as that of the simple harmonic oscillator.   In this case the particle, if perturbed, oscillates harmonically about the circle $r = a$, so the circular orbit is a stable one.   On the other hand, if the coefficient of $x$ is negative, the motion is nonoscillatory, and the result is that $x$ eventually increases exponentially with time; the orbit is unstable.   (If the coefficient of $x$ is zero, then higher terms in the expansion must be included in order to determine the stability.)   Hence we can state that a circular orbit of radius $a$ is stable if the force function $f(r)$ satisfies the inequality

$$f(a) + \frac{a}{3} f'(a) < 0 \qquad (5.53)$$

In particular, if the radial force function is a power law, namely,

$$f(r) = -cr^n$$

then the condition for stability reads

$$-ca^n - \frac{a}{3} cna^{n-1} < 0$$

which reduces to

$$n > -3$$

Thus the inverse-square law ($n = -2$) gives stable circular orbits, as does the law of direct distance ($n = 1$). The latter case is that of the two-dimensional harmonic oscillator. For the inverse fourth power ($n = -4$) circular orbits are unstable. It can be shown that circular orbits are also unstable for the inverse cube law of force ($n = -3$). To show this it is necessary to include terms of higher power than one in the radial equation.

## 5.14. Apsides and Apsidal Angles for Nearly Circular Orbits

An *apsis*, or *apse*, is a point in an orbit at which the radius vector assumes an extreme value (maximum or minimum). The perihelion and aphelion points are the apsides of planetary orbits. The angle swept out by the radius vector between two consecutive apsides is called the *apsidal angle*. Thus the apsidal angle is $\pi$ for elliptic orbits under the inverse square law of force.

In the case of motion in a nearly circular orbit, we have seen that $r$ oscillates about the circle $r = a$ (if the orbit is stable). From Equation (5.52) it follows that the period $\tau_r$ of this oscillation is given by

$$\tau_r = 2\pi \sqrt{\frac{m}{-\left[\frac{3}{a} f(a) + f'(a)\right]}}$$

The apsidal angle in this case is just the amount by which the polar angle $\theta$ increases during the time that $r$ oscillates from a minimum value to the succeeding maximum value. This time is clearly $\frac{1}{2}\tau_r$. Now $\dot\theta = h/r^2$, therefore $\dot\theta$ remains approximately constant, and we can write

$$\dot\theta \simeq \frac{h}{a^2} = \left[-\frac{f(a)}{ma}\right]^{1/2}$$

The last step above follows from Equation (5.50). Hence the apsidal angle is given by

$$\psi = \frac{1}{2}\tau_r\dot\theta = \pi\left[3 + a\frac{f'(a)}{f(a)}\right]^{-1/2} \tag{5.54}$$

Thus for the power law of force $f(r) = -cr^n$, we obtain

$$\psi = \pi(3 + n)^{-1/2}$$

The apsidal angle is independent of the size of the orbit in this case. The orbit is *re-entrant*, or repetitive, in the case of the inverse-square law ($n = -2$) for which $\psi = \pi$ and also in the case of the linear law ($n = 1$) for which $\psi = \pi/2$. If, however, say $n = 2$, then $\psi = \pi/\sqrt{5}$ which is an irrational multiple of $\pi$, and so the motion does not repeat itself.

If the law of force departs slightly from the inverse-square law, then the apsides will either advance or regress steadily, depending on whether the apsidal angle is slightly greater or slightly less than $\pi$.   (See Figure 5.10.)

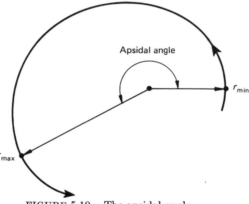

FIGURE 5.10    The apsidal angle.

Let us suppose, for example, that the force is of the form

$$f(r) = -\frac{k}{r^2} - \frac{\epsilon}{r^4} \tag{5.55}$$

where $\epsilon$ is very small.   (This is the form of the force function in the plane of a ring, as shown in Example 2, Section 5.3.)   The apsidal angle, from Equation (5.54), is

$$\psi = \pi\left(3 + a\frac{2ka^{-3} + 4\epsilon a^{-5}}{-ka^{-2} - \epsilon a^{-4}}\right)^{-1/2}$$

$$= \pi\left(\frac{1 - \epsilon k^{-1}a^{-2}}{1 + \epsilon k^{-1}a^{-2}}\right)^{-1/2}$$

$$\simeq \pi\left(1 + \frac{\epsilon}{ka^2}\right) \tag{5.56}$$

In the last step above we have neglected powers of the quantity $\epsilon/ka^2$ higher than one.   We see that the apsides advance if $\epsilon$ is positive, whereas they regress if $\epsilon$ is negative.

For a given planet, the gravitational perturbation owing to the other planets in the solar system is indeed approximated by a term of the form $\epsilon/r^4$ in Equation (5.55). The cumulative effect of one planet may be considered to be approximately the same as if that planet were smeared out into a ring, Section 5.3. For the innermost planet, Mercury, the calculated perturbations are such as to cause an advance of Mercury's perihelion of 531 sec of arc per century. The observed advance is 574 sec per century. The discrepancy of 43 sec per century is apparently explained by Einstein's general theory of relativity.

The gravitational field near the earth departs slightly from the inverse-square law. This is due to the fact that the earth is not quite a true sphere. As a result, the perigee of an artificial satellite whose orbit lies near the earth's equatorial plane will advance steadily in the direction of the satellite's motion. The observation of this advance is, in fact, one method of accurately determining the shape of the earth. Such observations have shown that the earth is slightly pear-shaped. In addition to causing an advance of the perigee of an orbiting satellite, the earth's oblateness also causes the plane of the orbit to precess if the orbit is not in the plane of the earth's equator.


## DRILL EXERCISES

5.1   By equating the gravitational force to the centripetal force, show that the period of revolution of a planet moving in a circular orbit is proportional to the 3/2 power of the radius of the orbit.

5.2   Find the radius required for a synchronous (24 hr) satellite going in a circular orbit around the earth.

5.3   Compute the mass of the earth from the fact that the period of the moon's revolution around the earth is 27.3 days, and the radius of the moon's orbit, assumed circular, is $3.84 \times 10^5$ km.


## PROBLEMS

5.4   Verify that the gravitational field on the inside of a thin spherical shell is zero (a) by finding the field directly, and (b) by calculating the potential and showing that it is constant over the interior of the shell.

5.5   Show that a particle dropped into a straight hole drilled through the earth, passing through the center, would execute simple harmonic motion. Find the period of oscillation, and show that it depends only on the density of the earth, not the size. Assume the earth to be a uniform solid sphere.

5.6   A particle slides through a smooth straight tube that passes obliquely through the earth. Show that the motion is simple harmonic with the same period as that of Problem 5.5. Neglect any effects of rotation.

5.7 If the solar system was embedded in a uniform dust cloud of density $\rho$, what would be the law of force on a planet a distance $r$ from the center of the sun?

5.8 A particle moving in a central field describes the spiral orbit $r = r_0 e^{k\theta}$. Show that the force law is inverse-cube and that $\theta$ varies logarithmically with $t$.

5.9 A particle moves in an inverse-cube field of force. Show that, in addition to the exponential spiral orbit of Problem 5.8, there are two other possible types of orbit, and give their equations.

5.10 The orbit of a particle moving in a central field is a circle passing through the origin, namely $r = r_0 \cos \theta$. Show that the force law is inverse-fifth power.

5.11 A particle moves in a spiral orbit given by $r = a\theta$. If $\theta$ increases linearly with $t$, is the force a central field? If not, determine how $\theta$ would have to vary with $t$ for a central force.

5.12 A rocket ship is initially going in a circular orbit close to the earth. It is desired to place the ship into a new orbit such that the apogee distance is equal to the radius of the moon's orbit around the earth. If a single rocket thrust is used to accomplish this, determine the ratio of the final and initial speeds. Assume that the radius of the original circular orbit is $\frac{1}{60}$ the distance to the moon. Second, calculate the apogee distance if the speed ratio is 1 percent too great. This problem illustrates the extreme accuracy needed to achieve a circumlunar orbit.

5.13 Compute the period of Halley's comet from the data given in the text, Section 5.9. Find also the comet's speed at perihelion and aphelion.

5.14 A comet is first seen at a distance of $d$ astronomical units from the sun and it is traveling with a speed of $q$ times the earth's speed. Show that the orbit of the comet is hyperbolic, parabolic, or elliptic, depending on whether the quantity $q^2 d$ is greater than, equal to, or less than 2, respectively.

5.15 A particle moves in an elliptic orbit in an inverse-square force field. Prove that the product of the minimum and maximum speeds is equal to $(2\pi a/\tau)^2$ where $a$ is the semimajor axis and $\tau$ is the periodic time.

5.16 Prove the statement made in Section 5.10 that the time average of the potential energy of a particle describing an elliptical orbit, in the inverse-square force field $f(r) = -k/r^2$, is $-k/a$ where $a$ is the semimajor axis of the ellipse.

5.17 Find the apsidal angle for nearly circular orbits in a central field for which the law of force is

$$f(r) = -k \frac{e^{-br}}{r^2}$$

5.18 If the solar system was embedded in a uniform dust cloud (Problem 5.7) what would be the apsidal angle of a planet be for motion in a nearly circular orbit? This was once suggested as a possible explanation for the advance of the perihelion of mercury.

5.19   Show that the radial differential equation of motion of a particle in a central field, Equation (5.49), is the same as that of a particle undergoing rectilinear motion in an "effective potential" $U(r)$ given by

$$U(r) = V(r) + \frac{1}{2}\frac{mh^2}{r^2}$$

in which the true force $f(r) = -dV(r)/dr$. Make a rough plot of $U(r)$ for the case of a stable circular orbit, say $V(r) = -k/r$, and for an unstable one, say $V(r) = -k/r^3$.

5.20   Show that the stability condition for a circular orbit of radius $a$ is equivalent to the condition that $d^2U/dr^2 > 0$ for $r = a$ where $U(r)$ is the "effective potential" defined in the previous problem.

5.21   Find the condition for which circular orbits are stable if the force function in a central field is of the form

$$f(r) = -\frac{k}{r^2} - \frac{\epsilon}{r^4}$$

5.22   Show that a circular orbit of radius $r$ is stable in Problem 5.17 if $r$ is less than $b^{-1}$.

5.23   A comet is going in a parabolic orbit lying in the plane of the earth's orbit. Regarding the earth's orbit as circular of radius $a$, show that the points where the comet intersects the earth's orbit are given by

$$\cos\theta = -1 + \frac{2p}{a}$$

where $p$ is the perihelion distance of the comet defined at $\theta = 0$.

5.24   Use the result of the above problem to show that the time interval that the comet remains inside the earth's orbit is the fraction

$$\frac{2^{1/2}}{3\pi}\left(\frac{2p}{a}+1\right)\left(1-\frac{p}{a}\right)^{1/2}$$

of a year, and that the maximum value of this time interval is $2/3\pi$ year, or about 11 weeks.

5.25   In advanced texts on potential theory it is shown that the potential energy of a particle of mass $m$ in the gravitational field of an oblate spheroid, like the earth, is approximately

$$V(r) = -\frac{k}{r}\left(1+\frac{\epsilon}{r^2}\right)$$

where $r$ refers to distances in the equatorial plane, $k = GMm$ as before, and $\epsilon = (\frac{2}{5})R\Delta R$ in which $R$ is the equatorial radius and $\Delta R$ is the difference between the equatorial and polar radii. From this, find the apsidal angle for a satellite moving in a nearly circular orbit in the equatorial plane of the earth where $R = 4000$ mi, $\Delta R = 13$ mi.

5.26   According to the special theory of relativity, a particle moving in a central field with potential energy $V(r)$ will describe the same orbit that a particle with a potential energy

$$V(r) - \frac{[E - V(r)]^2}{2m_0 c^2}$$

would describe according to nonrelativistic mechanics. Here $E$ is the total energy, $m_0$ is the rest mass of the particle, and $c$ is the speed of light. From this, find the apsidal angle for motion in an inverse-square force field, $V(r) = -k/r$.

    5.27    An asteroid is observed to have a speed $v_a$ when it is a distance $r_a$ from the sun, and its direction of motion makes an angle $\varphi$ with the radius vector from the sun. Show that the major axis of the elliptical orbit of the asteroid makes an angle

$$\cot^{-1}\left(\tan\varphi - \frac{2r_e v_e^2}{r_a v_a^2} \csc 2\varphi\right)$$

with the initial radius vector of the asteroid, where $r_e$ and $v_e$ are the earth's orbital radius and speed, respectively.

# 6. Dynamics
# of Systems
# of Many Particles

In studying a system or collection of many free particles, we shall be mainly interested in the general features of the motion of such a system.

## 6.1. Center of Mass and Linear Momentum

Our general system consists of $n$ particles of masses $m_1, m_2, \ldots, m_n$ whose position vectors are, respectively, $\mathbf{r}_1, \mathbf{r}_2, \ldots, \mathbf{r}_n$. We define the *center of mass* of the system as the point whose position vector $\mathbf{r}_{cm}$ (Figure 6.1) is given by

$$\mathbf{r}_{cm} = \frac{m_1 \mathbf{r}_1 + m_2 \mathbf{r}_2 + \cdots + m_n \mathbf{r}_n}{m_1 + m_2 + \cdots + m_n} = \frac{\Sigma m_i \mathbf{r}_i}{m} \tag{6.1}$$

where $m = \Sigma m_i$ is the total mass of the system. The above definition is clearly equivalent to the three equations

$$x_{cm} = \frac{\Sigma m_i x_i}{m} \qquad y_{cm} = \frac{\Sigma m_i y_i}{m} \qquad z_{cm} = \frac{\Sigma m_i z_i}{m}$$

We define the *linear momentum* $\mathbf{p}$ of the system as the vector sum of the momenta of the individual particles, namely,

$$\mathbf{p} = \Sigma \mathbf{p}_i = \Sigma m_i \mathbf{v}_i \tag{6.2}$$

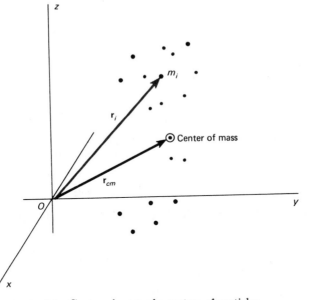

FIGURE 6.1   Center of mass of a system of particles.

From Equation (6.1), by differentiating with respect to the time $t$, it follows that

$$\mathbf{p} = \Sigma m_i \mathbf{v}_i = m\mathbf{v}_{cm} \tag{6.3}$$

that is, the linear momentum of a system of particles is equal to the velocity of the center of mass multiplied by the total mass of the system.

Suppose now that there are external forces $\mathbf{F}_1, \mathbf{F}_2, \ldots, \mathbf{F}_i, \ldots, \mathbf{F}_n$, acting on the respective particles. In addition, there may be internal forces of interaction between any two particles of the system. We shall denote these internal forces by $\mathbf{F}_{ij}$, meaning the force exerted on particle $i$ by particle $j$, with the understanding that $\mathbf{F}_{ii} = 0$. The equation of motion of particle $i$ is then

$$\mathbf{F}_i + \sum_{j=1}^{n} \mathbf{F}_{ij} = m_i \ddot{\mathbf{r}}_i = \dot{\mathbf{p}}_i \tag{6.4}$$

where $\mathbf{F}_i$ means the total external force acting on particle $i$. The second term in the above equation represents the vector sum of all the internal forces exerted on particle $i$ by all other particles of the system. Adding Equation (6.4) for the $n$ particles, we have

$$\sum_{i=1}^{n} \mathbf{F}_i + \sum_{i=1}^{n} \sum_{j=1}^{n} \mathbf{F}_{ij} = \sum_{i=1}^{n} \dot{\mathbf{p}}_i \tag{6.5}$$

In the double summation above, for every force $\mathbf{F}_{ij}$ there is also a force $\mathbf{F}_{ji}$, and these two forces are equal and opposite

$$\mathbf{F}_{ij} = -\mathbf{F}_{ji} \tag{6.6}$$

from the law of action and reaction, Newton's third law. Consequently, the internal forces cancel in pairs, and the double sum vanishes. We can therefore write Equation (6.5) in the following way:

$$\Sigma \mathbf{F}_i = \Sigma \dot{\mathbf{p}}_i = \dot{\mathbf{p}} = m\mathbf{a}_{cm} \tag{6.7}$$

In words: *The acceleration of the center of mass of a system of particles is the same as that of a single particle having a mass equal to the total mass of the system and acted upon by the sum of the external forces.*

Consider, for example, a swarm of particles moving in a *uniform* gravitational field. Then, since $\mathbf{F}_i = m_i \mathbf{g}$ for each particle,

$$\Sigma \mathbf{F}_i = \Sigma m_i \mathbf{g} = m\mathbf{g}$$

The last step follows from the fact that $\mathbf{g}$ is constant. Hence

$$\mathbf{a}_{cm} = \mathbf{g} \tag{6.8}$$

This is the same as the equation for a single particle or projectile. Thus the center of mass of the shrapnel from an artillery shell that has burst in mid-air will follow the same parabolic path that the shell would have taken had it not burst.

In the special case in which there are *no* external forces acting on a system (or if $\Sigma \mathbf{F}_i = 0$), then $\mathbf{a}_{cm} = 0$ and $\mathbf{v}_{cm} = \text{constant}$. Thus the linear momentum of the system remains constant:

$$\Sigma \mathbf{p}_i = \mathbf{p} = m\mathbf{v}_{cm} = \text{constant} \tag{6.9}$$

This is the *principle of conservation of linear momentum*. In Newtonian mechanics the constancy of the linear momentum of an isolated system is directly related to, and is in fact a consequence of, the third law. But even in those cases in which the forces between particles do not directly obey the law of action and reaction, such as the magnetic forces between moving charges, the principle of conservation of linear momentum still holds when due account is taken of the total linear momentum of the particles and the electromagnetic field.[1]

### 6.2. Angular Momentum of a System

We previously stated that the angular momentum of a single particle is defined as the cross product $\mathbf{r} \times m\mathbf{v}$. The angular momentum $\mathbf{L}$ of a system of particles is defined accordingly, as the vector sum of the individual angular momenta, namely

[1] See, for example, W. T. Scott, *The Physics of Electricity and Magnetism*, 2d ed., John Wiley and Sons, Inc., New York, 1966.

$$L = \sum_{i=1}^{n} (\mathbf{r}_i \times m_i \mathbf{v}_i)$$

Let us calculate the time derivative of the angular momentum. Using the rule for differentiating the cross product, we find

$$\frac{d\mathbf{L}}{dt} = \sum_{i=1}^{n} (\mathbf{v}_i \times m_i \mathbf{v}_i) + \sum_{i=1}^{n} (\mathbf{r}_i \times m_i \mathbf{a}_i) \tag{6.10}$$

Now the first term on the right vanishes, because $\mathbf{v}_i \times \mathbf{v}_i = 0$ and, since $m_i \mathbf{a}_i$ is equal to the total force acting on particle $i$, we can write

$$\frac{d\mathbf{L}}{dt} = \sum_{i=1}^{n} \left[ \mathbf{r}_i \times \left( \mathbf{F}_i + \sum_{j=1}^{n} \mathbf{F}_{ij} \right) \right]$$

$$= \sum_{i=1}^{n} \mathbf{r}_i \times \mathbf{F}_i + \sum_{i=1}^{n} \sum_{j=1}^{n} \mathbf{r}_i \times \mathbf{F}_{ij} \tag{6.11}$$

where, as in Section 6.1, $\mathbf{F}_i$ denotes the total external force on particle $i$, and $\mathbf{F}_{ij}$ denotes the (internal) force exerted on particle $i$ by any other particle $j$. Now the double summation on the right consists of pairs of terms of the form

$$(\mathbf{r}_i \times \mathbf{F}_{ij}) + (\mathbf{r}_j \times \mathbf{F}_{ji}) \tag{6.12}$$

Denoting the vector displacement of particle $j$, relative to particle $i$ by $\mathbf{r}_{ij}$, we see from the triangle shown in Figure 6.2 that

$$\mathbf{r}_{ij} = \mathbf{r}_j - \mathbf{r}_i \tag{6.13}$$

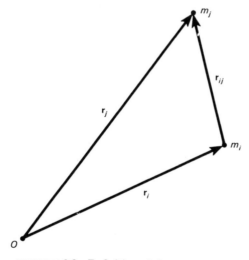

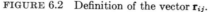

FIGURE 6.2   Definition of the vector $\mathbf{r}_{ij}$.

Therefore, since $\mathbf{F}_{ji} = -\mathbf{F}_{ij}$, the expression (6.12) reduces to

$$-\mathbf{r}_{ij} \times \mathbf{F}_{ij} \tag{6.14}$$

which clearly vanishes if the internal forces are central, that is, if they act along the lines connecting pairs of particle. Hence the double sum in Equation (6.11) vanishes. Now the cross product $\mathbf{r}_i \times \mathbf{F}_i$ is the moment of the external force $\mathbf{F}_i$. The sum $\sum \mathbf{r}_i \times \mathbf{F}_i$ is therefore the total moment of all the external forces acting on the system. If we denote the total external moment by $\mathbf{N}$, the Equation (6.11) takes the form

$$\frac{d\mathbf{L}}{dt} = \mathbf{N} \tag{6.15}$$

That is, *the time rate of change of the angular momentum of a system is equal to the total moment of all the external forces acting on the system.*

If a system is isolated, then $\mathbf{N} = 0$, and the angular momentum remains constant in both magnitude and direction:

$$\mathbf{L} = \sum \mathbf{r}_i \times m_i \mathbf{v}_i = \text{constant} \tag{6.16}$$

This is a statement of the *principle of conservation of angular momentum.* It is a generalization for a single particle in a central field. Like the constancy of linear momentum discussed in the preceding section, the angular momentum of an isolated system is also constant in the case of a system of moving charges when the angular momentum of the electromagnetic field is considered.[2]

### 6.3.   Kinetic Energy of a System of Particles

The total kinetic energy $T$ of a system of particles is given by the sum of the individual energies, namely,

$$T = \sum \tfrac{1}{2} m_i v_i^2 = \sum \tfrac{1}{2} m_i (\mathbf{v}_i \cdot \mathbf{v}_i) \tag{6.17}$$

As shown in Figure 6.3, we can express each position vector $\mathbf{r}_i$ in the form

$$\mathbf{r}_i = \mathbf{r}_{cm} + \bar{\mathbf{r}}_i \tag{6.18}$$

where $\bar{\mathbf{r}}_i$ is the position of particle $i$ relative to the center of mass. Taking the derivative with respect to $t$, we have

$$\mathbf{v}_i = \mathbf{v}_{cm} + \bar{\mathbf{v}}_i \tag{6.19}$$

Here $\mathbf{v}_{cm}$ is the velocity of the center of mass and $\bar{\mathbf{v}}_i$ is the velocity of particle $i$ relative to the center of mass. The expression for $T$ can therefore be written

[2] See footnote 1.

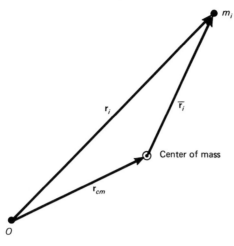

FIGURE 6.3  Definition of $\bar{\mathbf{r}}_i$.

$$T = \Sigma \tfrac{1}{2} m_i (\mathbf{v}_{cm} + \bar{\mathbf{v}}_i) \cdot (\mathbf{v}_{cm} + \bar{\mathbf{v}}_i)$$
$$= \Sigma \tfrac{1}{2} m_i v_{cm}{}^2 + \Sigma m_i (\mathbf{v}_{cm} \cdot \bar{\mathbf{v}}_i) + \Sigma \tfrac{1}{2} m_i \bar{v}_i{}^2$$
$$= \tfrac{1}{2} v_{cm}{}^2 \Sigma m_i + \mathbf{v}_c \cdot \Sigma m_i \bar{\mathbf{v}}_i + \Sigma \tfrac{1}{2} m_i \bar{v}_i{}^2$$

Now, from Equation (6.18), we have

$$\Sigma m_i \bar{\mathbf{r}}_i = \Sigma m_i (\mathbf{r}_i - \mathbf{r}_{cm}) = \Sigma m_i \mathbf{r}_i - m \mathbf{r}_{cm} = 0$$

Similarly, we obtain

$$\Sigma m_i \bar{\mathbf{v}}_i = 0$$

Therefore the expression for the kinetic energy reduces to

$$T = \tfrac{1}{2} m v_{cm}{}^2 + \Sigma \tfrac{1}{2} m_i \bar{v}_i{}^2 \qquad (6.20)$$

Thus the total kinetic energy of a system of particles is given by the sum of the kinetic energy of translation of the center of mass (the first term on the right) plus the kinetic energy of motion of the individual particles relative to the center of mass (the last term).   This separation of kinetic energy into its parts is convenient, for example, in molecular physics.   Thus, for a molecule, the total kinetic energy consists of translational energy of the whole molecule plus the energy of vibration and rotation within the molecule.

## 6.4.  Motion of Two Interacting Bodies.
## The Reduced Mass

Let us consider the motion of a system consisting of two bodies (treated as particles) that interact with one another by a central force.  We shall assume the system is isolated, and hence the center of mass moves with constant velocity. For simplicity, we shall take the center of mass as the origin. We have then

$$m_1 \bar{\mathbf{r}}_1 + m_2 \bar{\mathbf{r}}_2 = 0 \tag{6.21}$$

where, as shown in Figure 6.4, the vectors $\bar{\mathbf{r}}_1$ and $\bar{\mathbf{r}}_2$ represent the positions of the particles $m_1$ and $m_2$, respectively, relative to the center of mass. Now, if $\mathbf{R}$ is the position vector of particle 1 relative to particle 2, then

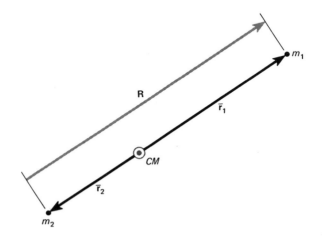

FIGURE 6.4   The two-body problem.

$$\mathbf{R} = \bar{\mathbf{r}}_1 - \bar{\mathbf{r}}_2 = \bar{\mathbf{r}}_1 \left( 1 + \frac{m_1}{m_2} \right) \tag{6.22}$$

The last step follows from Equation (6.21).

The differential equation of motion of particle 1 relative to the center of mass is

$$m_1 \frac{d^2 \bar{\mathbf{r}}_1}{dt^2} = \mathbf{F}_1 = f(\mathbf{R}) \frac{\mathbf{R}}{\mathbf{R}} \tag{6.23}$$

in which $f(\mathbf{R})$ is the magnitude of the mutual force between the two particles. By using Equation (6.22), we can write

$$\mu \frac{d^2 \mathbf{R}}{dt^2} = f(\mathbf{R}) \frac{\mathbf{R}}{\mathbf{R}} \tag{6.24}$$

where

$$\mu = \frac{m_1 m_2}{m_1 + m_2} \tag{6.25}$$

The quantity $\mu$ is called the *reduced mass*. The new equation of motion, (6.24), gives the motion of particle 1 relative to particle 2. This equation is precisely the same as the ordinary equation of motion of a single particle of mass $\mu$ moving in a central field of force given by $f(\mathbf{R})$. Thus the fact that $m_2$ is moving relative to the center of mass is automatically accounted for by replacing $m_1$ by the reduced mass $\mu$. If the bodies are of equal mass $m$, then $\mu = m/2$. On the other hand, if $m_2$ is very much greater than $m_1$, so that $m_1/m_2$ is very small, then $\mu$ is nearly equal to $m_1$.

For two bodies attracting one another by gravitation, we have

$$f(\mathrm{R}) = -\frac{Gm_1m_2}{\mathrm{R}^2} \tag{6.26}$$

In this case the equation of motion is

$$\mu\ddot{\mathbf{R}} = -\frac{Gm_1m_2}{\mathrm{R}^2}\left(\frac{\mathbf{R}}{\mathrm{R}}\right) \tag{6.27}$$

This is the same as the equation of a single particle in an inverse-square central field (as treated in Chapter 5). Since the choice of subscripts is arbitrary, we conclude that either particle describes a central conic about the other as a focus. Thus, regarding the earth and the moon as an isolated system, the moon. describes an ellipse with the center of the earth as a focus, and the earth describes an ellipse with the center of the moon as a focus.

### 6.5. Collisions

Whenever two bodies undergo a collision, the force that either exerts on the other during the contact is an internal force, if the two bodies are regarded together as a single system. The total linear momentum is therefore unchanged. We can therefore write

$$\mathbf{p}_1 + \mathbf{p}_2 = \mathbf{p}_1' + \mathbf{p}_2' \tag{6.28}$$

or, equivalently

$$m_1\mathbf{v}_1 + m_2\mathbf{v}_2 = m_1\mathbf{v}_1' + m_2\mathbf{v}_2' \tag{6.29}$$

The subscripts 1 and 2 refer to the two bodies, and the primes indicate the respective momenta and velocities *after* the collision. The above equations are quite general. They apply to any two bodies regardless of their shapes, rigidity, and so on.

With regard to the energy balance, we can write

$$\frac{p_1^2}{2m_1} + \frac{p_2^2}{2m_2} = \frac{p_1'^2}{2m_1} + \frac{p_2'^2}{2m_2} + Q \tag{6.30}$$

or

$$\tfrac{1}{2}m_1v_1^2 + \tfrac{1}{2}m_2v_2^2 = \tfrac{1}{2}m_1v_1'^2 + \tfrac{1}{2}m_2v_2'^2 + Q \tag{6.31}$$

Here the quantity $Q$ is introduced to indicate the net energy loss, or gain, that occurs as a result of the collision.

In the case of a perfectly elastic collision, there is no change in the total kinetic energy, so that $Q = 0$. If there is an energy loss, then $Q$ is positive.

This is called an *endoergic* collision. It may happen that there is an energy gain. This would occur, for example, if an explosive was present on one of the bodies at the point of contact. In this case $Q$ is negative, and the collision is called *exoergic*.

The study of collisions is of particular importance in atomic and nuclear physics. Here the bodies involved may be atoms, nuclei, or various elementary particles, such as electrons, protons, and so on.

### Direct Collisions

Let us consider the special case of a head-on collision of two bodies, or particles, in which the motion takes place entirely on a single straight line, as shown in Figure 6.5. In this case the momentum balance equation, Equation (6.29), can be written without the use of vector notation as

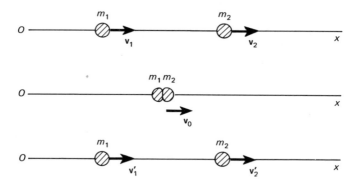

FIGURE 6.5   Head-on collision of two particles.

$$m_1v_1 + m_2v_2 = m_1v_1' + m_2v_2' \tag{6.32}$$

The direction along the line of motion is given by the signs of the $v$'s.

In order to compute the values of the velocities after the collision, given the values before the collision, we can use the above momentum equation together with the energy balance equation, Equation (6.31), if we know the value of $Q$. It is often convenient in this kind of problem to introduce another parameter $\epsilon$ called the *coefficient of restitution*. This quantity is defined as the ratio of the speed of separation $v'$ to the speed of approach $v$. In our notation, $\epsilon$ may be written as

$$\epsilon = \frac{|v_2' - v_1'|}{|v_2 - v_1|} = \frac{v'}{v} \tag{6.33}$$

The numerical value of $\epsilon$ depends primarily on the composition and physical makeup of the two bodies. It is easy to verify that in a perfectly elastic collision, the value of $\epsilon = 1$. To do this, we set $Q = 0$ in Equation (6.31), and solve it together with Equation (6.32), for the final velocities.

In the case of a perfectly inelastic collision, the two bodies stick together after colliding, so that $\epsilon = 0$.   For most real bodies, $\epsilon$ has a value somewhere between the two extremes of 0 and 1.   For ivory billiard balls, it is about 0.95. The value of the coefficient of restitution may also depend on the speed of approach.   This is particularly evident in the case of a silicone compound known under a trade name as "silly putty."   A ball of this material bounces when it strikes a hard surface at high speed, but at low speeds it acts like ordinary putty.

We can calculate the values of the final velocities from Equation (6.32) together with the definition of the coefficient of restitution, Equation (6.33). The result is

$$v_1' = \frac{(m_1 - \epsilon m_2)v_1 + (m_2 + \epsilon m_2)v_2}{m_1 + m_2}$$

$$v_2' = \frac{(m_1 + \epsilon m_1)v_1 + (m_2 - \epsilon m_1)v_2}{m_1 + m_2} \tag{7.}$$

Taking the inelastic case by setting $\epsilon = 0$, we find, as we should, that $v_1' = v_2'$, that is, there is no rebound.   On the other hand, in the special case that the bodies are of equal mass, $m_1 = m_2$, and are perfectly elastic, $\epsilon = 1$, then we obtain

$$v_1' = v_2$$
$$v_2' = v_1$$

The two bodies, therefore, just *exchange* their velocities as a result of the collision.

In the general case of a direct nonelastic collision, it is easily verified that the energy loss $Q$ is related to the coefficient of restitution by the equation

$$Q = \tfrac{1}{2}\mu v^2(1 - \epsilon^2)$$

in which $\mu = m_1 m_2/(m_1 + m_2)$ is the reduced mass, and $v = |v_2 - v_1|$ is the relative speed before impact.   The derivation is left as an exercise.

### 6.6.   Oblique Collisions and Scattering.   Comparison of Laboratory and Center-of-Mass Coordinates

We now turn our attention to the more general case of collisions in which the motion is not confined to a single straight line.   Here the vectorial form of the momentum equations, Equations (6.28) and (6.29), must be employed. Let us study the special case of a particle of mass $m_1$ with initial velocity $v_1$ (the incident particle) that strikes a particle of mass $m_2$ that is initially at rest (the target particle).   This is a typical problem found in nuclear physics. The momentum equations in this case are

$$\mathbf{p}_1 = \mathbf{p}_1' + \mathbf{p}_2' \tag{6.35}$$
$$m_1\mathbf{v}_1 = m_1\mathbf{v}_1' + m_2\mathbf{v}_2' \tag{6.36}$$

The energy balance condition is

$$\frac{p_1^2}{2m_1} = \frac{p_1'^2}{2m_1} + \frac{p_2'^2}{2m_2} + Q \tag{6.37}$$

or

$$\tfrac{1}{2}m_1v_1^2 = \tfrac{1}{2}m_1v_1'^2 + \tfrac{1}{2}m_2v_2'^2 + Q \tag{6.38}$$

Here, as before, the primes indicate the velocities and momenta after the collision, and $Q$ represents the net energy that is lost or gained as a result of the impact. The quantity $Q$ is of fundamental importance in atomic and nuclear physics, since it represents the energy released or absorbed in atomic and nuclear collisions. In many cases the target particle is broken up or changed by the collision. In such cases, the particles that leave the collision are different from the particles that enter. This is easily taken into account by assigning different masses, say $m_3$ and $m_4$ to the particles leaving the collision. In any case, the law of conservation of linear momentum is always assumed to be valid. According to the theory of relativity, however, the mass of a particle varies with speed in a definite way which we shall study in a later chapter. For now, we can state that the momentum conservation expressed by Equation (6.28) is relativistically correct, by assuming that mass is a function of speed.

### Center-of-Mass Coordinates

Theoretical calculations in nuclear physics are often done in terms of quantities referred to a coordinate system in which the center of mass of the colliding particles is at rest. On the other hand, the experimental observations on scattering of particles are carried out in terms of the laboratory coordinates. It is of interest, therefore, to consider briefly the problem of conversion from one coordinate system to the other.

The velocity vectors in the laboratory system and in the center-of-mass system are illustrated diagramatically in Figure 6.6. In the figure, $\varphi_1$ is the angle of deflection of the incident particle after it strikes the target particle and $\varphi_2$ is the angle that the line of motion of the target particle makes with the line of motion of the incident particle. Both $\varphi_1$ and $\varphi_2$ are measured in the laboratory system. In the center-of-mass system, since the center of mass must lie on the line joining the two particles at all times, both particles approach the center of mass, collide, and recede from the center of mass in opposite directions. The angle $\theta$ denotes the angle of deflection of the incident particle in the center-of-mass system as indicated.

From the definition of the center of mass, the linear momentum in the center-of-mass system is zero both before and after the collision. Hence we can write

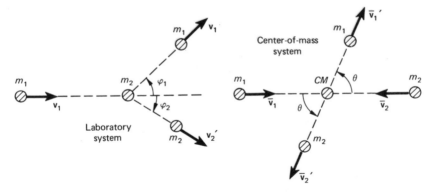

FIGURE 6.6   Comparison of laboratory and center-of-mass coordinates.

$$\bar{\mathbf{p}}_1 + \bar{\mathbf{p}}_2 = 0 \tag{6.39}$$
$$\bar{\mathbf{p}}_1' + \bar{\mathbf{p}}_2' = 0 \tag{6.40}$$

The bars are used to indicate that the quantity in question is referred to the center-of-mass system.   The energy balance equation reads

$$\frac{\bar{p}_1^2}{2m_1} + \frac{\bar{p}_2^2}{2m_2} = \frac{\bar{p}_1'^2}{2m_1} + \frac{\bar{p}_2'^2}{2m_2} + Q \tag{6.41}$$

We can eliminate $\bar{p}_2$ and $\bar{p}_2'$ from the energy equation by using the momentum relations.   The result, which is conveniently expressed in terms of the reduced mass, is

$$\frac{\bar{p}_1^2}{2\mu} = \frac{\bar{p}_1'^2}{2\mu} + Q \tag{6.42}$$

The momentum relations, Equations (6.39) and (6.40), expressed in terms of velocities, read

$$m_1\bar{\mathbf{v}}_1 + m_2\bar{\mathbf{v}}_2 = 0 \tag{6.43}$$
$$m_1\bar{\mathbf{v}}_1' + m_2\bar{\mathbf{v}}_2' = 0 \tag{6.44}$$

The velocity of the center of mass is

$$\mathbf{v}_{cm} = \frac{m_1\mathbf{v}_1}{m_1 + m_2} \tag{6.45}$$

Hence we have

$$\bar{\mathbf{v}}_1 = \mathbf{v}_1 - \mathbf{v}_{cm} = \frac{m_2\mathbf{v}_1}{m_1 + m_2} \tag{6.46}$$

The relationships among the velocity vectors $\mathbf{v}_{cm}$, $\mathbf{v}_1'$, and $\bar{\mathbf{v}}_1'$ are shown in Figure 6.7.   From the figure, we see that

$$v_1' \sin \varphi_1 = \bar{v}_1' \sin \theta \tag{6.47}$$
$$v_1' \cos \varphi_1 = \bar{v}_1' \cos \theta + v_{cm} \tag{6.48}$$

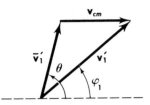

FIGURE 6.7   Relationships between the velocity vectors in the laboratory
system and the center-of-mass system.

Hence, by dividing, we find the equation connecting the scattering angles to be
expressible in the form

$$\tan \varphi_1 = \frac{\sin \theta}{\gamma + \cos \theta} \tag{6.49}$$

in which $\gamma$ is a numerical parameter whose value is given by

$$\gamma = \frac{v_{cm}}{\bar{v}_1'} = \frac{m_1 v_1}{\bar{v}_1'(m_1 + m_2)} \tag{6.50}$$

The last step follows from Equation (6.45).

Now we can readily calculate the value of $\bar{v}_1'$ in terms of the initial energy
of the incident particle from the energy equation, Equation (6.42). This
gives us the necessary information to find $\gamma$ and thus determine the relationship between the scattering angles. For example, in the case of a perfectly
elastic collision, $Q = 0$, we find from the energy equation that $\bar{p}_1 = \bar{p}_1'$, or
$\bar{v}_1 = \bar{v}_1'$. This result, together with Equation (6.46) yields the value

$$\gamma = \frac{m_1}{m_2} \tag{6.51}$$

for an elastic collision.

Two special cases of such elastic collisions are instructive to consider.
First, if the mass $m_2$ of the target particle is very much greater than the mass
$m_1$ of the incident particle, then $\gamma$ is very small. Hence $\tan \varphi_1 \approx \tan \theta$, or
$\varphi_1 \approx \theta$. That is, the scattering angles as seen in the laboratory and in the
center-of-mass systems are nearly equal.

The second special case is that of equal masses of the incident and target
particles $m_1 = m_2$. In this case $\gamma = 1$, and the scattering relation reduces to

$$\tan \varphi_1 = \frac{\sin \theta}{1 + \cos \theta} = \tan \frac{\theta}{2}$$

$$\varphi_1 = \frac{\theta}{2}$$

That is, the angle of deflection in the laboratory system is just half that in
the center-of-mass system. Furthermore, since the angle of deflection of the
target particle is $\pi - \theta$ in the center-of-mass system, as shown in Figure 6.6,
then the same angle in the laboratory system is $(\pi - \theta)/2$. Therefore the

two particles leave the point of impact at right angles to each other as seen in the laboratory system.

In the general case of nonelastic collisions, it is left as a problem to show that $\gamma$ is expressible as

$$\gamma = \frac{m_1}{m_2}\left[1 - \frac{Q}{T}\left(1 + \frac{m_1}{m_2}\right)\right]^{-1/2} \tag{6.52}$$

in which $T$ is the kinetic energy of the incident particle as measured in the laboratory system.

### 6.7.  Impulse in Collisions

Forces of extremely short duration in time, such as those exerted by bodies undergoing collisions, are called *impulsive forces*. If we confine our attention to one body, or particle, the differential equation of motion, as we know, is

$$\frac{d(m\mathbf{v})}{dt} = \mathbf{F} \tag{6.53}$$

or, in differential form

$$d(m\mathbf{v}) = \mathbf{F}\,dt \tag{6.54}$$

Let us take the time integral over the interval $t = t_1$ to $t = t_2$.  This is the time during which the force is considered to act.  Then we have

$$\Delta(m\mathbf{v}) = \int_{t_1}^{t_2} \mathbf{F}\,dt \tag{6.55}$$

The time integral of the force is the impulse, previously defined in Section 2.8. It is customarily denoted by the symbol $\hat{\mathbf{P}}$.  The above equation is, accordingly, expressed as

$$\Delta(m\mathbf{v}) = \hat{\mathbf{P}} \tag{6.56}$$

We can think of an *ideal impulse* as produced by a force that tends to infinity but lasts for a time interval which approaches zero in such a way that the integral $\int \mathbf{F}\,dt$ remains finite.  Such an ideal impulse would produce an instantaneous change in the momentum and velocity of a body without producing any displacement

### *Relationship between Impulse and Coefficient of Restitution*

Let us apply the concept of impulse to the case of the direct collision of two spherical bodies (treated in Section 6.5).  We shall divide the impulse into two parts, namely, the impulse of compression, $\hat{\mathbf{P}}_c$, and the impulse of restitution, $\hat{P}_r$.  We are concerned only with components along the line of centers. Therefore, for the compression we can write

$$m_1 v_0 - m_1 v_1 = \hat{P}_c \qquad (6.57)$$
$$m_2 v_0 - m_2 v_2 = -\hat{P}_c \qquad (6.58)$$

where $v_0$ is the common velocity of both particles at the instant their relative speed is zero.  Similarly, for the restitution, we have

$$m_1 v_1' - m_1 v_0 = \hat{P}_r \qquad (6.59)$$
$$m_2 v_2' - m_2 v_0 = -\hat{P}_r \qquad (6.60)$$

Upon eliminating $v_0$ from Equations (6.57) and (6.58) and also from Equations (6.59) and (6.60), we obtain the following pair of equations

$$m_1 m_2 (v_2 - v_1) = \hat{P}_c (m_1 + m_2)$$
$$m_1 m_2 (v_1' - v_2') = \hat{P}_r (m_1 + m_2)$$

Division of the second equation by the first yields the relation

$$\frac{v_2' - v_1'}{v_1 - v_2} = \frac{\hat{P}_r}{\hat{P}_c} \qquad (6.61)$$

But the left-hand side is just the definition of the coefficient of restitution $\epsilon$. Hence we have

$$\epsilon = \frac{\hat{P}_r}{\hat{P}_c} \qquad (6.62)$$

The coefficient of restitution is thus equal to the ratio of the impulse of restitution to the impulse of compression.

### 6.8.  Motion of a Body with Variable Mass. Rocket Motion

In the case of a body whose mass changes with time, it is necessary to use care in setting up the differential equations of motion.  The concept of impulse can be helpful in this type of problem.

Consider the general case of the motion of a body with changing mass. Let $\mathbf{F}_{ext}$ denote the external force acting on the body at a given time, and let $\Delta m$ denote the increment of the mass of the body that occurs in a short time interval $\Delta t$.  Then $\mathbf{F}_{ext}\Delta t$ is the impulse delivered by the external force, and we have

$$\mathbf{F}_{ext}\Delta t = (\mathbf{p}_{\text{total}})_{t+\Delta t} - (\mathbf{p}_{\text{total}})_t$$

for the change in the total linear momentum of the system.  Hence if $\mathbf{v}$ denotes the velocity of the body and $\mathbf{V}$ the velocity of the mass increment $\Delta m$ relative to the body, then we can write

$$\mathbf{F}_{ext}\Delta t = (m + \Delta m)(\mathbf{v} + \Delta \mathbf{v}) - [m\mathbf{v} + \Delta m(\mathbf{v} + \mathbf{V})]$$

This reduces to

$$\mathbf{F}_{ext}\Delta t = m\Delta \mathbf{v} + \Delta m \Delta \mathbf{v} - \mathbf{V}\Delta m$$

or, by dividing by $\Delta t$, we can write

$$\mathbf{F}_{ext} = (m + \Delta m)\frac{\Delta \mathbf{v}}{\Delta t} - \mathbf{V}\frac{\Delta m}{\Delta t}$$

Thus, in the limit as $\Delta t$ approaches zero, we have the general equation

$$\mathbf{F}_{ext} = m\dot{\mathbf{v}} - \mathbf{V}\dot{m} \qquad (6.63)$$

Here the force $\mathbf{F}_{ext}$ may represent gravity, air resistance, and so on.   In the case of rockets, the term $\mathbf{V}\dot{m}$ represents the thrust.

Let us apply the equation to two special cases.   First, suppose that a body is moving through a fog or mist so that it collects mass as it goes.   In this case the initial velocity of the accumulated matter is zero.   Hence $\mathbf{V} = -\mathbf{v}$, and we get

$$\mathbf{F}_{ext} = m\dot{\mathbf{v}} + \mathbf{v}\dot{m} = \frac{d(m\mathbf{v})}{dt} \qquad (6.64)$$

for the equation of motion.   It applies *only* if the initial velocity of the matter that is being swept up is zero.   Otherwise the general equation (6.63) must be used.

For the second case, consider the motion of a rocket.   In this instance the sign of $\dot{m}$ is negative, because the rocket is losing mass in the form of ejected fuel.   Hence $\mathbf{V}\dot{m}$ is opposite to the direction of $\mathbf{V}$, the relative velocity of the ejected fuel.   For simplicity we shall solve the equation of motion for the case in which the external force $\mathbf{F}_{ext}$ is zero.   Then we have

$$m\dot{\mathbf{v}} = \mathbf{V}\dot{m} \qquad (6.65)$$

We can now separate the variables and integrate to find $\mathbf{v}$ as follows:

$$\int d\mathbf{v} = \int \frac{\mathbf{V}\,dm}{m}$$

If it is assumed that $\mathbf{V}$ is constant, then we can integrate between limits to find the *speed* as a function of $m$:

$$\int_{v_0}^{v} dv = -V\int_{m_0}^{m} \frac{dm}{m}$$

$$v = v_0 + V\ln\frac{m_0}{m}$$

Here $m_0$ is the initial mass of the rocket plus unburned fuel, $m$ is the mass at any time, and $V$ is the speed of the ejected fuel relative to the rocket. Owing to the nature of the logarithmic function, it is necessary to have a large fuel to payload ratio in order to attain the large speeds needed for satellite launching.

## DRILL EXERCISES

6.1    A system consists of three particles, each of unit mass, with instantaneous positions and velocities as follows:

$$\mathbf{r}_1 = \mathbf{i} + \mathbf{j} + \mathbf{k} \qquad \mathbf{v}_1 = -\mathbf{i}$$
$$\mathbf{r}_2 = \mathbf{i} + \mathbf{k} \qquad \mathbf{v}_2 = 2\mathbf{j}$$
$$\mathbf{r}_3 = \mathbf{k} \qquad \mathbf{v}_3 = \mathbf{i} + \mathbf{j} + \mathbf{k}$$

Calculate the following quantities:
   (a)  The instantaneous position of the center of mass
   (b)  The velocity of the center of mass
   (c)  The linear momentum of the system
   (d)  The angular momentum of the system about the origin
   (e)  The kinetic energy

6.2    A gun of mass $m$ fires a bullet of mass $\gamma m$ where $\gamma$ is a small fraction. If $v_0$ is the speed of the bullet just as it leaves the gun barrel, what is the recoil speed of the gun?

## PROBLEMS

6.3    A block of wood of mass $m$ rests on a horizontal surface. A bullet of mass $\gamma m$ is fired horizontally with speed $v_0$ into the block, coming to rest in the block. What fraction of the original kinetic energy of the bullet is lost as heat in the block, immediately upon impact? If $\mu$ is the coefficient of sliding friction of the block, how far will it slide before coming to rest?

6.4    An artillery shell is fired at angle of elevation of 45° with an initial speed $v_0$. At the uppermost part of the trajectory the shell bursts into two equal fragments, one of which moves directly downward, relative to the ground, with initial speed $\sqrt{2}v_0$. What is the direction and speed of the other fragment immediately after the burst.

6.5    Three particles of equal mass lie on a straight line along which the particles move. Initially the particles are located at the points $-1$, $0$, and $+1$, and their velocities are $4v_0$, $2v_0$, and $v_0$, respectively. Find the final velocities of the particles assuming that all collisions are perfectly elastic.

6.6    A ball is dropped from a height $h$ onto a horizontal pavement. If the coefficient of restitution is $\epsilon$, show that the total vertical distance the ball goes before the rebounds cease is $h(1 + \epsilon^2)/(1 - \epsilon^2)$, and find the total length of time that the ball bounces.

6.7    Consider the earth and the moon as an isolated system, and show that each describes an ellipse about their common center of mass.

6.8  Show that the kinetic energy of a two-particle system is $\frac{1}{2}mv_{cm}^2 + \frac{1}{2}\mu v^2$ where $m = m_1 + m_2$, v is the relative speed, and $\mu$ is the reduced mass.

6.9  If two bodies undergo a direct collision, show that the loss in kinetic energy is equal to

$$\tfrac{1}{2}\mu v^2(1 - \epsilon^2)$$

where $\mu$ is the reduced mass, v is the relative speed before impact, and $\epsilon$ is the coefficient of restitution.

6.10  A moving particle of mass $m_1$ collides elastically with a target particle of mass $m_2$ which is initially at rest. If the collision is head-on, show that the incident particle loses a fraction $4\mu/m$ of its original kinetic energy where $\mu$ is the reduced mass and $m = m_1 + m_2$.

6.11  Show that the angular momentum of a two-particle system is

$$\mathbf{r}_{cm} \times m\mathbf{V}_{cm} + \mathbf{R} \times \mu\mathbf{v}$$

where $m = m_1 + m_2$, $\mu$ is the reduced mass, $\mathbf{R}$ is the relative position vector, and v is the relative velocity of the two particles.

6.12  Show that the constant $c$ in Equation (5.58) for the period of a planet around the sun should be $2\pi[G(M + m)]^{-1/2}$ rather than $2\pi(GM)^{-1/2}$ where $M$ is the mass of the sun and $m$ is the mass of the planet.

6.13  A proton of mass $m_p$ with initial velocity $\mathbf{v}_0$ collides with a helium atom, mass $4m_p$, that is initially at rest. If the proton leaves the point of impact at an angle of 45° with its original line of motion, find the final velocities of each particle. Assume that the collision is perfectly elastic.

6.14  Work the above problem for the case that the collision is inelastic and that $Q$ is equal to $\frac{1}{4}$ of the initial energy of the proton.

6.15  Referring to Problem 6.13, find the scattering angle of the proton in the center-of-mass system.

6.16  Find the scattering angle of the proton in the center-of-mass system for Problem 6.14.

6.17  A particle of mass $m$ with initial momentum $p_1$ collides with a particle of equal mass at rest. If the magnitudes of the final momenta of the two particles are $p_1'$ and $p_2'$, respectively, show that the energy loss of the collision is given by

$$Q = \frac{p_1'p_2'}{m} \cos\psi$$

where $\psi$ is the angle between the paths of the two particles after colliding.

6.18  Find the equation of motion for a rocket fired vertically upward, assuming $g$ is constant. Find the ratio of fuel to payload to achieve a final speed equal to the escape speed $v_e$ from the earth if the speed of the exhaust gas is $kv_e$ where $k$ is a given constant, and the fuel burning rate is $\dot{m}$. Compute the numerical value of the fuel-payload ratio for $k = \frac{1}{4}$, and $\dot{m}$ is equal to 1 percent of the initial mass of fuel, per second.

6.19   Find the differential equation of motion of a raindrop falling through a mist collecting mass as it falls.   Assume that the drop remains spherical and that the rate of accretion is proportional to the cross-sectional area of the drop multiplied by the speed of fall.   Show that if the drop starts from rest when it is infinitely small, then the acceleration is constant and equal to $g/7$.

6.20   A uniform heavy chain of length $a$ hangs initially with a part of length $b$ hanging over the edge of a table.   The remaining part, of length $a$-$b$, is coiled up at the edge of the table.   If the chain is released, show that the speed of the chain when the last link leaves the end of the table is $[2g(a^3 - b^3)/3a^2]^{1/2}$.

6.21   A rocket traveling through the atmosphere experiences a linear air resistance $-k\mathbf{v}$.   Find the differential equation of motion when all other external forces are negligible.   Integrate the equation and show that if the rocket starts from rest, the final speed is given by $v = V\alpha[1 - (m/m_0)^{1/\alpha}]$ where $V$ is the relative speed of the exhaust fuel, $\alpha = |\dot{m}/k| = $ constant, and $m_0$ is the initial mass of the rocket plus fuel, $m$ being the final mass of the rocket.

# 7. Mechanics of Rigid Bodies. Planar Motion

A rigid body may be regarded as a system of particles whose *relative* positions are fixed, or, in other words, the distance between any two particles is constant. This definition of a rigid body is idealized. In the first place, as pointed out in the definition of a particle, there are no true particles in nature. Secondly, real extended bodies are not strictly rigid; they become more or less deformed (stretched, compressed, or bent) when external forces are applied. We shall for the present, however, neglect such deformations.

## 7.1. Center of Mass of a Rigid Body

We have already defined the center of mass (Section 6.1) of a system of particles as the point $(x_{cm}, y_{cm}, z_{cm})$ where

$$x_{cm} = \frac{\Sigma x_i m_i}{\Sigma m_i} \qquad y_{cm} = \frac{\Sigma y_i m_i}{\Sigma m_i} \qquad z_{cm} = \frac{\Sigma z_i m_i}{\Sigma m_i}$$

For a rigid extended body, we can replace the summation by an integration over the volume of the body, namely,

$$x_{cm} = \frac{\int_v \rho x \, dv}{\int_v \rho \, dv} \qquad y_{cm} = \frac{\int_v \rho y \, dv}{\int_v \rho \, dv} \qquad z_{cm} = \frac{\int_v \rho z \, dv}{\int_v \rho \, dv} \qquad (7.1)$$

where $\rho$ is the density, and $dv$ is the element of volume.

If a rigid body is in the form of a thin shell, the equations for the center of mass become

$$x_{cm} = \frac{\int_s \rho x \, ds}{\int_s \rho \, ds} \qquad y_{cm} = \frac{\int_s \rho y \, ds}{\int_s \rho \, ds} \qquad z_{cm} = \frac{\int_s \rho z \, ds}{\int_s \rho \, ds} \qquad (7.2)$$

where $ds$ is the element of area, and $\rho$ is the mass per unit area, the integration extending over the area of the body.

Similarly, if the body is in the form of a thin wire, we have

$$x_{cm} = \frac{\int_l \rho x \, dl}{\int_l \rho \, dl} \qquad y_{cm} = \frac{\int_l \rho y \, dl}{\int_l \rho \, dl} \qquad z_{cm} = \frac{\int_l \rho z \, dl}{\int_l \rho \, dl} \qquad (7.3)$$

In this case $\rho$ is the mass per unit length, and $dl$ is the element of length.

For uniform homogeneous bodies, the density factors $\rho$ are constant in each case and therefore may be canceled out in each equation above.

If a body is composite, that is, if it consists of two or more parts whose centers of mass are known, then it is clear, from the definition of the center of mass, that we can write

$$x_{cm} = \frac{x_1 m_1 + x_2 m_2 + \cdots}{m_1 + m_2 + \cdots} \qquad (7.4)$$

with similar equations for $y_{cm}$ and $z_{cm}$.   Hence $(x_1, y_1, z_1)$ is the center of mass of the part $m_1$, and so on.

### Symmetry Considerations

If a body possesses symmetry, it is possible to take advantage of that symmetry in locating the center of mass.   Thus, if the body has a plane of symmetry, that is, if each particle $m_i$ has a mirror image of itself $m_i'$ relative to some plane, then the center of mass lies in that plane.   To prove this, let us suppose that the $xy$ plane is a plane of symmetry.   We have then

$$z_{cm} = \frac{\Sigma(z_i m_i + z_i' m_i')}{\Sigma(m_i + m_i')}$$

But $m_i = m_i'$ and $z_i = -z_i'$.   Hence the terms in the numerator cancel in pairs, and so $z_{cm} = 0$; that is, the center of mass lies in the $xy$ plane.

Similarly, if the body has a line of symmetry, it is easy to show that the center of mass lies on that line.   The proof is left as an exercise.

### Solid Hemisphere

To find the center of mass of a solid homogeneous hemisphere of radius $a$, we know from symmetry that the center of mass lies on the radius that is normal to the plane face.   Choosing coordinate axes as shown in Figure 7.1, we have that the center of mass lies on the $z$ axis.   To calculate $z_{cm}$ we use a circular element of volume of thickness $dz$ and radius $(a^2 - z^2)^{1/2}$, as shown. Thus

$$dv = \pi(a^2 - z^2)\,dz$$

Therefore

$$z_{cm} = \frac{\int_0^a \rho\pi z(a^2 - z^2)\,dz}{\int_0^a \rho\pi(a^2 - z^2)\,dz} = \frac{3}{8}\,a \qquad (7.5)$$

### *Hemispherical Shell*

For a hemispherical shell of radius $a$ we use the same axes as in the previous problem (Figure 7.1).   Again, from symmetry, the center of mass is

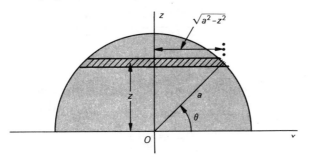

**FIGURE 7.1**   Coordinates for calculating the center of mass of a hemisphere.

located on the $z$ axis.   For our element of surface we choose a circular strip of width $a\,d\theta$.   Hence we can write

$$ds = 2\pi(a^2 - z^2)^{1/2}a\,d\theta$$

But $\theta = \sin^{-1}(z/a)$, so $d\theta = (a^2 - z^2)^{-1/2}\,dz$.   Therefore

$$ds = 2\pi a\,dz$$

The location of the center of mass is accordingly given by

$$z_{cm} = \frac{\int_0^a \rho 2\pi a z\,dz}{\int_0^a \rho 2\pi a\,dz} = \frac{1}{2}\,a \qquad (7.6)$$

### *Semicircle*

To find the center of mass of a thin wire bent into the form of a semicircle of radius $a$, we use axes as shown in Figure 7.2.   We have

$$dl = a\,d\theta$$

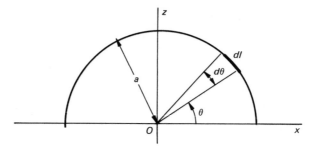

FIGURE 7.2   Coordinates for calculating the center of mass of a
semicircular wire.

and

$$z = a \sin \theta$$

Hence

$$z_{cm} = \frac{\int_0^\pi \rho(a \sin \theta)a \, d\theta}{\int_0^\pi \rho a \, d\theta} = \frac{2}{\pi} a \qquad (7.7)$$

*Semicircular Lamina*

In the case of a uniform semicircular lamina, the center of mass is on
the $z$ axis (Figure 7.2).   It is left as a problem to show that

$$z_{cm} = \frac{4}{3\pi} a \qquad (7.8)$$


## 7.2.   Some Theorems on
## Static Equilibrium of a Rigid Body

We have found (Section 6.1) that the acceleration of the center of mass
of a system is equal to the vector sum of the external forces divided by the
mass.   In particular, if the system is a rigid body, and if the sum of all the
external forces vanishes

$$\mathbf{F}_1 + \mathbf{F}_2 + \cdots = 0 \qquad (7.9)$$

then the center of mass, if initially at rest, will remain at rest.   Thus Equation
(7.9) expresses the condition for *translational equilibrium* of a rigid body.

Similarly, the vanishing of the total moment of all the applied forces

$$\mathbf{r}_1 \times \mathbf{F}_1 + \mathbf{r}_2 \times \mathbf{F}_2 + \cdots = 0 \qquad (7.10)$$

means that the angular momentum of the body does not change (Section 6.2).
This is the condition for *rotational equilibrium* of a rigid body—that is, the
condition that the body, if initially at rest, will not start to rotate.   Equations
(7.9) and (7.10) together constitute the necessary conditions for complete
equilibrium of a rigid body.

### Equilibrium in a Uniform Gravitational Field

Let us consider a rigid body in a uniform gravitational field, say at the surface of the earth. Since the sum of the gravitational forces is equal to $m\mathbf{g}$ where $m$ is the mass of the body, we can write the condition for translational equilibrium as

$$\mathbf{F}_1 + \mathbf{F}_2 + \cdots + m\mathbf{g} = 0 \tag{7.11}$$

where $\mathbf{F}_1$, $\mathbf{F}_2$, and so on, are all the external forces other than gravity.

Similarly, the condition for rotational equilibrium may be written

$$\mathbf{r}_1 \times \mathbf{F}_1 + \mathbf{r}_2 \times \mathbf{F}_2 + \cdots + \sum_i \mathbf{r}_i \times m_i\mathbf{g} = 0 \tag{7.12}$$

But $\mathbf{g}$ is a constant vector, so we can write

$$\sum_i \mathbf{r}_i \times m_i\mathbf{g} = \left(\sum_i m_i\mathbf{r}_i\right) \times \mathbf{g} = m\mathbf{r}_{cm} \times \mathbf{g} = \mathbf{r}_{cm} \times m\mathbf{g} \tag{7.13}$$

The above equation states that the moment of the force of gravity about any point is the same as that of a single force $m\mathbf{g}$ acting at the center of mass.[1] The equation for rotational equilibrium then becomes

$$\mathbf{r}_1 \times \mathbf{F}_1 + \mathbf{r}_2 \times \mathbf{F}_2 + \cdots + \mathbf{r}_{cm} \times m\mathbf{g} = 0 \tag{7.14}$$

### Equilibrium Under Coplanar Forces

If the lines of action of a set of forces acting on a rigid body are coplanar, that is, if they all lie in a plane, then we can write $\mathbf{F}_1 = \mathbf{i}X_1 + \mathbf{j}Y_1$, and so on. The component forms of the equations of equilibrium, Equations (7.9) and (7.10), (which the student will recall from elementary physics) are then

Translational equilibrium:
$$X_1 + X_2 + \cdots = 0 \qquad Y_1 + Y_2 + \cdots = 0 \tag{7.15}$$
Rotational equilibrium:
$$x_1Y_1 - y_1X_1 + x_2Y_2 - y_2X_2 + \cdots = 0 \tag{7.16}$$

### 7.3. Rotation of a Rigid Body About a Fixed Axis. Moment of Inertia

The simplest type of rigid-body motion, other than pure translation, is that in which the body is constrained to rotate about a fixed axis. Let us choose the $z$ axis of an appropriate coordinate system as the axis of rotation.

---

[1] The apparent center of gravitational force is called the *center of gravity*. In a uniform gravitational field such as we are considering, the center of mass and the center of gravity coincide.

The path of a representative particle $m_i$ located at the point $(x_i, y_i, z_i)$ is then a circle of radius $(x_i^2 + y_i^2)^{1/2} = R_i$ centered on the $z$ axis. A representative cross section parallel to the $xy$ plane is shown in Figure **7.3**.

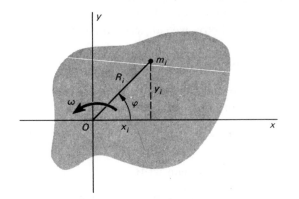

**FIGURE 7.3**   Cross section of a rigid body that is rotating about the $z$ axis.

The speed $v_i$ of particle $i$ is given by

$$v_i = R_i\omega = (x_i^2 + y_i^2)^{1/2}\omega \tag{7.17}$$

where $\omega$ is the angular speed of rotation. From a study of the figure, we see that the velocity has components as follows:

$$\dot{x}_i = -v_i \sin \varphi = -\omega y_i \tag{7.18}$$
$$\dot{y}_i = v_i \cos \varphi = \omega x_i \tag{7.19}$$
$$\dot{z}_i = 0 \tag{7.20}$$

where $\varphi$ is defined as shown in the figure. The above equations can also be obtained by taking the components of

$$\mathbf{v}_i = \boldsymbol{\omega} \times \mathbf{r}_i \tag{7.21}$$

where $\boldsymbol{\omega} = \mathbf{k}\omega$.

Let us calculate the kinetic energy of rotation of the body. We have

$$T = \sum_i \tfrac{1}{2}m_i v_i^2 = \tfrac{1}{2}\left(\sum_i m_i R_i^2\right)\omega^2 = \tfrac{1}{2}I\omega^2 \tag{7.22}$$

where

$$I = \sum_i m_i R_i^2 = \sum_i m_i(x_i^2 + y_i^2) \tag{7.23}$$

The quantity $I$, defined by the above equation, is of particular importance in the study of the motion of rigid bodies. It is called the *moment of inertia*.

To show how the moment of inertia further enters the picture, let us next calculate the angular momentum about the axis of rotation. Since the angular momentum of a single particle is, by definition, $\mathbf{r}_i \times m_i\mathbf{v}_i$, the $z$ component is

$$m_i(x_i\dot{y}_i - y_i\dot{x}_i) = m_i(x_i^2 + y_i^2)\omega = m_i R_i^2\omega \tag{7.24}$$

where we have made use of Equations (7.18) and (7.19).   The total $z$ component of the angular momentum, which we shall call $L$, is then given by summing over all the particles, namely,

$$L = \sum_i m_i R_i^2 \omega = I\omega \tag{7.25}$$

In Section 6.2 we found that the rate of change of angular momentum for any system is equal to the total moment of the external forces.   For a body constrained to rotate about a fixed axis, we have

$$N = \frac{dL}{dt} = \frac{d(I\omega)}{dt} \tag{7.26}$$

where $N$ is the total moment of all the applied forces about the axis of rotation (the component of $\mathbf{N}$ along the axis).   If the body is rigid, then $I$ is constant, and we can write

$$N = I\frac{d\omega}{dt} \tag{7.27}$$

The analogy between the equations for translation and for rotation about a fixed axis is shown below:

|   *Translation*   |   |   *Rotation*   |   |
|---|---|---|---|
| Linear momentum | $p = mv$ | Angular momentum | $L = I\omega$ |
| Force | $F = m\dot{v}$ | Torque | $N = I\dot{\omega}$ |
| Kinetic energy | $T = \frac{1}{2}mv^2$ | Kinetic energy | $T = \frac{1}{2}I\omega^2$ |

Thus the moment of inertia is analogous to mass; it is a measure of the rotational inertia of a body relative to some fixed axis of rotation, just as mass is a measure of translational inertia of a body.

## 7.4.   Calculation of the Moment of Inertia

In actual calculations of the moment of inertia $\Sigma m R^2$ for extended bodies, we can replace the summation by an integration over the body, just as we did in calculation of the center of mass.   Thus we may write

$$I = \int R^2 \, dm \tag{7.28}$$

where $dm$, the element of mass, is given by a density factor multiplied by an appropriate differential (volume, area, or length).   It is important to remember that $R$ is the perpendicular distance from the element of mass to the axis of rotation.

In the case of a composite body, it is clear, from the definition of the moment of inertia, that we may write

$$I = I_1 + I_2 + \cdots \tag{7.29}$$

where $I_1$, $I_2$, etc., are the moments of inertia of the various parts about the particular axis chosen.

Let us calculate the moments of inertia for some important special cases.

### Thin Rod

For a thin uniform rod of length $a$ and mass $m$, we have, for an axis perpendicular to the rod at one end [Figure 7.4(a)],

$$I = \int_0^a x^2 \rho \, dx = \tfrac{1}{3}\rho a^3 = \tfrac{1}{3}ma^2 \tag{7.30}$$

The last step follows from the fact that $m = \rho a$.

If the axis is taken at the center of the rod [Figure 7.4(b)], we have

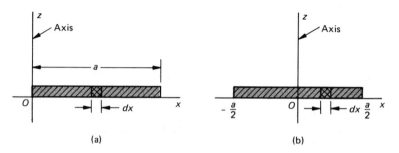

(a)                                                  (b)

FIGURE 7.4    Coordinates for calculating the moment of inertia of a rod<br>(a) about one end (b) about the center.

$$I = \int_{-a/2}^{a/2} x^2 \rho \, dx = \tfrac{1}{12}\rho a^3 = \tfrac{1}{12}ma^2 \tag{7.31}$$

### Hoop or Cylindrical Shell

In the case of a thin circular hoop or cylindrical shell, for the central or symmetry axis all particles lie at the same distance from the axis. Thus

$$I = ma^2 \tag{7.32}$$

where $a$ is the radius, and $m$ is the mass.

### Circular Disc or Cylinder

To calculate the moment of inertia of a uniform circular disc of radius $a$ and mass $m$, we shall use polar coordinates. The element of mass, a thin ring of radius $r$ and thickness $dr$, is given by

$$dm = \rho 2\pi r \, dr$$

where $\rho$ is the mass per unit area. The moment of inertia about an axis through the center of the disc normal to the plane faces (Figure 7.5) is obtained as follows:

$$I = \int_0^a \rho(r^2)(2\pi r \, dr) = 2\pi\rho \, \frac{a^4}{4} = \frac{1}{2} \, ma^2 \qquad (7.33)$$

The last step results from the relation $m = \rho\pi a^2$.

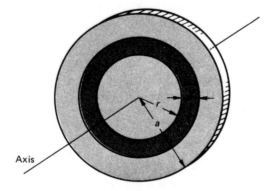

FIGURE 7.5   Coordinates for finding the moment of inertia of a disc.

Clearly, Equation (7.33) also applies to a uniform right-circular cylinder of radius $a$ and mass $m$, the axis being the central axis of the cylinder.

### Sphere

Let us find the moment of inertia of a uniform solid sphere of radius $a$ and mass $m$ about an axis (the $z$ axis) passing through the center. We shall divide the sphere into thin circular discs, as shown in Figure 7.6. The moment

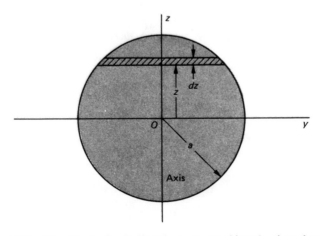

FIGURE 7.6   Coordinates for finding the moment of inertia of a sphere about the $z$ axis.

of inertia of a representative disc of radius $y$, from Equation (7.33), is $\frac{1}{2}y^2\, dm$. But $dm = \rho\pi y^2\, dz$, hence

$$I = \int_{-a}^{a} \frac{1}{2}\pi\rho y^4\, dz = \int_{-a}^{a} \frac{1}{2}\pi\rho(a^2 - z^2)^2\, dz = \frac{8}{15}\pi\rho a^5 \qquad (7.34)$$

The last step above should be filled in by the student.  Since the mass $m$ is given by

$$m = \frac{4}{3}\pi a^3 \rho$$

we have

$$I = \frac{2}{5}ma^2 \qquad (7.35)$$

### Spherical Shell

The moment of inertia of a thin uniform spherical shell can be found very simply by application of Equation (7.34).  If we differentiate with respect to $a$, namely,

$$dI = \frac{8}{3}\pi\rho a^4\, da$$

the result is the moment of inertia of a shell of thickness $da$ and radius $a$.  The mass of the shell is $4\pi a^2 \rho\, da$.  Hence we can write

$$I = \frac{2}{3}ma^2 \qquad (7.36)$$

for the moment of inertia of a thin shell of radius $a$ and mass $m$.  The student should verify the above result by direct integration.

### Perpendicular-Axis Theorem

Consider a rigid body which is in the form of a thin plane lamina of any shape.  Let us place the lamina in the $xy$ plane (Figure 7.7).  The moment of inertia about the $z$ axis is given by

$$I_z = \sum_i m_i(x_i^2 + y_i^2) = \sum_i m_i x_i^2 + \sum_i m_i y_i^2$$

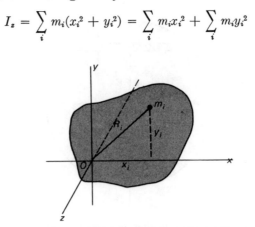

FIGURE 7.7  The perpendicular axis theorem.

But the sum $\sum_i m_i x_i^2$ is just the moment of inertia $I_y$ about the $y$ axis, because $z_i$ is zero for all particles.   Similarly, $\sum_i m_i y_i^2$ is the moment of inertia $I_x$ about the $x$ axis.   The above equation can therefore be written

$$I_z = I_x + I_y \tag{7.37}$$

This is the perpendicular-axis theorem.   In words: The moment of inertia of any plane lamina about an axis normal to the plane of the lamina is equal to the sum of the moments of inertia about any two mutually perpendicular axes passing through the given axis and lying in the plane of the lamina.

As an example of the use of this theorem, let us consider a thin circular disc in the $xy$ plane (Figure 7.8).   From Equation (7.33) we have

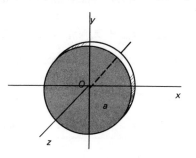

FIGURE 7.8

$$I_z = \tfrac{1}{2}ma^2 = I_x + I_y$$

In this case, however, we know from symmetry that $I_x = I_y$.   Therefore we must have

$$I_x = I_y = \tfrac{1}{4}ma^2 \tag{7.38}$$

for the moment of inertia about any axis in the plane of the disc passing through the center.   Equation (7.38) can also be obtained by direct integration.

### Parallel-Axis Theorem

Consider the equation for the moment of inertia about some axis, say the $z$ axis,

$$I = \sum_i m_i(x_i^2 + y_i^2)$$

Now we can express $x_i$ and $y_i$ in terms of the coordinates of the center of mass $(x_{cm}, y_{cm}, z_{cm})$ and the coordinates *relative* to the center of mass $(\bar{x}_i, \bar{y}_i, \bar{z}_i)$ (Figure 7.9) as follows:

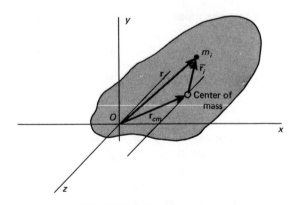

FIGURE 7.9  Parallel axis theorem.

$$x_i = x_{cm} + \bar{x}_i \quad y_i = y_{cm} + \bar{y}_i \tag{7.39}$$

We have, therefore, after substituting and collecting terms,

$$I = \sum_i m_i(\bar{x}_i{}^2 + \bar{y}_i{}^2) + \sum_i m_i(x_{cm}{}^2 + y_{cm}{}^2)$$

$$+ 2x_{cm}\sum_i m_i\bar{x}_i + 2y_{cm}\sum_i m_i\bar{y}_i \tag{7.40}$$

The first sum on the right is just the moment of inertia about an axis parallel to the $z$ axis and passing through the center of mass.  We shall call it $I_{cm}$. The second sum is clearly equal to the mass of the body multiplied by the square of the distance between the center of mass and the $z$ axis.  Let us call this distance $l$.  That is, $l^2 = x_{cm}{}^2 + y_{cm}{}^2$.

Now, from the definition of the center of mass,

$$\sum_i m_i\bar{x}_i = \sum_i m_i\bar{y}_i = 0$$

Hence, the last two sums on the right of Equation (7.40) vanish.  The final result may be written

$$I = I_{cm} + ml^2 \tag{7.41}$$

This is the *parallel-axis* theorem.  It is applicable to any rigid body, solid as well as laminar.  The theorem states, in effect, that the moment of inertia of a rigid body about any axis is equal to the moment of inertia about a parallel axis passing through the center of mass plus the product of the mass of the body and the square of the distance between the two axes.

Applying the above theorem to a circular disc, we have, from Equations (7.33) and (7.41),

$$I = \tfrac{1}{2}ma^2 + ma^2 = \tfrac{3}{2}ma^2 \tag{7.42}$$

for the moment of inertia of a uniform circular disc about an axis perpendicular to the plane of the disc and passing through the edge.   Furthermore, from Equations (7.38) and (7.41), we find

$$I = \tfrac{1}{4}ma^2 + ma^2 = \tfrac{5}{4}ma^2 \tag{7.43}$$

for the moment of inertia about an axis in the plane of the disc and tangent to the edge.

### Radius of Gyration

For some purposes it is convenient to express the moment of inertia of a rigid body in terms of a distance $k$ called the *radius of gyration*, where $k$ is defined by the equation

$$I = mk^2 \qquad \text{or} \qquad k = \sqrt{\frac{I}{m}} \tag{7.44}$$

For example, we find for the radius of gyration of a thin rod about an axis passing through one end [refer to Equation (7.30)]

$$k = \sqrt{\frac{\tfrac{1}{3}ma^2}{m}} = \frac{a}{\sqrt{3}}$$

Moments of inertia for various objects can be tabulated simply by listing the squares of their radii of gyration, as in Appendix VI.

## 7.5.   The Physical Pendulum

A rigid body which is free to swing under its own weight about a fixed horizontal axis of rotation is known as a *physical pendulum* or *compound pendulum*.   A physical pendulum is shown in Figure 7.10, where $O$ represents the location of the axis of rotation, and $CM$ is the center of mass.   The distance between $O$ and $CM$ is $l$, as shown.

Denoting the angle between the line $OCM$ and the vertical line $OA$ by $\theta$, the moment of the gravitational force (acting at $CM$) about the axis of rotation is of magnitude

$$mgl \sin \theta$$

The fundamental equation of motion $N = I\dot{\omega}$ then takes the form $-mgl \sin \theta = I\ddot{\theta}$ or

$$\ddot{\theta} + \frac{mgl}{I} \sin \theta = 0 \tag{7.45}$$

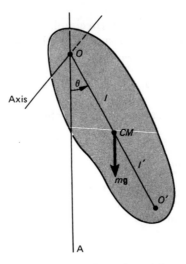

FIGURE 7.10   The physical pendulum.

The above equation is identical in form to the equation of motion of a simple pendulum.   For small oscillations, as in the case of the simple pendulum, we can replace $\sin \theta$ by $\theta$:

$$\ddot{\theta} + \frac{mgl}{I} \theta = 0 \tag{7.46}$$

The solution is

$$\theta = \theta_0 \cos (2\pi f t + \epsilon) \tag{7.47}$$

where $\theta_0$ is the amplitude and $\epsilon$ is a phase angle.   The frequency of oscillation $f$ is given by

$$f = \frac{1}{2\pi} \sqrt{\frac{mgl}{I}} \tag{7.48}$$

The period $T$ is therefore given by

$$T = \frac{1}{f} = 2\pi \sqrt{\frac{I}{mgl}} \tag{7.49}$$

(To avoid confusion, we shall not use a specific symbol to designate the angular frequency $2\pi f$.)   We can also express the period in terms of the radius of gyration $k$, namely,

$$T = 2\pi \sqrt{\frac{k^2}{gl}} \tag{7.50}$$

Thus the period is the same as that of a simple pendulum of length $k^2/l$.

As an example, a thin uniform rod of length $a$ swinging as a physical pendulum about one end ($k^2 = a^2/3$) has a period

$$T = 2\pi \sqrt{\frac{2a}{3g}}$$

### Center of Oscillation

By use of the parallel-axis theorem, we can express the radius of gyration $k$ in terms of the radius of gyration about the center of mass $k_{cm}$, as follows:

$$I = I_{cm} + ml^2$$

or

$$mk^2 = mk_{cm}^2 + ml^2$$

Canceling the $m$'s, we get

$$k^2 = k_{cm}^2 + l^2 \tag{7.51}$$

Equation (7.50) can therefore be written as

$$T = 2\pi \sqrt{\frac{k_{cm}^2 + l^2}{gl}} \tag{7.52}$$

Suppose that the axis of rotation of a physical pendulum is shifted to a different position $O'$ at a distance $l'$ from the center of mass, as shown in Figure 7.10. The period of oscillation $T'$ about this new axis is given by

$$T' = 2\pi \sqrt{\frac{k_{cm}^2 + l'^2}{gl'}}$$

It follows that the periods of oscillation about $O$ and about $O'$ will be equal, provided

$$\frac{k_{cm}^2 + l^2}{l} = \frac{k_{cm}^2 + l'^2}{l'}$$

The above equation readily reduces to

$$ll' = k_{cm}^2 \tag{7.53}$$

The point $O'$, related to $O$ by the above equation, is called the *center of oscillation* for the point $O$. It is clear that $O$ is also the center of oscillation for $O'$. Thus, for a rod of length $a$ swinging about one end, we have $k_{cm}^2 = a^2/12$ and $l = a/2$. Hence, from Equation (7.53), $l' = a/6$, and so the rod will have the same period when swinging about an axis located a distance $a/6$ from the center as it does for an axis passing through one end.

### 7.6.   A General Theorem Concerning Angular Momentum

In order to study the more general case of rigid-body motion, that in which the axis of rotation is *not* fixed, we need to develop a fundamental theorem about angular momentum. In Section 6.2 we showed that the time rate of change of angular momentum of any system is equal to the applied torque:

$$\frac{d\mathbf{L}}{dt} = \mathbf{N} \qquad\qquad (7.54)$$

or, explicitly

$$\frac{d}{dt} \sum_i (\mathbf{r}_i \times m_i\mathbf{v}_i) = \sum_i (\mathbf{r}_i \times \mathbf{F}_i) \qquad\qquad (7.55)$$

In the above equation all quantities are referred to some inertial coordinate system.

Let us now introduce the center of mass by expressing the position vector of each particle $\mathbf{r}_i$ in terms of the position of the center of mass $\mathbf{r}_{cm}$ and the position vector of particle $i$ relative to the center of mass $\bar{\mathbf{r}}_i$ (as in Section 6.3), namely,

$$\mathbf{r}_i = \mathbf{r}_{cm} + \bar{\mathbf{r}}_i$$

and

$$\mathbf{v}_i = \mathbf{v}_{cm} + \bar{\mathbf{v}}_i$$

Equation (7.55) then becomes

$$\frac{d}{dt} \sum_i [(\mathbf{r}_{cm} + \bar{\mathbf{r}}_i) \times m_i(\mathbf{v}_{cm} + \bar{\mathbf{v}}_i)] = \sum_i (\mathbf{r}_{cm} + \bar{\mathbf{r}}_i) \times \mathbf{F}_i \qquad (7.56)$$

Upon expanding and using the fact that $\Sigma m_i\bar{\mathbf{r}}_i$ and $\Sigma m_i\bar{\mathbf{v}}_i$ both vanish, we find that Equation (7.56) reduces to

$$\mathbf{r}_{cm} \times \sum_i m_i\mathbf{a}_{cm} + \frac{d}{dt} \sum_i \bar{\mathbf{r}}_i \times m_i\bar{\mathbf{v}}_i = \mathbf{r}_{cm} \times \sum_i \mathbf{F}_i + \sum_i \bar{\mathbf{r}}_i \times \mathbf{F}_i \quad (7.57)$$

where $\mathbf{a}_{cm} = \dot{\mathbf{v}}_{cm}$.

In Section 6.1 we showed that the translation of the center of mass of any system of particles obeys the equation

$$\sum_i \mathbf{F}_i = \sum_i m_i\mathbf{a}_i = m\mathbf{a}_{cm} \qquad\qquad (7.58)$$

Consequently, the first term on the left of Equation (7.57) cancels the first term on the right.  The final result is

$$\frac{d}{dt} \sum_i \bar{\mathbf{r}}_i \times m_i\bar{\mathbf{v}}_i = \sum_i \bar{\mathbf{r}}_i \times \mathbf{F}_i \qquad\qquad (7.59)$$

The sum on the left in the above equation is just the angular momentum of the system about the center of mass, and the sum on the right is the total moment of the external forces about the center of mass. Calling these quantities $\bar{\mathbf{L}}$ and $\bar{\mathbf{N}}$, respectively, we have

$$\frac{d\bar{\mathbf{L}}}{dt} = \bar{\mathbf{N}} \tag{7.60}$$

This important result states that the time rate of change of angular momentum about the center of mass of any system is equal to the total moment of the external forces about the center of mass. This is true even if the center of mass is accelerating. If we choose any point *other* than the center of mass as a reference point, then that point must be at rest in an inertial coordinate system (except for certain special cases which we shall not attempt to discuss). An example of the use of the above theorem is given in Section 7.8.

### 7.7.   Laminar Motion of a Rigid Body

If the motion of a body is such that all particles move parallel to some fixed plane, then that motion is called *laminar*. In laminar motion the axis of rotation may change position, but it does not change in direction. Rotation about a fixed axis is a special case of laminar motion. The rolling of a cylinder on a plane surface is another example of laminar motion.

If a body undergoes a laminar displacement, that displacement can be specified as follows: Choose some reference point of the body, for example, the center of mass. The reference point undergoes some displacement $\Delta\mathbf{r}$. In addition, the body rotates about the reference point through some angle $\Delta\varphi$. Clearly, any laminar displacement can be so specified. Consequently, laminar motion can be specified by giving the translational velocity of a convenient reference point together with the angular velocity.

The fundamental equation governing translation of a rigid body is

$$\mathbf{F} = m\ddot{\mathbf{r}}_{cm} = m\dot{\mathbf{v}}_{cm} = m\mathbf{a}_{cm} \tag{7.61}$$

where $\mathbf{F}$ represents the sum of all the external forces acting on the body, $m$ is the mass, and $\mathbf{a}_{cm}$ is the acceleration of the center of mass.

Application of Equation (7.25) to the case of laminar motion of a rigid body yields

$$\bar{L} = I_{cm}\omega \tag{7.62}$$

for the magnitude of the angular momentum about an axis $C$ passing through the center of mass where $\omega$ is the angular speed of rotation about that axis. The fundamental equation governing the rotation of the body, Equation (7.60), then becomes

$$\frac{d\bar{L}}{dt} = I_{cm}\dot{\omega} = \bar{N} \tag{7.63}$$

where $\bar{N}$ is the total moment of the applied forces about the axis $C$.

### 7.8.  Body Rolling Down an Inclined Plane

As an illustration of laminar motion, we shall study the motion of a round object (cylinder, ball, and so on) rolling down an inclined plane.  As shown in Figure 7.11, there are three forces acting on the body.  These are

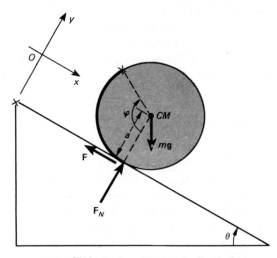

FIGURE 7.11   Body rolling an inclined plane.

(1) the downward force of gravity, (2) the normal reaction of the plane $F_N$, and (3) the frictional force parallel to the plane $F$.  Choosing axes as shown, the component equations of the translation of the center of mass are

$$m\ddot{x}_{cm} = mg \sin \theta - F \tag{7.64}$$
$$m\ddot{y}_{cm} = -mg \cos \theta + F_N \tag{7.65}$$

where $\theta$ is the inclination of the plane to the horizontal.  Since the body remains in contact with the plane, we have

$$y_{cm} = \text{constant}$$

Hence

$$\ddot{y}_{cm} = 0$$

Therefore, from Equation (7.65),

$$F_N = mg \cos \theta \tag{7.66}$$

The only force which exerts a moment about the center of mass is the frictional force $F$.  The magnitude of this moment is $Fa$ where $a$ is the radius of the body.  Hence the rotational equation, Equation (7.63), becomes

$$I_{cm}\dot{\omega} = Fa \tag{7.67}$$

To discuss the problem further, we need to make some assumptions regarding the contact between the plane and the body.  We shall solve the equations of motion for two cases.

### Motion with No Slipping

If the contact is perfectly rough so that no slipping can occur, we have the following relations:

$$\begin{aligned}
x_{cm} &= a\varphi \\
\dot{x}_{cm} &= a\dot{\varphi} = a\omega \\
\ddot{x}_{cm} &= a\ddot{\varphi} = a\dot{\omega}
\end{aligned} \tag{7.68}$$

where $\varphi$ is the angle of rotation.  Equation (7.67) can then be written

$$\frac{I_{cm}}{a^2}\ddot{x}_{cm} = F \tag{7.69}$$

Substituting the above value for $F$ into Equation (7.64) yields

$$m\ddot{x}_{cm} = mg\sin\theta - \frac{I_{cm}}{a^2}\ddot{x}_{cm}$$

Solving for $\ddot{x}_{cm}$, we find

$$\ddot{x}_{cm} = \frac{mg\sin\theta}{m + (I_{cm}/a^2)} = \frac{g\sin\theta}{1 + (k_{cm}^2/a^2)} \tag{7.70}$$

where $k_{cm}$ is the radius of gyration about the center of mass.  The body therefore rolls down the plane with constant linear acceleration and also with constant angular acceleration by virtue of Equation (7.68).

For example, the acceleration of a uniform cylinder ($k_{cm}^2 = a^2/2$) is

$$\frac{g\sin\theta}{1 + \frac{1}{2}} = \frac{2}{3}g\sin\theta$$

whereas that of a uniform sphere ($k_{cm}^2 = 2a^2/5$) is

$$\frac{g\sin\theta}{1 + \frac{2}{5}} = \frac{5}{7}g\sin\theta$$

### Energy Considerations

The above results can also be obtained from energy considerations.  In a uniform gravitational field the potential energy $V$ of a rigid body is given by the sum of the potential energies of the individual particles, namely,

$$V = \Sigma(m_i g z_i) = mg z_{cm}$$

where $z_{cm}$ is the vertical distance of the center of mass from some (arbitrary) reference plane.  Now if the forces, other than gravity, acting on the body do no work, then the motion is conservative, and we can write

$$T + V = T + mgz_{cm} = E = \text{constant}$$

where $T$ is the kinetic energy.

In the case of the body rolling down the inclined plane, Figure 7.11, the kinetic energy of translation is $\frac{1}{2}m\dot{x}_{cm}^2$ and that of rotation is $\frac{1}{2}I_{cm}\omega^2$, so the energy equation reads

$$\tfrac{1}{2}m\dot{x}_{cm}^2 + \tfrac{1}{2}I_{cm}\omega^2 + mgz_{cm} = E$$

But $\omega = \dot{x}_{cm}/a$ and $z_{cm} = -x_{cm}\sin\theta$. Hence

$$\frac{1}{2}\,m\dot{x}_{cm}^2 + \frac{1}{2}\,mk_{cm}{}^2\frac{\dot{x}_{cm}^2}{a^2} - mgx_{cm}\sin\theta = E$$

In the case of pure rolling motion the frictional force does not affect the energy equation, since this force is perpendicular to the displacement and consequently does no work. Hence $E$ is constant.

Differentiating with respect to $t$ and collecting terms yields

$$m\dot{x}_{cm}\ddot{x}_{cm}\left(1 + \frac{k_{cm}{}^2}{a^2}\right) - mg\dot{x}_{cm}\sin\theta = 0$$

Canceling the common factor $\dot{x}_{cm}$ (assuming, of course, that $\dot{x}_{cm} \neq 0$) and solving for $\ddot{x}_{cm}$, we find the same result as that obtained previously using forces and moments.


### Occurrence of Slipping

Let us now consider the case in which the contact with the plane is not perfectly rough but has a certain coefficient of sliding friction $\mu$. If slipping occurs, then the magnitude of the frictional force $\mathbf{F}$ is given by

$$F = F_{\max} = \mu F_N = \mu mg\cos\theta \tag{7.71}$$

The equation of translation, Equation (7.64), then becomes

$$m\ddot{x}_{cm} = mg\sin\theta - \mu mg\cos\theta \tag{7.72}$$

and the rotational equation, Equation (7.67), is

$$I_{cm}\dot{\omega} = \mu mga\cos\theta \tag{7.73}$$

From Equation (7.72) we see that again the center of mass undergoes constant acceleration:

$$\ddot{x}_{cm} = g(\sin\theta - \mu\cos\theta) \tag{7.74}$$

and, at the same time, the angular acceleration is constant:

$$\dot{\omega} = \frac{\mu mga\cos\theta}{I_{cm}} = \frac{\mu ga\cos\theta}{k_{cm}{}^2} \tag{7.75}$$

Let us integrate these two equations with respect to $t$, assuming that the body starts from rest, that is, at $t = 0$, $\dot{x}_{cm} = 0$, $\dot{\varphi} = 0$. We obtain

$$\dot{x}_{cm} = g(\sin\theta - \mu\cos\theta)t$$
$$\omega = \dot{\varphi} = g(\mu a\cos\theta/k_{cm}{}^2)t \tag{7.76}$$

Consequently, the linear speed and the angular speed have a constant ratio, and we can write

$$\dot{x}_{cm} = \gamma a \omega$$

where

$$\gamma = \frac{\sin \theta - \mu \cos \theta}{\mu a^2 \cos \theta / k_{cm}^2} = \frac{k_{cm}^2}{a^2} \left( \frac{\tan \theta}{\mu} - 1 \right) \tag{7.77}$$

Now $a\omega$ cannot be greater than $\dot{x}_{cm}$, so $\gamma$ cannot be less than unity. The limiting case, that for which we have pure rolling, is given by $\dot{x}_{cm} = a\omega$, that is,

$$\gamma = 1$$

Solving for $\mu$ in Equation (7.77) with $\gamma = 1$, we find that the critical value of $\mu$ is given by

$$\mu_{crit} = \frac{\tan \theta}{1 + (a/k_{cm})^2} \tag{7.78}$$

If $\mu$ is greater than that given above, then the body rolls without slipping.

For example, if a ball is placed on a 45° plane, it will roll without slipping provided $\mu$ is greater than $\tan 45°/(1 + \frac{5}{2})$ or $\frac{2}{7}$.

### 7.9.   Motion of a Rigid Body Under an Impulsive Force

In the previous chapter we introduced the concept of an impulsive force acting on a particle.   We found that the effect of such a force, or impulse, is to produce a sudden change in the velocity of the particle.   In this section we shall extend the impulse concept to the case of laminar motion of an extended rigid body.

*Free Motion*

Suppose a body is free to move in a plane and is subjected to an impulse $\hat{\mathbf{P}}$. Then, according to the general theory discussed in Section 7.7, we have both the translation and the rotation of the body to consider.

First, the translation is given by the general formula

$$\mathbf{F} = m\dot{\mathbf{v}}_{cm}$$

If $\mathbf{F}$ is an impulsive type of force, we have

$$\int \mathbf{F} \, dt = \hat{\mathbf{P}} = m\Delta\mathbf{v}_{cm}$$

Thus the result of the impulse is to cause the velocity of the center of mass to change by the amount

$$\Delta \mathbf{v}_{cm} = \frac{\hat{\mathbf{P}}}{m} \tag{7.79}$$

Secondly, the rotation of the body is governed by the equation

$$N = \dot{L} = I_{cm}\dot{\omega}$$

We can integrate with respect to the time $t$ and obtain the following relation

$$\int N \, dt = I_{cm}\Delta\omega \tag{7.80}$$

We call the integral $\int N dt$ the *rotational impulse*. Let us use the symbol $\hat{L}$ to designate it. The effect of a rotational impulse, then, is to change the angular velocity of the body by the amount

$$\Delta\omega = \frac{\hat{L}}{I_{cm}} \tag{7.81}$$

Now if the primary impulse $\hat{\mathbf{P}}$ is applied to the body in such a manner that its line of action is a distance $b$ from the center of mass, then the moment $N = Fb$. Consequently,

$$\hat{L} = \hat{P}b \tag{7.82}$$

We can then express the change in the angular velocity produced by an impulse as

$$\Delta\omega = \frac{\hat{P}b}{I_{cm}} \tag{7.83}$$

To summarize: *the effect of an impulse on a rigid body that is free to move in laminar motion is (1) to produce a sudden change in the velocity of the center of mass—the translational effect, and (2) to produce a sudden change in the angular velocity of the body—the rotational effect.*

### Constrained Motion

In the event that a body subjected to an impulse is not free, but is constrained to rotate about a fixed axis, we need only consider the rotation condition $N = I\dot{\omega}$. Thus

$$\int N \, dt = \hat{L} = I\Delta\omega$$

In the above equation, $I$ is the moment of inertia about the fixed axis of rotation, and $N$ is the moment about that axis. In this case the rotational impulse $\hat{L}$ that is produced by a single primary impulse $\hat{P}$ whose line of action is a distance $b$ from the axis of rotation is also given by

$$\hat{L} = \hat{P}b$$

so that

$$\Delta\omega = \frac{\hat{P}b}{I} \tag{7.84}$$

is the change in the angular velocity about the fixed axis of rotation.

*Effect of Several Simultaneous Impulses*

If a number of different impulses are applied to a rigid body at the same instant, the resulting change in the velocity of the center of mass and the angular velocity of the body are obtained by properly adding the impulses and the moments, respectively. Thus, the translational effect of several impulses is obtained by vector addition of the individual impulses, so that Equation (7.79) becomes

$$\Delta \mathbf{v}_{cm} = \frac{\hat{\mathbf{P}}_1 + \hat{\mathbf{P}}_2 + \cdots}{m} \tag{7.85}$$

Similarly, for the rotational effect, Equation (7.83) is modified to read

$$\Delta \omega = \frac{\hat{P}_1 b_1 + \hat{P}_2 b_2 + \cdots}{I_{cm}} \tag{7.86}$$

In the case of a body that is constrained to rotate about a fixed axis, there is a secondary impulse due to the reaction of the axis on the body whenever an external impulse is applied. The motion is then determined by the sum of all impulses according to the above equations.

## EXAMPLES

### 1. *Impulse Applied to a Free Rod*

As an illustration of the above theory, consider a rod that is free to slide on a smooth horizontal surface. Let an impulse $\hat{\mathbf{P}}$ be applied to the rod a distance $b$ from the center of mass and in a direction at right angles to the length of the rod, as shown in Figure 7.12.

If the rod is initially motionless, then the equations of translation and rotation are, respectively,

$$\mathbf{V}_{cm} = \frac{\hat{\mathbf{P}}}{m} \tag{7.87}$$

$$\omega = \frac{\hat{P}b}{I_{cm}} \tag{7.88}$$

In particular, if the rod is uniform and of length $2a$, then $I_{cm} = ma^2/3$. Accordingly

$$\omega = \hat{P}\,\frac{3b}{ma^2} \tag{7.89}$$

and thus the velocity imparted to the mass center is the same regardless of the point of application of the impulse, whereas the angular velocity acquired by

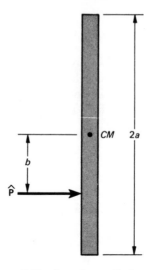

FIGURE 7.12    Impulse applied to a free rod.

the rod depends on the location of the applied impulse.   We see also, that the final kinetic energy of the rod is

$$T = \frac{1}{2}\,mv_{cm}{}^2 + \frac{1}{2}\,I_{cm}\omega^2 = \frac{\hat{P}^2}{2m} + \frac{3\hat{P}^2}{2m}\left(\frac{b}{a}\right)^2$$

This clearly depends on the point at which the impulse is applied.

### 2.   Impulse Applied to a Rod Constrained to Rotate About a Fixed Axis

Let us next consider the case in which the same rod is constrained to rotate about a fixed axis.   Suppose the axis $O$ is located at one end of the rod as indicated in Figure 7.13.   In this case we find the rotational equation for the motion to be

$$\hat{L} = \hat{P}(a + b) = I_0\omega \tag{7.90}$$

Then, since $I_0 = (\frac{4}{3})ma^2$, we obtain

$$\omega = \hat{P}\,\frac{3(a + b)}{4ma^2} \tag{7.91}$$

for the angular velocity imparted to the rod.   Now, since the rod is rotating about $O$, the center of mass is moving. Its speed is

$$v_{cm} = a\omega$$

or

$$v_{cm} = \hat{P}\,\frac{3(a + b)}{4ma} \tag{7.92}$$

We note that this is *not* equal to $\hat{P}/m$.   At first sight this result may seem to

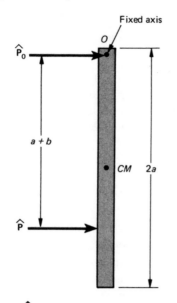

FIGURE 7.13 Impulse $\hat{\mathbf{P}}$ applied to a rod that is constrained to rotate about one end. The reactive impulse at the axis is $\hat{\mathbf{P}}_0$.

be in contradiction to the general equation for translation, Equation (7.79). Actually, there is no contradiction, because there is also another impulse that acts on the rod at the same time as the primary impulse. This second impulse is the reactive impulse exerted on the rod by the axis at $O$. Let us call this reactive impulse $\hat{\mathbf{P}}_0$. The total impulse applied to the rod is then the vector sum $\hat{\mathbf{P}} + \hat{\mathbf{P}}_0$. The velocity acquired by the center of mass is accordingly

$$\mathbf{v}_{cm} = \frac{\hat{\mathbf{P}} + \hat{\mathbf{P}}_0}{m} \tag{7.93}$$

We can now calculate the value of $\hat{\mathbf{P}}_0$ by using the value of $\mathbf{v}_{cm}$ given by Equation (7.92). Thus

$$\hat{\mathbf{P}} \frac{3(a + b)}{4ma} = \frac{\hat{\mathbf{P}} + \hat{\mathbf{P}}_0}{m}$$

which yields

$$\hat{\mathbf{P}}_0 = \hat{\mathbf{P}} \frac{3b - a}{4a} \tag{7.94}$$

for the impulse delivered to the rod by the constraining axis. From the law of action and reaction, the impulse exerted by the rod on the axis at $O$ is $-\hat{\mathbf{P}}_0$.

It should be noted that the reactive impulse vanishes if the point of application of the primary impulse is properly chosen. This point is called the *center of percussion*. In the case of the above rod this point is such that $b = a/3$.

### 7.10.   Collisions of Rigid Bodies

In problems involving collisions of extended rigid bodies, the forces that the bodies exert on each other during contact are always equal and opposite. Therefore the principles of conservation of linear and angular momentum are valid.   The concepts of linear and rotational impulse are often helpful in such problems.

## EXAMPLE

*Collision of a Ball and a Rod*

Consider, as an example, the impact of a ball of mass $m'$ with a uniform rod of length $2a$ and mass $m$.   Let us suppose that the rod is initially at rest on a smooth horizontal surface, as above, and that the point of impact is a distance $b$ from the center of the rod as shown in Figure 7.14.   Equations

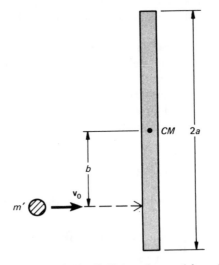

FIGURE 7.14   Collision of a particle and a rod.

(7.87) and (7.88) give the motion of the rod after the impact in terms of the impulse $\hat{\mathbf{P}}$ delivered to the rod by the ball.   We know also that the impulse received by the ball as a result of the impact is $-\hat{\mathbf{P}}$.   Hence we can write the equations for translation as

$$\hat{\mathbf{P}} = m\mathbf{v}_{cm} \qquad (7.95)$$
$$-\hat{\mathbf{P}} = m'(\mathbf{v}_1 - \mathbf{v}_0) \qquad (7.96)$$

in which $\mathbf{v}_{cm}$ is the velocity of the center of mass of the rod after the impact,

$\mathbf{v}_0$ is the initial velocity of the ball before impact, and $\mathbf{v}_1$ is the final velocity of the ball. The two equations of translation together imply conservation of linear momentum, for, upon eliminating $\hat{\mathbf{P}}$, we have

$$m'\mathbf{v}_0 = m'\mathbf{v}_1 + m\mathbf{v}_{cm} \tag{7.97}$$

In order to determine the rotation of the rod after the impact, we can make use of the principle of conservation of angular momentum. The initial angular momentum of the ball, about the center of mass, is $bm'v_0$, and the final angular momentum is $bm'v_1$. For the rod, the initial angular momentum is zero, and the final angular momentum is $I_{cm}\omega$. Hence

$$bm'v_0 = bm'v_1 + I_{cm}\omega \tag{7.98}$$

The above translational and rotational equations do not give us enough information to find all three velocity parameters of the final motion, namely $v_1$, $v_{cm}$, and $\omega$. In order to calculate the final motion completely, we need another equation. This can be the energy balance equation

$$\tfrac{1}{2}m'v_0{}^2 = \tfrac{1}{2}m'v_1{}^2 + \tfrac{1}{2}mv_{cm}^2 + \tfrac{1}{2}I_{cm}\omega^2 + Q \tag{7.99}$$

in which $Q$ is the energy loss. Alternatively, we can use the equation for the coefficient of restitution

$$\epsilon = \frac{\text{speed of separation}}{\text{speed of approach}}$$

In the problem under discussion, we have

$$\text{speed of approach} = v_0$$

To find the speed of separation, we need to know the speed of the rod at the point of impact. This is given by the sum of the translational speed of the center of mass and the rotational speed of that point relative to the center. Thus the speed of the impact point immediately after the collision is just $v_{cm} + b\omega$. Accordingly, we can write

$$\text{speed of separation} = v_{cm} + b\omega - v_1$$

Therefore

$$\epsilon v_0 = v_{cm} + b\omega - v_1$$

We now have enough equations to solve for the final motion. Setting $I_{cm} = ma^2/3$, we obtain

$$v_{cm} = v_0(\epsilon + 1)\left(\frac{m}{m'} + 3\frac{b^2}{a^2} + 1\right)^{-1}$$

$$v_1 = v_0 - \frac{m}{m'}v_{cm} \tag{7.100}$$

$$\omega = v_{cm}\left(\frac{3b}{a^2}\right)$$

The reader should verify this result.

## DRILL EXERCISES

7.1   Find the center of mass of the following:
(a) A thin uniform wire bent into a "U," each section being of the same length $a$
(b) The area bounded by the parabola $y = ax^2$ and the line $y = b$
(c) The volume bounded by the paraboloid of revolution $z = a(x^2 + y^2)$ and the plane $z = b$
(d) A solid uniform right circular cone of height $h$

7.2   A solid uniform sphere of radius $a$ contains a spherical cavity of radius $b$ centered a distance $c$ from the center of the sphere, where $a > b + c$. Find the center of mass.

7.3   Find the moments of inertia of each of the figures in Exercise 7.1 about their symmetry axes.

7.4   Find the moment of inertia of the sphere in Exercise 7.2 about an axis passing through the centers of the sphere and the cavity.

7.5   Show that the moment of inertia of a solid uniform octant of a sphere of radius $a$ is $(2/5)ma^2$ about an axis along one of the straight edges. [*Note*:   This is the same formula as that for a solid sphere of the same radius.]


## PROBLEMS

7.6   A uniform wire bent into a semicircle hangs on a rough peg.   The line joining the ends of the wire makes an angle $\theta$ with the horizontal, and the wire is just on the verge of slipping.   What is the coefficient of friction between the wire and the peg?

7.7   A solid uniform hemisphere rests in limiting equilibrium against a vertical wall.   The rounded side of the hemisphere is in contact with the wall and the floor.   If the coefficient of friction $\mu$ is the same for the wall and the floor, find the angle between the plane face of the hemisphere and the floor.

7.8   A uniform hemispherical shell rests in limiting equilibrium on a rough inclined plane of inclination $\theta$.   The rounded side of the shell is in contact with the plane, and the coefficient of friction is $\mu$.   Find the inclination of the shell.

7.9   Given that a set of forces $F_1$, $F_2$, . . . acting on a rigid body is (a) in translational equilibrium and (b) in rotational equilibrium about some point $O$.   Prove that the set of forces is also in rotational equilibrium about any other point $O'$.

7.10  Show that the moments of inertia of a solid uniform rectangular parallelepiped, elliptic cylinder, and ellipsoid are, respectively, $(m/3)(a^2 + b^2)$, $(m/4)(a^2 + b^2)$, and $(m/5)(a^2 + b^2)$, where $m$ is the mass, and $2a$ and $2b$ are the principal diameters of the solid at right angles to the axis of rotation, the axis being through the center in each case.

7.11  A circular hoop of radius $a$ swings as a physical pendulum about a point on the circumference. Find the period of oscillation if the axis of rotation is (a) normal to the plane of the hoop, and (b) in the plane of the hoop.

7.12  Show that the period of a physical pendulum is equal to $2\pi(d/g)^{1/2}$, where $d$ is the distance between the point of suspension $O$ and the center of oscillation $O'$.

7.13  A uniform solid ball has a few turns of light string wound around it. If the end of the string is held steady, and the ball is allowed to fall under gravity, what is the acceleration of the center of the ball?

7.14  Two men are holding the ends of a uniform plank of length $l$ and mass $m$. Show that if one man suddenly lets go, the load supported by the other man suddenly drops from $mg/2$ to $mg/4$. Show also that the initial downward acceleration of the free end is $\frac{3}{2}g$.

7.15  A uniform solid ball contains a hollow spherical cavity at its center, the radius of the cavity being $\frac{1}{2}$ the radius of the ball. Show that the acceleration of the ball rolling down a rough inclined plane is just $\frac{98}{101}$ of that of a uniform solid ball with no cavity. [*Note:* This suggests a method for non-destructive testing.]

7.16  Two weights of mass $m_1$ and $m_2$ are tied to the ends of a light inextensible cord. The cord passes over a pulley of radius $a$ and moment of inertia $I$. Find the accelerations of the weights, assuming $m_1 > m_2$ and neglecting friction in the axle of the pulley.

7.17  A uniform right-circular cylinder of radius $a$ is balanced on the top of a perfectly rough fixed cylinder of radius $b(b > a)$, the axes of the two cylinders being parallel. If the balance is slightly disturbed, find the point at which the rolling cylinder leaves the fixed one.

7.18  A ladder leans against a smooth vertical wall. If the floor is also smooth, and the initial angle between the floor and the ladder is $\theta_0$, show that the ladder, in sliding down, will lose contact with the wall when the angle between the floor and the ladder is $\sin^{-1}\left(\frac{2}{3}\sin\theta_0\right)$.

7.19  A long uniform rod of length $l$ stands vertically on a rough floor. The rod is slightly disturbed and falls to the floor. (a) Find the horizontal and vertical components of the reaction at the floor as functions of the angle $\theta$ between the rod and the vertical at any instant. (b) Find also the angle at which the rod begins to slip and in what direction the slipping occurs. Let $\mu$ be the coefficient of friction between the rod and the floor.

7.20  A billiard ball of radius $a$ is initially spinning about a horizontal axis with angular speed $\omega_0$, and with zero forward speed. If the coefficient of friction between the ball and the billiard table is $\mu$, find the distance the ball travels before slipping ceases to occur.

7.21  A ball is initially projected, without rotation, at a speed $v_0$ up a rough inclined plane of inclination $\theta$ and coefficient of friction $\mu$. Find the position of the ball as a function of time, and determine the position of the ball when pure rolling begins. Assume that $\mu$ is greater than $\frac{2}{7}\tan\theta$.

7.22  (a) A uniform circular hoop rests on a smooth horizontal surface. If it is struck tangentially at a point on the circumference, about what point does the hoop begin to rotate?  (b) Find the height at which a billiard ball should be struck so that it will roll without slipping.

7.23  Show that the center of oscillation of a physical pendulum is also the center of percussion for an impulse applied at the axis of rotation.

7.24  A ballistic pendulum is made of a long plank of length $l$ and mass $m$. It is free to swing about one end $O$, and is initially at rest in a vertical position. A bullet of mass $m'$ is fired horizontally into the pendulum at a distance $l'$ from $O$, the bullet coming to rest in the plank.  If the resulting amplitude of oscillation of the pendulum is $\theta_0$, find the speed of the bullet.

7.25  Two uniform rods $AB$ and $BC$ of equal mass $m$ and equal length $l$ are smoothly joined at $B$.  The system is initially at rest on a smooth horizontal surface, the points $A$, $B$, and $C$ lying in a straight line.  If an impulse $\hat{P}$ is applied at $A$ at right angles to the rod, find the initial motion of the system. [*Hint:* Isolate the rods.]

7.26  Work the above problem for the case in which rods are initially at right angles to each other.

# 8. Motion of Rigid Bodies in Three Dimensions

In the motion of a rigid body constrained either to rotate about a fixed axis or to move parallel to a fixed plane, the direction of the axis of rotation does not change. In the more general cases of rigid-body motion, the direction of the axis of rotation varies. The situation here is considerably more complicated. In fact, even in the case of a body on which no external forces whatever are acting, the motion is not simple.

## 8.1. Angular Momentum of a Rigid Body. Products of Inertia

Because of the fact that angular momentum is of capital importance in the study of the dynamics of rigid bodies, we shall begin with a derivation of the general expression for angular momentum of a rigid body. As defined in Section 6.2, the angular momentum **L** of any system of particles is the vector sum of the individual angular momenta of all the particles, namely,

$$\mathbf{L} = \sum_i (\mathbf{r}_i \times m\mathbf{v}_i)$$

In this chapter we shall be concerned with the vectorial character of angular momentum and its relation to the fundamental equation of rotational motion

217

$$\mathbf{N} = \frac{d\mathbf{L}}{dt}$$

in which $\mathbf{N}$ is the applied torque.   The conditions under which the above equation is valid were discussed in Section 7.6 of the preceding chapter.

We shall first calculate the angular momentum of a rigid body which is rotating about a single fixed point.   In this case we can imagine a coordinate system fixed in the body with origin $O$ at the fixed point (Figure 8.1).

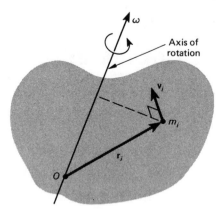

FIGURE 8.1   The velocity vector $\mathbf{v}_i$ of a representative particle of a rigid body rotating about a given axis defined by the angular velocity vector $\boldsymbol{\omega}$.

Referring to Section 1.26, we know that the velocity $\mathbf{v}_i$ of any constituent particle of the body is expressible as a cross product

$$\mathbf{v}_i = \boldsymbol{\omega} \times \mathbf{r}_i$$

where $\boldsymbol{\omega}$ is the angular velocity of the body, and $\mathbf{r}_i$ is the position vector of the particle.   Consequently, for all particles, the total angular momentum is

$$\mathbf{L} = \sum_i [m_i \mathbf{r}_i \times (\boldsymbol{\omega} \times \mathbf{r}_i)] \tag{8.1}$$

Now the $x$ component of the triple cross product

$$\mathbf{r}_i \times (\boldsymbol{\omega} \times \mathbf{r}_i)$$

is given by

$$[\mathbf{r}_i \times (\boldsymbol{\omega} \times \mathbf{r}_i)]_x = \omega_x(y_i^2 + z_i^2) - \omega_y x_i y_i - \omega_z x_i z_i \tag{8.2}$$

as may easily be shown by expansion of the determinant form of the cross product.   (The student should do this as an exercise.)

The $x$ component of the angular momentum is therefore

$$\begin{aligned} L_x &= \Sigma m_i [\omega_x(y_i^2 + z_i^2) - \omega_y x_i y_i - \omega_z x_i z_i] \\ &= \omega_x \Sigma m_i(y_i^2 + z_i^2) - \omega_y \Sigma m_i x_i y_i - \omega_z \Sigma m_i x_i z_i \end{aligned} \tag{8.3}$$

Analogous expressions hold for $L_y$ and $L_z$.

In computing the angular momentum for an extended rigid body, the summations are replaced by integrations over the volume, as before.  We introduce the following abbreviations:

$$I_{xx} = \Sigma(y_i^2 + z_i^2)m_i = \int(y^2 + z^2)\,dm \qquad \textit{(moment of inertia about x axis)}$$
$$I_{yy} = \Sigma(z_i^2 + x_i^2)m_i = \int(z^2 + x^2)\,dm \qquad \textit{(moment of inertia about y axis)}$$
$$I_{zz} = \Sigma(x_i^2 + y_i^2)m_i = \int(x^2 + y^2)\,dm \qquad \textit{(moment of inertia about z axis)}$$
$$I_{yx} = I_{xy} = -\Sigma x_i y_i m_i = -\int xy\,dm \qquad \textit{(xy product of inertia)}$$
$$I_{zx} = I_{xz} = -\Sigma x_i z_i m_i = -\int xz\,dm \qquad \textit{(xz product of inertia)}$$
$$I_{zy} = I_{yz} = -\Sigma z_i y_i m_i = -\int zy\,dm \qquad \textit{(zy product of inertia)}$$

We have already calculated the moments of inertia for a number of simple cases in the preceding chapter.  The products of inertia are found by a similar type of calculation.

Using the above notation, we can express the angular momentum as

$$\mathbf{L} = \mathbf{i}L_x + \mathbf{j}L_y + \mathbf{k}L_z$$
$$= \mathbf{i}(I_{xx}\omega_x + I_{xy}\omega_y + I_{xz}\omega_z)$$
$$+ \mathbf{j}(I_{yx}\omega_x + I_{yy}\omega_y + I_{yz}\omega_z) + \mathbf{k}(I_{zx}\omega_x + I_{zy}\omega_y + I_{zz}\omega_z) \quad (8.4)$$

It is apparent that the angular momentum vector $\mathbf{L}$ is not always in the same direction as the axis of rotation or the angular velocity vector $\boldsymbol{\omega}$.

### EXAMPLE

A body of arbitrary shape rotates about the $z$ axis.  Find the angular momentum $\mathbf{L}$.  Since, in this case, $\omega_x = \omega_y = 0$, and $\omega_z = \omega$, then we have immediately

$$\mathbf{L} = \mathbf{i}I_{xz}\omega + \mathbf{j}I_{yz}\omega + \mathbf{k}I_{zz}\omega$$

In particular, if either of the products of inertia $I_{xz}$ or $I_{yz}$ is not zero, then $\mathbf{L}$ has a component perpendicular to $\omega$, and so the angular momentum is not in the same direction as the axis of rotation. See the example at the end of the following section.

### 8.2.   Use of Matrices in Rigid Body Dynamics. The Inertia Tensor

We now see that the rotational properties of a rigid body about a given point require a set of nine quantities $I_{xx}$, $I_{xy}$, . . . in order to be completely specified.  There are many other examples that occur in which such sets of quantities are required for the complete description of some physical property at a point.  Such sets are called *tensors*, provided they obey certain transformation rules which we shall not attempt to discuss here.  The set defined above is known as the *inertia tensor* of the body.

Consider the general expression for the angular momentum, Equation (8.4). In matrix notation this equation reads

$$\begin{bmatrix} L_x \\ L_y \\ L_z \end{bmatrix} = \begin{bmatrix} I_{xx} & I_{xy} & I_{xz} \\ I_{yx} & I_{yy} & I_{yz} \\ I_{zx} & I_{zy} & I_{zz} \end{bmatrix} \begin{bmatrix} \omega_x \\ \omega_y \\ \omega_z \end{bmatrix} \tag{8.5}$$

Here, as in the treatment of coordinate transformations in Section 1.15, vectors are represented by column matrices. The $3 \times 3$ matrix involving the moments and products of inertia embodies a complete characterization of a rigid body with respect to its rotational properties. This matrix is a particular way of representing the inertia tensor.

Let us introduce a single symbol $\mathbf{I}$ for the inertia tensor. Then the angular momentum is expressed as

$$\mathbf{L} = \mathbf{I}\boldsymbol{\omega} \tag{8.6}$$

in which it is understood that the vectors $\mathbf{L}$ and $\boldsymbol{\omega}$ are column matrices.

### Principal Axes

A considerable simplification in the mathematical formulas for rigid body motion results if a coordinate system is employed such that the products of inertia all vanish. The axes of such a coordinate system are said to be the principal axes for the body at the point $O$, the origin of the coordinate system in question.

If the coordinate axes are principal axes of the body, then the inertia tensor takes the diagonal form

$$\mathbf{I} = \begin{bmatrix} I_{xx} & 0 & 0 \\ 0 & I_{yy} & 0 \\ 0 & 0 & I_{zz} \end{bmatrix} \tag{8.7}$$

In particular, the angular momentum becomes

$$\mathbf{L} = \mathbf{i}I_{xx}\omega_x + \mathbf{j}I_{yy}\omega_y + \mathbf{k}I_{zz}\omega_z \tag{8.8}$$

when principal axes are used. In this case the three moments of inertia are said to be the *principal moments* of the body at the point $O$.

Let us investigate the question of finding principal axes. First, if the body possesses some sort of symmetry, then it is usually possible to choose a coordinate system by inspection such that the products of inertia each consist of two parts of equal magnitude and opposite sign and therefore vanish. For example, the symmetrical plane laminar body shown in Figure 8.2 has, as principal axes at point $O$, the coordinate axes shown.

A body does not necessarily have to be symmetric in order that the products of inertia vanish, however. Consider, for example, a plane lamina of any shape (Figure 8.3). If the $xy$ plane is the plane of the lamina, then $z = 0$ and both $I_{yz}$ and $I_{xz}$ vanish. Now, relative to any given origin in the plane of the lamina, it is easy to prove that there always exists a set of axes such that the integral $\int xy \, dm$ vanishes.   To show this, we observe that the integral changes sign as the $Oxy$ system is rotated through an angle of 90°, because the lamina

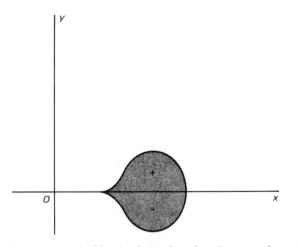

FIGURE 8.2   A symmetrical lamina located so that the $xy$ product of inertia is zero.

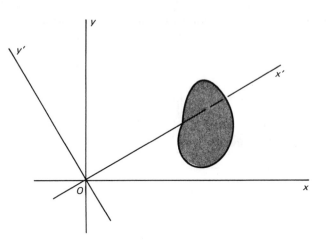

FIGURE 8.3   Rotated axes.

passes from one quadrant to the next, as shown.   Consequently, the integral must vanish for some angle of rotation bétween 0° and 90°.   This angle defines a set of coordinate axes for which all products of inertia vanish.   This is, by definition, a set of principal axes.

It can be shown in a similar way that for any rigid body there always exists a set of principal axes at any given point.   A general method of finding principal axes will be discussed in the next section.

Suppose that a body is rotating about a principal axis, say the $z$ axis. Then $\omega_x = \omega_y = 0$, and $\omega_z = \omega$. The expression for the angular momentum reduces to one term

$$\mathbf{L} = \mathbf{k} I_{zz} \omega$$

whereas the angular velocity is

$$\boldsymbol{\omega} = \mathbf{k} \omega$$

In this case the angular momentum vector is parallel to the angular velocity vector or the axis of rotation. We have therefore the following important fact: **L** *is either in the same direction as the axis of rotation, or is not, depending on whether the axis of rotation is, or is not, a principal axis.*

### Dynamic Balancing

The above rule finds application in the case of a rotating device such as a flywheel or fan blade. If the device is *statically balanced*, the center of mass lies on the axis of rotation. To be *dynamically balanced* the axis of rotation must also be a principal axis so that, as the body rotates, the angular momentum vector **L** will lie along the axis. Otherwise, if the rotational axis is not a principal one, the angular momentum vector varies in direction: it describes a cone as the body rotates (Figure 8.4). Then, since $d\mathbf{L}/dt$ is equal to the applied torque, there must be a torque exerted on the body. The direction of this torque is at right angles to the axis. The result is a reaction on the bearings. Thus in the case of a dynamically unbalanced rotator, there may be violent vibration and wobbling, even if the rotator is statically balanced.

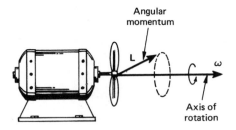

FIGURE 8.4   A rotating fan blade. The angular momentum vector **L** describes a cone about the axis of rotation when the blade is not dynamically balanced.

### EXAMPLE

A thin rod of length $l$ and mass $m$ is constrained to rotate with constant angular velocity $\boldsymbol{\omega}$ about an axis passing through the center making an angle $\alpha$ with the rod. Find **L**. We choose principal axes fixed on the rod as shown in Figure 8.5. Then we have

$$I_{xx} = I_{yy} = \frac{ml^2}{12}$$

and all other moments and products of inertia are zero.   Since the axis of rotation lies in the $yz$ plane, the components of $\boldsymbol{\omega}$ are

$$\omega_x = 0$$
$$\omega_y = \omega \sin \alpha$$
$$\omega_z = \omega \cos \alpha$$

The angular momentum vector is therefore

$$\mathbf{L} = \mathbf{j}\,\frac{ml^2}{12}\,\omega \sin \alpha$$

Thus $\mathbf{L}$ remains in the $y$ direction, as shown in the figure, and rotates with the body around $\boldsymbol{\omega}$.   (It is easy to verify that $\mathbf{r} \times m\mathbf{v}$ for each part of the rod is along $y$.)   In particular, if $\alpha = 90°$ then $\mathbf{L}$ and $\boldsymbol{\omega}$ point in the same direction, namely, the direction of the $y$ axis.   Otherwise they are in different directions.

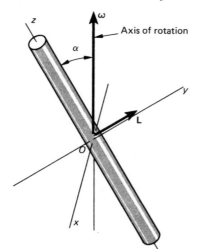

FIGURE 8.5   Rigid rod constrained to rotate about an oblique axis passing through the center.

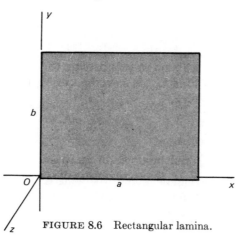

FIGURE 8.6   Rectangular lamina.

## 8.3.  Determination of Principal Axes

Evidently, the general problem of finding the principal axes of a rigid body is equivalent to the mathematical problem of diagonalizing a $3 \times 3$ matrix.   From matrix theory, it is known that any symmetric square matrix can be diagonalized.   In our case $I_{xy} = I_{yx}$, and similarly for the other pairs. Hence the matrix is symmetric, and so there must exist a set of principal axes at any point.

As shown in Appendix V, the diagonalization is accomplished by finding the roots of the secular equation

$$|\mathbf{I} - \lambda\mathbf{1}| = 0$$

where $\mathbf{1}$ is the unit matrix.   Explicitly, this equation reads

$$\begin{vmatrix} I_{xx} - \lambda & I_{xy} & I_{xz} \\ I_{yx} & I_{yy} - \lambda & I_{yz} \\ I_{zx} & I_{zy} & I_{zz} - \lambda \end{vmatrix} = 0 \tag{8.9}$$

It is a cubic in $\lambda$, namely

$$-\lambda^3 + A\lambda^2 + B\lambda + C = 0$$

in which $A$, $B$, and $C$ are simple functions of the $I$'s.   The three roots, $\lambda_1$, $\lambda_2$, and $\lambda_3$ are the three principal moments of inertia.

In order to find the orientation of the principal axes, we make use of the physical fact that when the body is rotating about one of its principal axes, the angular momentum vector is in the same direction as the angular velocity vector.   Let the direction angles of one of the principal axes be $\alpha$, $\beta$, and $\gamma$, and let the body rotate with angular velocity $\omega$ about this axis.   The angular momentum is then given by

$$\mathbf{L} = \lambda\boldsymbol{\omega} = \mathbf{I}\boldsymbol{\omega}$$

in which $\lambda$ is one of the three roots $\lambda_1$, $\lambda_2$, or $\lambda_3$.   Explicitly, the above equation reads

$$\begin{bmatrix} \lambda\omega\cos\alpha \\ \lambda\omega\cos\beta \\ \lambda\omega\cos\gamma \end{bmatrix} = \begin{bmatrix} I_{xx} & I_{xy} & I_{xz} \\ I_{yx} & I_{yy} & I_{yz} \\ I_{zx} & I_{zy} & I_{zz} \end{bmatrix} \begin{bmatrix} \omega\cos\alpha \\ \omega\cos\beta \\ \omega\cos\gamma \end{bmatrix} \tag{8.10}$$

In turn, this equation is equivalent to the three scalar equations

$$\begin{aligned} (I_{xx} - \lambda)\cos\alpha + I_{xy}\cos\beta + I_{xz}\cos\gamma &= 0 \\ I_{yx}\cos\alpha + (I_{yy} - \lambda)\cos\beta + I_{yz}\cos\gamma &= 0 \\ I_{zx}\cos\alpha + I_{zy}\cos\beta + (I_{zz} - \lambda)\cos\gamma &= \end{aligned} \tag{8.11}$$

in which the common factor $\omega$ has been cancelled out.   Thus the direction cosines of the principal axes can be found by solving the above equations.   The roots are not independent, for they must clearly satisfy the condition

$$\cos^2\alpha + \cos^2\beta + \cos^2\gamma = 1$$

*Determination of Principal Axes when One Is Known*

In many instances a body possesses some kind of symmetry so that at least one principal axis can be found by inspection.   If such is the case, then the other two principal axes can be determined as follows.

Suppose the $z$ axis is known to be a principal axis at the origin in a suitable coordinate system.   Then, by definition

$$I_{zx} = I_{zy} = 0$$

The other two principal axes must lie in the $xy$ plane.   The first two of Equations (8.11) now reduce to

$$(I_{xx} - \lambda) \cos \alpha + I_{xy} \cos \beta = 0$$
$$I_{xy} \cos \alpha + (I_{yy} - \lambda) \cos \beta = 0$$

Let $\theta$ denote the angle between the $x$ axis and one of the unknown principal axes.   Then $\tan \theta = \cos \beta / \cos \alpha$. Elimination of $\lambda$ between the two equations yields

$$I_{xy}(\tan^2 \theta - 1) = (I_{yy} - I_{xx}) \tan \theta$$

from which $\theta$ can be found.   In this application it is helpful to use the trigonometric identity $\tan 2\theta = 2 \tan \theta / (1 - \tan^2 \theta)$.   This gives

$$\tan 2\theta = \frac{2I_{xy}}{I_{xx} - I_{yy}} \tag{8.12}$$

There are two values of $\theta$ lying between $\pi/2$ and $-\pi/2$ that satisfy the above equation, and these give the directions of the two principal axes in the $xy$ plane.

### EXAMPLES

1. Find the directions of the principal axes in the plane of a rectangular lamina, of sides $a$ and $b$, at a corner, Figure 8.6.   From the previous chapter we have

$$I_{xx} = \tfrac{1}{3}mb^2$$
$$I_{yy} = \tfrac{1}{3}ma^2$$
$$I_{zz} = I_{xx} + I_{yy} = \tfrac{1}{3}m(a^2 + b^2)$$

The last equation follows from the perpendicular axis theorem.   Since $z = 0$ for all points of the plate, the two products of inertia involving $z$ must vanish:

$$I_{xz} = I_{yz} = 0$$

Finally, the $xy$ product of inertia is given by

$$I_{xy} = -\int xy \, dm = -\int_0^b \int_0^a xy\rho \, dx \, dy = -\rho \frac{a^2b^2}{4}$$

in which $\rho$ is the mass per unit area. Furthermore, since

$$m = \rho ab$$

we have

$$I_{xy} = -\tfrac{1}{4}mab$$

for the $xy$ product of inertia of the plate. The application of Equation (8.12) then gives

$$\tan 2\theta = \frac{-2(mab/4)}{(mb^2/3) - (ma^2/3)} = \frac{3ab}{2(a^2 - b^2)}$$

or

$$\theta = \frac{1}{2}\tan^{-1}\left[\frac{3ab}{2(a^2 - b^2)}\right]$$

for the directions of the principal axes.

2. Find the principal moments of inertia of a square plate about a corner. Here Equation (8.9) reads

$$\begin{vmatrix} \tfrac{1}{3}ml^2 - \lambda & -\tfrac{1}{4}ml^2 & 0 \\ -\tfrac{1}{4}ml^2 & \tfrac{1}{3}ml^2 - \lambda & 0 \\ 0 & 0 & \tfrac{2}{3}ml^2 - \lambda \end{vmatrix} = 0$$

or

$$[(\tfrac{1}{3}ml^2 - \lambda)^2 - (\tfrac{1}{4}ml^2)^2](\tfrac{2}{3}ml^2 - \lambda) = 0$$

The second factor gives

$$\lambda = \tfrac{2}{3}ml^2$$

for one of the principal moments. The first factor gives

$$\tfrac{1}{3}ml^2 - \lambda = \pm\tfrac{1}{4}ml^2$$

or

$$\lambda = \tfrac{7}{12}ml^2$$

and

$$\lambda = \tfrac{1}{12}ml^2$$

These three values of $\lambda$ are the three principal moments.

3. Find the directions of the principal axes for the above problem. Equations (8.11) give

$$(\tfrac{1}{3}ml^2 - \lambda)\cos\alpha - \tfrac{1}{4}ml^2\cos\beta = 0$$
$$-\tfrac{1}{4}ml^2\cos\alpha + (\tfrac{1}{3}ml^2 - \lambda)\cos\beta = 0$$
$$(\tfrac{2}{3}ml^2 - \lambda)\cos\gamma = 0$$

From the last equation we see that $\gamma = 90°$ is one root. If we set $\lambda$ equal to $\tfrac{1}{12}ml^2$, the first equation becomes

$$\cos\alpha - \cos\beta = 0$$

This, together with Equation (9.65) gives

$$2 \cos^2 \alpha = 1$$

or, taking the positive root, we have $\alpha = 45°$ for one principal axis. The other is given by taking the negative root, that is $\alpha = 135°$. Thus one principal axis is along the diagonal, the other perpendicular to the diagonal and in the plane of the plate, and the third principal axis is normal to the plate.

### 8.4.   Rotational Kinetic Energy of a Rigid Body

Let us calculate the kinetic energy of a rigid body which is turning about a fixed point with angular velocity $\boldsymbol{\omega}$. For the velocity $\mathbf{v}_i$ of a representative particle $i$, we have, as in our calculation of angular momentum

$$\mathbf{v}_i = \boldsymbol{\omega} \times \mathbf{r}_i$$

where $\mathbf{r}_i$ is the position vector of the particle relative to the fixed point. The kinetic energy $T$ is therefore given by the summation

$$T = \sum_i \tfrac{1}{2} m_i \mathbf{v}_i \cdot \mathbf{v}_i = \tfrac{1}{2} \sum_i [(\boldsymbol{\omega} \times \mathbf{r}_i) \cdot (m_i \mathbf{v}_i)]$$

Now in the triple scalar product we can exchange the dot and the cross. (See Section 1.14.)   Hence

$$T = \tfrac{1}{2} \sum_i [\boldsymbol{\omega} \cdot (\mathbf{r}_i \times m_i \mathbf{v}_i)] = \tfrac{1}{2} \boldsymbol{\omega} \cdot \sum_i (\mathbf{r}_i \times m_i \mathbf{v}_i)$$

But $\sum_i (\mathbf{r}_i \times m_i \mathbf{v}_i)$ is, by definition, the angular momentum $\mathbf{L}$. Thus we can write

$$T = \tfrac{1}{2} \boldsymbol{\omega} \cdot \mathbf{L} \tag{8.13}$$

The above equation gives the rotational kinetic energy of a rigid body in terms of the angular velocity $\boldsymbol{\omega}$ and the angular momentum $\mathbf{L}$. It is analogous with the equation $T = \tfrac{1}{2} \mathbf{v} \cdot \mathbf{p}$ for the kinetic energy of translation of a particle or system in which $\mathbf{v}$ is the velocity of the center of mass and $\mathbf{p}$ is the linear momentum. The total kinetic energy of a moving and rotating body is the sum $\tfrac{1}{2} \boldsymbol{\omega} \cdot \mathbf{L} + \tfrac{1}{2} \mathbf{v} \cdot \mathbf{p}$.

By employing matrix notation for the angular momentum, we have

$$T = \tfrac{1}{2} \boldsymbol{\omega}^T \mathsf{I} \boldsymbol{\omega} \tag{8.14}$$

Here the row matrix $\boldsymbol{\omega}^T$ is the transpose matrix of the column matrix $\boldsymbol{\omega}$, and $\mathsf{I}$ is the inertia tensor. In explicit form

$$T = \tfrac{1}{2}[\omega_x \omega_y \omega_z] \begin{bmatrix} I_{xx} & I_{xy} & I_{xz} \\ I_{yx} & I_{yy} & I_{yz} \\ I_{zx} & I_{zy} & I_{zz} \end{bmatrix} \begin{bmatrix} \omega_x \\ \omega_y \\ \omega_z \end{bmatrix} \tag{8.15}$$

or

$$T = \tfrac{1}{2}(I_{xx}\omega_x{}^2 + I_{yy}\omega_y{}^2 + I_{zz}\omega_z{}^2 + 2I_{yz}\omega_y\omega_z + 2I_{zx}\omega_z\omega_x + 2I_{xy}\omega_x\omega_y) \tag{8.16}$$

## EXAMPLES

1. Find the inertia tensor for a square plate of side $l$ and mass $m$ in a coordinate system $Oxyz$ where $O$ is at one corner and the $x$ and $y$ axes are along the two edges. Utilizing the results of the example in Section 8.3, we have $I_{xx} = I_{yy} = ml^2/3$, $I_{zz} = 2ml^2/3$, $I_{xy} = -ml^2/4$, $I_{xz} = I_{yz} = 0$. Hence the inertia tensor is

$$\mathbf{I} = \begin{bmatrix} ml^2/3 & -ml^2/4 & 0 \\ -ml^2/4 & ml^2/3 & 0 \\ 0 & 0 & 2ml^2/3 \end{bmatrix} = \frac{ml^2}{3} \begin{bmatrix} 1 & -\tfrac{3}{4} & 0 \\ -\tfrac{3}{4} & 1 & 0 \\ 0 & 0 & 2 \end{bmatrix}$$

2. Find the angular momentum of the above plate when it is rotating about a diagonal. In this case, the angular velocity vector can be expressed as the column matrix

$$\boldsymbol{\omega} = \begin{bmatrix} \omega_x \\ \omega_y \\ \omega_z \end{bmatrix} = \begin{bmatrix} \omega/\sqrt{2} \\ \omega/\sqrt{2} \\ 0 \end{bmatrix} = \frac{\omega}{\sqrt{2}} \begin{bmatrix} 1 \\ 1 \\ 0 \end{bmatrix}$$

Consequently, the angular momentum is

$$\mathbf{L} = \mathbf{I}\boldsymbol{\omega} = \frac{ml^2\omega}{3\sqrt{2}} \begin{bmatrix} 1 & -\tfrac{3}{4} & 0 \\ -\tfrac{3}{4} & 1 & 0 \\ 0 & 0 & 2 \end{bmatrix} \begin{bmatrix} 1 \\ 1 \\ 0 \end{bmatrix}$$

$$= \frac{ml^2\omega}{3\sqrt{2}} \begin{bmatrix} \tfrac{1}{4} \\ \tfrac{1}{4} \\ 0 \end{bmatrix} = \frac{ml^2\omega}{12\sqrt{2}} \begin{bmatrix} 1 \\ 1 \\ 0 \end{bmatrix}$$

3. Find the kinetic energy of rotation in the above problem. Using the above results, we have

$$T = \frac{1}{2}\boldsymbol{\omega}^{\mathrm{T}}\mathbf{I}\boldsymbol{\omega} = \frac{1}{2}\boldsymbol{\omega}^{\mathrm{T}}\mathbf{L} = \frac{ml^2\omega^2}{24}[1 \quad 1 \quad 0] \begin{bmatrix} 1 \\ 1 \\ 0 \end{bmatrix}$$

$$= \frac{ml^2\omega^2}{12}$$

### 8.5.   Moment of Inertia of a Rigid Body About an Arbitrary Axis.   The Momental Ellipsoid

Let us apply the fundamental definition

$$I = \Sigma m_i \text{R}_i{}^2$$

to find the moment of inertia of a rigid body about any axis.   In the above formula $\text{R}_i$ is the perpendicular distance from the representative particle $m_i$ to the axis, as shown in Figure 8.7.

Suppose we designate the direction of the axis of rotation by the unit vector $\mathbf{n}$.   Then

$$\text{R}_i = |\mathbf{r}_i \times \mathbf{n}|$$

where

$$\mathbf{r}_i = \mathbf{i}x_i + \mathbf{j}y_i + \mathbf{k}z_i$$

is the position vector of the $i$th particle.   Further, let $\cos \alpha$, $\cos \beta$, and $\cos \gamma$ be the direction cosines of the axis, that is

$$\mathbf{n} = \mathbf{i} \cos \alpha + \mathbf{j} \cos \beta + \mathbf{k} \cos \gamma$$

Then we have

$$\text{R}_i{}^2 = |\mathbf{r}_i \times \mathbf{n}|^2 = (y_i \cos \gamma - z_i \cos \beta)^2 \\ + (z_i \cos \alpha - x_i \cos \gamma)^2 + (x_i \cos \beta - y_i \cos \alpha)^2$$

Upon rearranging terms, we find

$$\text{R}_i{}^2 = (y_i{}^2 + z_i{}^2)\cos^2 \alpha + (z_i{}^2 + x_i{}^2)\cos^2 \beta + (x_i{}^2 + y_i{}^2)\cos^2 \gamma \\ - 2 \, y_i z_i \cos \gamma \cos \beta - 2 \, z_i x_i \cos \alpha \cos \gamma - 2 \, x_i y_i \cos \alpha \cos \beta$$

The moment of inertia $I = \Sigma m_i \text{R}_i{}^2$ can thus be expressed as

$$I = I_{xx}\cos^2 \alpha + I_{yy}\cos^2 \beta + I_{zz}\cos^2 \gamma \\ + 2 \, I_{yz}\cos \gamma \cos \beta + 2 \, I_{zx}\cos \alpha \cos \gamma + 2 \, I_{xy}\cos \alpha \cos \beta \quad (8.17)$$

The above formula gives the moment of inertia of a rigid body about any axis in terms of the direction cosines of that axis and the moments and products of inertia of the body in an arbitrary coordinate system which has its origin on the axis.   If the coordinate axes happen to be principal axes at the origin, then the products of inertia vanish, and the formula reduces to the simpler one

$$I = I_{xx}\cos^2 \alpha + I_{yy}\cos^2 \beta + I_{zz}\cos^2 \gamma \quad (8.18)$$

A useful abbreviation of the general formula is obtained by the use of the inertia tensor, namely,

$$I = \mathbf{n}^T \mathsf{I} \mathbf{n} \quad (8.19)$$

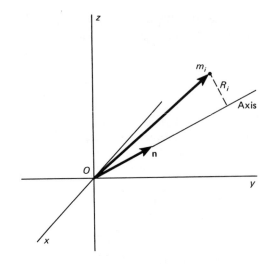

FIGURE 8.7   Definition of the unit vector **n** giving the direction of the axis
of rotation.

## The Momental Ellipsoid

A very instructive geometric interpretation of the general formula for the
moment of inertia may be obtained in the following way.   Consider an arbitrary axis of rotation about a given point $O$.   We define a point $P$ on the axis of
rotation such that the distance $OP$ is numerically equal to the reciprocal of
the square root of the moment of inertia about the axis:

$$OP = \frac{1}{\sqrt{I}}$$

Now let $x$, $y$, and $z$ be the coordinates of the point $P$, and let $\alpha$, $\beta$, and $\gamma$ be the
direction angles of the line $OP$.   Then we have $\cos \alpha = x/OP = x\sqrt{I}$,
$\cos \beta = y/OP = y\sqrt{I}$, $\cos \gamma = z/OP = z\sqrt{I}$, so that

$$\mathbf{n} = \begin{bmatrix} x\sqrt{I} \\ y\sqrt{I} \\ z\sqrt{I} \end{bmatrix} = \mathbf{r}\sqrt{I}$$

If we substitute into the general formula for the moment of inertia, Equation
(8.19), we obtain

$$\mathbf{r}^T \mathbf{I} \mathbf{r} = 1 \tag{8.20}$$

or, explicitly,

$$x^2 I_{xx} + y^2 I_{yy} + z^2 I_{zz} + 2yz I_{yz} + 2zx I_{zx} + 2xy I_{xy} = 1 \qquad (8.21)$$

The above equation is the equation of a surface, Figure 8.8.   It defines the locus of points $P$ as the direction of the axis $OP$ is varied.   Being of the second degree, it is the equation of a general quadric surface in three dimensions. Since $I$ is never zero for any extended body, the surface is bounded and must therefore be an ellipsoid.[1]   It is called the *momental ellipsoid of the body at the point O.*

If the coordinate axes are principal axes, the equation of the momental ellipsoid is

$$x^2 I_{xx} + y^2 I_{yy} + z^2 I_{zz} = 1 \qquad (8.22)$$

Thus, the principal axes of the body coincide with the principal axes of the momental ellipsoid.   Since there are always at least three principal axes for every ellipsoid, it follows that there always exist at least three principal axes for a body at a given point.

If two of the three principal moments of inertia are equal, then the ellipsoid of inertia is one of revolution.   If all three principal moments are equal

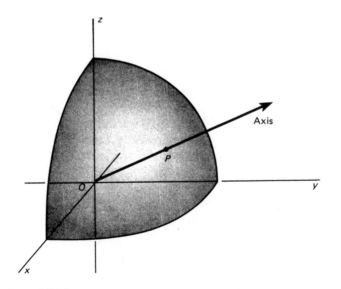

FIGURE 8.8   One octant of the momental ellipsoid   The distance $OP$ is equal to the reciprocal of the square root of the moment of inertia about the axis.

---

[1] In the case of an infinitely thin straight body, the moment of inertia about the axis of the body is zero.   The momental ellipsoid degenerates into a cylinder in this case.

at a point $O$, the momental ellipsoid is a sphere, and it follows that the moment of inertia is the same for any line passing through $O$, no matter what its direction may be.

## EXAMPLE

Find the equation of the momental ellipsoid of a rectangular lamina of sides $a$ and $b$ for coordinate origin at the center. From the moments of inertia, we have immediately

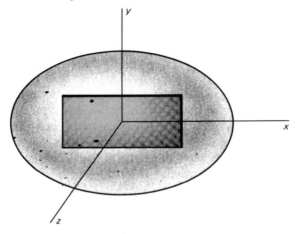

FIGURE 8.9    Momental ellipsoid of a rectangular block.

$$x^2\left(\frac{mb^2}{12}\right) + y^2\left(\frac{ma^2}{12}\right) + z^2\left(\frac{ma^2 + mb^2}{12}\right) = 1$$

for the equation of the momental ellipsoid. We note that the principal diameters of the ellipsoid are in the ratios

$$b^{-1} : a^{-1} : (a^2 + b^2)^{-1/2}$$

For instance, if $a/b = 2$, the ratios are $2 : 1 : 2/\sqrt{5}$. Thus the long diameter of the momental ellipsoid corresponds to the long axis of the lamina. The ellipsoid is shown in Figure 8.9..

## 8.6.  Euler's Equations of Motion of a Rigid Body

Consider the fundamental equation governing the rotation of a rigid body under the action of a torque

$$\mathbf{N} = \frac{d\mathbf{L}}{dt}$$

We have shown that $\mathbf{L}$ is most simply expressed if the coordinate axes are principal axes for the body, namely

$$\mathbf{L} = \mathbf{i}I_{xx}\omega_x + \mathbf{j}I_{yy}\omega_y + \mathbf{k}I_{zz}\omega_z$$

Here $I_{xx}$, $I_{yy}$, and $I_{zz}$ are the principal moments of inertia of the body at the origin of the coordinate system.   Now in order that the above expression for the angular momentum remains valid as the body rotates, the coordinate system must also rotate with the body.   Thus the angular velocity of the body and the angular velocity of the coordinate system are usually one and the same.   (There is an exception.   If two of the three principal moments of inertia are equal so that the momental ellipsoid is one of revolution, then the coordinate axes need not be fixed in the body to be principal axes.   This case will be considered later in Section 8.8.)

According to the theory of rotating coordinate systems developed in Chapter 4, the time rate of change of the angular momentum vector, when referred to a rotating system, is given by the formula

$$\frac{d\mathbf{L}}{dt} = \dot{\mathbf{L}} + \boldsymbol{\omega} \times \mathbf{L}$$

Thus, the equation of motion in the rotating system is

$$\mathbf{N} = \dot{\mathbf{L}} + \boldsymbol{\omega} \times \mathbf{L}$$

In rectangular components, the above equation reads

$$\begin{aligned}
N_x &= \dot{L}_x + (\boldsymbol{\omega} \times \mathbf{L})_x \\
N_y &= \dot{L}_y + (\boldsymbol{\omega} \times \mathbf{L})_y \\
N_z &= \dot{L}_z + (\boldsymbol{\omega} \times \mathbf{L})_z
\end{aligned} \tag{8.23}$$

or more explicitly

$$\begin{aligned}
N_x &= I_{xx}\dot{\omega}_x + \omega_y\omega_z(I_{zz} - I_{yy}) \\
N_y &= I_{yy}\dot{\omega}_y + \omega_z\omega_x(I_{xx} - I_{zz}) \\
N_z &= I_{zz}\dot{\omega}_z + \omega_x\omega_y(I_{yy} - I_{xx})
\end{aligned} \tag{8.24}$$

These are known as Euler's equations for the motion of a rigid body.   They are of fundamental importance in the theory of rotation of extended rigid bodies.

### Body Constrained to Rotate About a Fixed Axis

As an application of Euler's equations, we consider the special case of a rigid body that is constrained to rotate about a fixed axis with constant angular velocity.   Then

$$\dot{\omega}_x = \dot{\omega}_y = \dot{\omega}_z = 0$$

and Euler's equations then reduce to

$$N_x = \omega_y\omega_z(I_{zz} - I_{yy})$$
$$N_y = \omega_z\omega_x(I_{xx} - I_{zz}) \tag{8.25}$$
$$N_z = \omega_x\omega_y(I_{yy} - I_{xx})$$

These give the components of the torque exerted on the body by the constraining support.

In particular, if the axis of rotation is a principal axis, then two of the three components of $\omega$ are equal to zero. Consequently, all three components of the torque $N$ vanish. This agrees with the previous statement concerning dynamic balancing in Section 8.2.

## EXAMPLE

Calculate the torque that must be exerted on a thin rod to cause it to rotate with constant angular speed $\omega$ about an axis through the center at an angle $\alpha$ with the rod, as in the example on p. 222. Using the results of the example cited, we find that Euler's equations give the components of the torque as

$$N_x = \omega^2 \sin \alpha \cos \alpha \left(-\frac{ml^2}{12}\right)$$
$$N_y = 0$$
$$N_z = 0$$

Thus the torque vanishes if either the sine or the cosine vanishes, that is, if $\alpha = 0$ or $90°$. In either case the rod is rotating about a principal axis.

### 8.7. Free Rotation of a Rigid Body Under No Forces. Geometric Description of the Motion

Let us consider the case of a rigid body that is free to rotate in any direction about a certain point $O$. There are no torques acting on the body. This is the case of free rotation and is exemplified, for example, by a body supported on a smooth pivot at its center of mass. Another example is that of a rigid body moving freely under no forces or falling freely in a uniform gravitational field so that there are no torques. The point $O$ in this case is the center of mass.

With zero torque the angular momentum of the body, as seen from the outside, must remain constant in direction and magnitude according to the general principle of conservation of angular momentum. However, with respect to rotating axes fixed in the body, the direction of the angular momentum vector may change, although its magnitude must remain constant. This fact can be expressed by the equation

$$\mathbf{L} \cdot \mathbf{L} = \text{constant}$$

In terms of components referred to the principal axes of the body, the above equation reads

$$I_{xx}^2 \omega_x^2 + I_{yy}^2 \omega_y^2 + I_{zz}^2 \omega_z^2 = L^2 = \text{constant} \tag{8.26}$$

As the body rotates, the components of $\boldsymbol{\omega}$ may vary, but they must always satisfy the above equation.

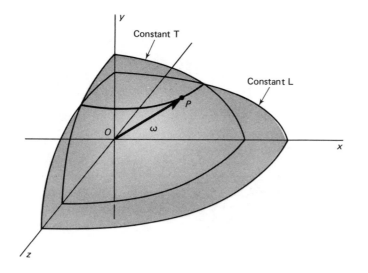

FIGURE 8.10  Intersecting ellipsoids of constant $L$ and constant $T$ for a rigid body undergoing free rotation.

A second relation is obtained by considering the kinetic energy of rotation. Again, since there is zero torque, the total rotational kinetic energy must remain constant. This may be expressed as

$$\boldsymbol{\omega} \cdot \mathbf{L} = 2T = \text{constant}$$

or, equivalently in terms of components,

$$I_{xx} \omega_x^2 + I_{yy} \omega_y^2 + I_{zz} \omega_z^2 = 2T = \text{constant} \tag{8.27}$$

We now see that the components of $\boldsymbol{\omega}$ must simultaneously satisfy two different equations expressing the constancy of kinetic energy and of magnitude of angular momentum. These are the equations of two ellipsoids whose principal axes coincide with the principal axes of the body. The first ellipsoid, Equation (8.26), has principal diameters in the ratios $I_{xx}^{-1} : I_{yy}^{-1} : I_{zz}^{-1}$. The

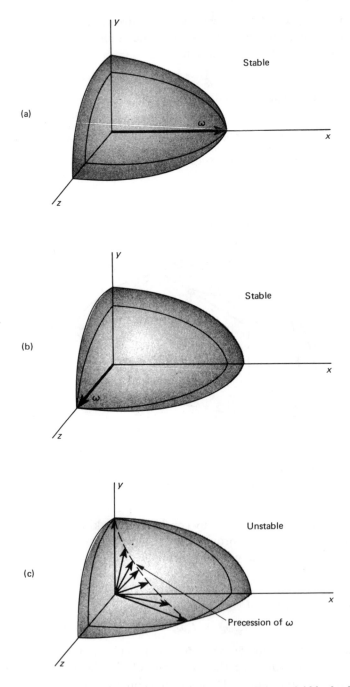

FIGURE 8.11  Ellipsoids of constant $L$ and constant $T$ for a rigid body that is rotating freely about the axis of (a) least, (b) greatest, and (c) intermediate moment of inertia.

second ellipsoid, Equation (8.27), has principal diameters in the ratios $I_{xx}^{-1/2} : I_{yy}^{-1/2} : I_{zz}^{-1/2}$. It is known as the *Poinsot ellipsoid* and is similar to the momental ellipsoid. As the body rotates, the extremity of the angular velocity vector thus describes a curve which is the intersection of the two ellipsoids. This is illustrated in Figure 8.10.

From the equations of the intersecting ellipsoids, it can be shown that in the case where the initial axis of rotation coincides with one of the principal axes of the body, then the curve of intersection diminishes to a point. In other words, the two ellipsoids just touch at a principal diameter, and the body rotates steadily about this axis. This is true, however, only if the initial rotation is about the axis of either the largest or the smallest moment of inertia. If it is about the intermediate axis, say the $y$ axis where $I_{xx} > I_{yy} > I_{zz}$, then the intersection of the two ellipsoids is not a point, but a curve that goes entirely around both, as illustrated in Figure 8.11. In this case the rotation is unstable since the axis of rotation precesses all around the body. These facts can easily be illustrated by tossing an oblong block, or a book, into the air.

### 8.8. Free Rotation of a Rigid Body with an Axis of Symmetry. Analytical Treatment

Although the geometric description of the motion of a rigid body given in the preceding section is helpful in visualizing free rotation under no torques, the method does not immediately give numerical values. We now proceed to augment that description with an analytical approach based on the direct integration of Euler's equations.

We shall solve Euler's equations for the special case in which the body possesses an axis of symmetry, so that two of the three principal moments of inertia are equal. (Actually, all that is required is that the momental ellipsoid have an axis of symmetry, not the body itself.)

Let us choose the $z$ axis as the axis of symmetry. We introduce the following notation:

$I_s = I_{zz}$ (*moment of inertia about the symmetry axis*)
$I = I_{xx} = I_{yy}$ (*moment about the axes normal to the symmetry axis*)

For the case of zero torque, Euler's equations then read

$$I\dot{\omega}_x + \omega_y\omega_z(I_s - I) = 0$$
$$I\dot{\omega}_y + \omega_z\omega_x(I - I_s) = 0 \qquad (8.28)$$
$$I_s\dot{\omega}_z = 0$$

From the last equation it follows that

$$\omega_z = \text{constant}$$

Let us now define a constant $\Omega$ as

$$\Omega = \omega_z \frac{I_s - I}{I} \tag{8.29}$$

Then the first two of Equations (8.28) may be written

$$\dot{\omega}_x + \Omega \omega_y = 0 \tag{8.30}$$
$$\dot{\omega}_y - \Omega \omega_x = 0 \tag{8.31}$$

To separate the variables in the above pair of equations, we differentiate the first with respect to $t$ and obtain

$$\ddot{\omega}_x + \Omega \dot{\omega}_y = 0$$

Upon solving for $\dot{\omega}_y$ and inserting the result into Equation (8.31), we find

$$\ddot{\omega}_x + \Omega^2 \omega_x = 0$$

This is the equation for simple harmonic motion. A solution is

$$\omega_x = \omega_1 \cos(\Omega t) \tag{8.32}$$

in which $\omega_1$ is a constant of integration. To find $\omega_y$, we differentiate the above equation with respect to $t$ and insert the result into Equation (8.30). We can then solve for $\omega_y$ to obtain

$$\omega_y = \omega_1 \sin(\Omega t) \tag{8.33}$$

Thus $\omega_x$ and $\omega_y$ vary harmonically in time with angular frequency $\Omega$, and their

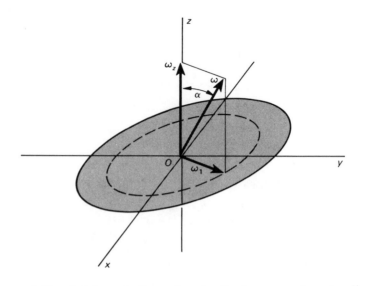

FIGURE 8.12   Angular velocity vectors for the free precession of a disc.

phases differ by $\pi/2$.   It follows that the projection of $\boldsymbol{\omega}$ on the $xy$ plane describes a circle of radius $\omega_1$ at the angular frequency $\Omega$.

We can summarize the above results as follows: In the free rotation of a rigid body with an axis of symmetry, the angular velocity vector describes a conical motion (precesses) about the symmetry axis.   It describes a surface called the *body cone*.   (See Figure 8.14.)   The angular frequency of this precession is the constant $\Omega$ defined by Equation (8.29).   Let $\alpha$ denote the angle between the symmetry axis ($z$ axis) and the axis of rotation (direction of $\boldsymbol{\omega}$) as shown in Figure 8.12.   Then we can express $\Omega$ as

$$\Omega = \left(\frac{I_s}{I} - 1\right) \omega \cos \alpha \tag{8.34}$$

giving the rate of precession of the angular velocity vector about the axis of symmetry.

*Description of the Rotation of a Rigid Body Relative to a Fixed Coordinate System.   The Eulerian Angles*

In the foregoing analysis of the free rotation of a rigid body, the precessional motion was relative to a coordinate system fixed in the body and rotating with it.   In order to describe the motion relative to an observer outside the body, we must use a fixed coordinate system.   In Figure 8.13 the coordinate system $OXYZ$ has a fixed orientation in space.   The primed system $Ox'y'z'$ is fixed in the body and rotates with it.   A third system $Oxyz$ is defined as follows: The $z$ axis coincides with the $z'$ axis or symmetry axis of the body, and the $x$ axis is the line of intersection of the $XY$ plane with the

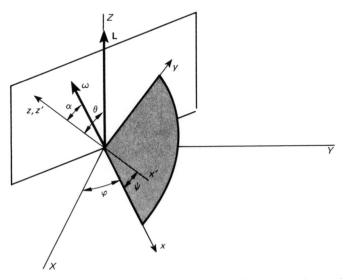

FIGURE 9.15   Diagram showing the relation of the Eulerian angles to the
fixed and rotating coordinate axes.

$x'y'$ plane. The angle between the $X$ axis and the $x$ axis is denoted by $\varphi$, and that between the $Z$ and $z$ axes by $\theta$. The turning of the body about the symmetry axis is determined by the angle between the $x$ axis and the $x'$ axis, denoted by $\psi$. The three angles $\varphi$, $\theta$, and $\psi$ are known as the *Eulerian angles*.

In the case in which there are no torques acting on the body, the angular momentum vector **L** is constant in magnitude and direction in the fixed system $OXYZ$. Let us choose the $Z$ axis as the direction of **L**. This is known as the *invariable line*. From the figure we see that the components of **L** in the $Oxyz$ system are

$$
\begin{aligned}
L_x &= 0 \\
L_y &= L \sin \theta \\
L_z &= L \cos \theta
\end{aligned}
\tag{8.35}
$$

We again restrict ourselves to the case of a body with an axis of symmetry (the $z$ axis) so that the momental ellipsoid is one of revolution, and consequently the $Oxyz$ axes are principal axes as well as the primed axes. We now have from the first of Equations (8.35) that $\omega_x = 0$. Hence $\boldsymbol{\omega}$ lies in the $yz$ plane. Let $\alpha$ denote the angle between the $z$ axis and the angular velocity $\boldsymbol{\omega}$. The components of $\boldsymbol{\omega}$ are then given by

$$
\begin{aligned}
\omega_x &= 0 \\
\omega_y &= \omega \sin \alpha \\
\omega_z &= \omega \cos \alpha
\end{aligned}
\tag{8.36}
$$

Consequently,

$$
\begin{aligned}
L_x &= I_{xx}\omega_x = 0 \\
L_y &= I_{yy}\omega_y = I\omega \sin \alpha \\
L_z &= I_{zz}\omega_z = I_s\omega \cos \alpha
\end{aligned}
\tag{8.37}
$$

It readily follows that

$$
\frac{L_y}{L_z} = \tan \theta = \frac{I}{I_s} \tan \alpha
\tag{8.38}
$$

giving the relation between the angles $\theta$ and $\alpha$.

According to the above result, $\theta$ is less than, or greater than $\alpha$, depending on whether $I$ is less than $I_s$ or greater than $I_s$, respectively. In other words, the angular momentum vector lies between the symmetry axis and the axis of rotation in the case of a flattened body ($I < I_s$), whereas in the case of an elongated body ($I > I_s$) the axis of rotation lies between the axis of symmetry and the angular momentum vector. The two cases are illustrated in Figure 8.14. In either case, as the body rotates, the axis of symmetry ($z$ axis) describes a conical motion or precesses about the constant angular momentum vector **L**. At the same time the axis of rotation ($\boldsymbol{\omega}$ vector) precesses about **L** with the same frequency. The surface traced out by $\boldsymbol{\omega}$ about **L** is called the *space cone*, Figure 8.14.

Referring to Figure 8.13, we see that the angular speed of rotation of the $yz$ plane about the $Z$ axis is equal to the time rate of change of the angle $\varphi$.

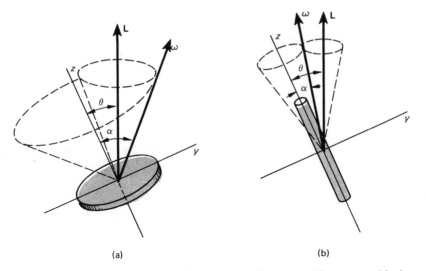

(a)                                                                (b)

FIGURE 8.14   Free rotation of (a) a disc and (b) a rod.   The space and body
cones are shown dotted.

Thus $\dot{\varphi}$ is the angular rate of precession of the axis of symmetry (and of the $\boldsymbol{\omega}$ vector) about the invariable line ($\mathbf{L}$ vector) as viewed from the outside. From a study of the figure, it is clear that the components of $\boldsymbol{\omega}$ are

$$\begin{aligned} \omega_x &= 0 \\ \omega_y &= \dot{\varphi} \sin \theta \\ \omega_z &= \dot{\varphi} \cos \theta + \dot{\psi} \end{aligned} \tag{8.39}$$

From the second of the above, and the second of Equations (8.36), we find

$$\dot{\varphi} = \omega \frac{\sin \alpha}{\sin \theta} \tag{8.40}$$

The above equation can be put into a somewhat more useful form by expressing $\theta$ as a function of $\alpha$ by means of Equation (8.38).   After some algebra, we obtain

$$\dot{\varphi} = \omega \left[ 1 + \left( \frac{I_s^{\,2}}{I^2} - 1 \right) \cos^2 \alpha \right]^{1/2} \tag{8.41}$$

for the rate of precession of the axis of symmetry about the invariable line.


## EXAMPLES


### 1. Free Precession of a Disc

As an example of the above theory, consider the case of a thin disc, or any symmetrical laminar body.   The perpendicular axis theorem gives

$$I_{xx} + I_{yy} = I_{zz}$$

or, in our present notation

$$2I = I_s$$

Consequently Equation (8.34) yields

$$\Omega = \left(\frac{2I}{I} - 1\right) \omega \cos \alpha = \omega \cos \alpha$$

for the rate of precession of the angular velocity vector about the symmetry axis, as seen in a rotating coordinate system fixed to the disc. If the disc is thick, then $I_s$ is not equal to $2I$, and the rate of precession is different from the above expression, depending on the value of the ratio $I/I_s$.

For the rate of precession of the symmetry axis about the **L** vector, or $z$ axis, as seen from the outside, Equation (8.41) gives

$$\dot{\varphi} = \omega(1 + 3 \cos^2 \alpha)^{1/2}$$

In particular, if $\alpha$ is quite small so that $\cos \alpha \approx 1$, then we have approximately

$$\Omega \approx \omega$$
$$\dot{\varphi} \approx 2\omega$$

Thus the symmetry axis precesses in space at just twice the angular speed of rotation. This precession is manifest as a wobbling motion.

### 2. Free Precession of the Earth

In the motion of the earth, it is known that the axis of rotation is very slightly inclined with respect to the geographic pole defining the axis of symmetry. The angle $\alpha$ is about 0.2 sec of arc (shown exaggerated in Figure 8.15.) It is also known that the ratio of the moments of inertia $I_s/I$ is about 1.00327 as determined from the earth's oblateness. From Equation (8.34) we have therefore

$$\Omega = 0.00327\omega$$

Then, since $\omega = 2\pi/\text{day}$, the period of the above precession is calculated to be

$$\frac{2\pi}{\Omega} = \frac{1}{0.00327} \text{ days} = 305 \text{ days}$$

The observed period of precession of the earth's axis of rotation about the pole is about 440 days. The disagreement between the observed and calculated values is attributed to the fact that the earth is not perfectly rigid.

With regard to the precession of the earth's symmetry axis as viewed from space, Equation (8.41) gives

$$\dot{\varphi} = 1.00327\omega$$

The associated period is then

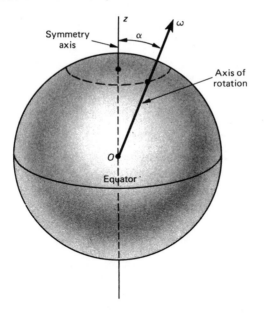

FIGURE 8.15   The earth with symmetry axis and the axis of rotation.   The
angle $\alpha$ is greatly exaggerated.

$$\frac{2\pi}{\dot{\varphi}} = \frac{2\pi}{\omega} \frac{1}{1.00327} \approx 0.997 \text{ day}$$

This free precession of the earth's axis in space is superimposed upon a very
much longer gyroscopic precession of 26,000 years, the latter resulting from the
fact that there is actually a torque exerted on the earth (because of its oblateness) by the sun and the moon.   The fact that the period of gyroscopic precession is so much longer than that of the free precession justifies the neglect of
the external torques in calculating the period of the free precession.

## 8.9.  Gyroscopic Precession.   Motion of a Top

In this section we shall study the motion of a symmetrical rigid body
which is free to turn about a fixed point and on which there is exerted a torque,
instead of no torque, as in the case of free precession.   The case is exemplified
by a simple gyroscope (or top).

The notation for our coordinate axes is shown in Figure 8.16(a).   For
clarity, only the $Z$, $y$, and $z$ axes are shown in Figure 8.16(b), the $x$ axis being
normal to the paper.   The origin $O$ is the fixed point about which the body
turns.

The torque about $O$ resulting from the weight is of magnitude $mgl \sin \theta$,
$l$ being the distance from $O$ to the center of mass $C$.   This torque is about the
$x$ axis, so that

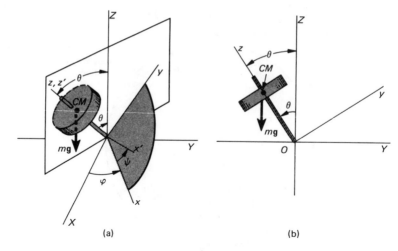

FIGURE 8.16   The simple gyroscope.

$$N_x = mgl \sin \theta$$
$$N_y = 0 \tag{8.42}$$
$$N_z = 0$$

Let us denote the angular velocity of the $Oxyz$ coordinate system by $\boldsymbol{\omega}$. In terms of the Eulerian angles, the components of $\boldsymbol{\omega}$ are clearly

$$\omega_x = \dot\theta$$
$$\omega_y = \dot\varphi \sin \theta \tag{8.43}$$
$$\omega_z = \dot\varphi \cos \theta$$

Thus the angular momentum of the spinning top has the following components

$$L_x = I_{xx}\omega_x = I\dot\theta$$
$$L_y = I_{yy}\omega_y = I\dot\varphi \sin \theta \tag{8.44}$$
$$L_z = I_{zz}(\omega_z + \dot\psi) = I_s(\dot\varphi \cos \theta + \dot\psi) = I_s S$$

Here we use the same notation for the moments of inertia as in the previous section. In the last equation we have abbreviated the quantity $\dot\varphi \cos \theta + \dot\psi$ by the letter $S$, called the spin.

The fundamental equation of motion, referred to our rotating coordinate system, is

$$\mathbf{N} = \dot{\mathbf{L}} + \boldsymbol{\omega} \times \mathbf{L}$$

Thus, in component form, we have the following equations of motion:

$$mgl \sin \theta = I\ddot\theta + I_s S\dot\varphi \sin \theta - I\dot\varphi^2 \cos \theta \sin \theta \tag{8.45}$$

$$0 = I \frac{d}{dt}(\dot\varphi \sin \theta) - I_s S\dot\theta + I\dot\theta\dot\varphi \cos \theta \tag{8.46}$$

$$0 = I_s \dot S \tag{8.47}$$

The last equation shows that $S$, the spin of the body about the symmetry axis, remains constant. Also, of course, the component of the angular momentum along that axis is constant

$$L_z = I_s S = \text{constant} \tag{8.48}$$

The second equation is then equivalent to

$$0 = \frac{d}{dt} (I\dot{\varphi} \sin^2 \theta + I_s S \cos \theta)$$

so that

$$I\dot{\varphi} \sin^2 \theta + I_s S \cos \theta = B = \text{constant} \tag{8.49}$$

### Steady Precession

Before proceeding with the integration of the remaining equations, we shall discuss an interesting special case, namely that of steady precession. This is the situation in which the axis of the gyroscope or top describes a right-circular cone about the vertical ($Z$ axis). In this case $\dot{\theta} = \ddot{\theta} = 0$, and Equation (8.45), after canceling the common factor $\sin \theta$, reduces to

$$mgl = I_s S\dot{\varphi} - I\dot{\varphi}^2 \cos \theta$$

or, solving for $S$, we find

$$S = \frac{mgl}{I_s\dot{\varphi}} + \frac{I}{I_s} \dot{\varphi} \cos \theta \tag{8.50}$$

as the condition for steady precession. Here $\dot{\varphi}$ is the angular frequency of the precession, that is, the angular frequency of the motion of the symmetry or spin axis about the vertical. In particular, if $\dot{\varphi}$ is very small, then $S$ is large. (This is the usual case for a top or gyroscope.) Then the second term on the right in Equation (8.50) may be ignored, and we may write approximately

$$\dot{\varphi} \simeq \frac{mgl}{I_s S} \tag{8.51}$$

which is the familiar result of elementary gyroscopic theory given in most general physics textbooks. Actually, since Equation (8.50) is a quadratic in $\dot{\varphi}$, there are two values of $\dot{\varphi}$ for a given value of $S$, but the above approximate value is the one that is usually observed.

### The Energy Equation and Nutation

If there are no frictional forces acting on the gyroscope to dissipate its energy, then the total energy $T + V$ remains constant:

$$\tfrac{1}{2}(I\omega_x^2 + I\omega_y^2 + I_s S^2) + mgl \cos \theta = E$$

or equivalently, in terms of the Eulerian angles,

$$\tfrac{1}{2}(I\dot{\theta}^2 + I\dot{\varphi}^2 \sin^2 \theta + I_s S^2) + mgl \cos \theta = E$$

From Equation (8.49), we can solve for $\dot{\varphi}$ and substitute into the above equation. The result is

$$\frac{1}{2} I \dot{\theta}^2 + \frac{(B - I_s S \cos \theta)^2}{2I \sin^2 \theta} + \frac{1}{2} I_s S^2 + mgl \cos \theta = E \qquad (8.52)$$

which is entirely in terms of $\theta$. This equation allows us, in principle, to find $\theta$ as a function of $t$ by integration. Let us make the substitution

$$u = \cos \theta$$

Then $\dot{u} = -(\sin \theta)\dot{\theta} = -(1 - u^2)^{1/2}\dot{\theta}$. We find

$$\dot{u}^2 = (1 - u^2)(2E - I_s S^2 - 2mglu)I^{-1} - (B - I_s Su)^2 I^{-2}$$

or

$$\dot{u}^2 = f(u)$$

from which $u$ (hence $\theta$) can be found as a function of $t$ by integration:

$$t = \int \frac{du}{\sqrt{f(u)}} \qquad (8.53)$$

Now $f(u)$ is a cubic polynomial, hence the integration can be carried out in terms of elliptic functions.

We need not actually perform the integration, however, to discuss the general properties of the motion. We see that $f(u)$ must be positive in order that $t$ be real. Hence the limits of the motion in $\theta$ are determined by the roots of the equation $f(u) = 0$. Since $\theta$ must lie between 0 and 90 degrees, then $u$ must take values between 0 and $+1$. A plot of $f(u)$ is shown in Figure 8.17 for the case in which there are two distinct roots $u_1$ and $u_2$ between 0 and $+1$. The corresponding values of $\theta$, namely $\theta_1$, and $\theta_2$ are then the limits of the vertical motion. The axis of the top oscillates back and forth

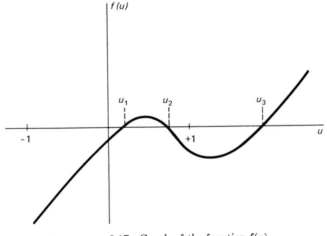

FIGURE 8.17   Graph of the function $f(u)$.

between these two values of $\theta$ as the top precesses about the vertical Figure 8.18.  This oscillation is called *nutation*.  If we have a double root, that is, if $u_1 = u_2$, then there is no nutation and the top precesses steadily.  The condition for a double root is, in fact, given by Equation (8.50).

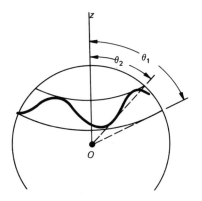

FIGURE 8.18   Illustrating the nutation of a simple gyroscope.

*Sleeping Top*

Anyone who has played with a top knows that, if it is spinning sufficiently fast and started in a vertical position, the axis of the top will remain fixed in the vertical direction, a condition called *sleeping*.  In terms of the above analysis, we see that sleeping must correspond to a double root at $u = +1$.  In this case, since $\theta = \dot\theta = 0$, $E = mgh + \frac{1}{2}I_sS^2$ and $B = I_sS$.  The equation

$$f(u) = 0$$

then becomes

$$(1 - u)^2 \left[ \frac{2mgh}{I} (1 + u) - \frac{(I_sS)^2}{I^2} \right] = 0$$

and we do, indeed, have a double root at $u = +1$.  Now setting the bracketed term in the above equation equal to zero gives us a third root $u_3$.  We find

$$u_3 = \frac{I_s^2 S^2}{2Imgh} - 1$$

If the root $u_3$ does not correspond to a physically possible value of $\theta$, that is, if $u_3$ is greater than 1, then the vertical sleeping motion will be stable.  This gives

$$S^2 > \frac{4Imgh}{I_s^2} \tag{8.54}$$

as the criterion for stability of the sleeping top.  If the top slows down through friction so that the above condition no longer holds, then it will begin to undergo nutation and will eventually topple over.

## DRILL EXERCISES

8.1   Write down the inertia tensor for a square plate of side $l$ and mass $m$ for a coordinate system with origin at the center of the plate, the $z$ axis being normal to the plate, and the $x$ and $y$ axes parallel to the edges.

8.2   Find the angular momentum and kinetic energy of the above plate when it is rotating about a diagonal, verifying Examples 2 and 3, Section 8.4.

8.3   A "rigid body" consists of six particles, each of mass $m$, fixed to the ends of three light rods of length $2a$, $2b$, and $2c$, respectively, the rods being held mutually perpendicular to one another at their mid points.   Show that a set of coordinate axes defined by the rods are principal axes, and write down the inertia tensor for the system in these axes.

8.4   Find the angular momentum and the kinetic energy of the above system when it is rotating with angular speed $\omega$ about an axis passing through the origin and the point $(a,b,c)$.

8.5   A uniform rectangular block spins about a long diagonal.   Find the inertia tensor for a coordinate system with origin at the center of the block and with axes normal to the faces.   Find also the angular momentum and the kinetic energy.   Find also the inertia tensor for axes with origin at one corner.

## PROBLEMS

8.6   Find the inertia tensor of a uniform triangular lamina $AOB$, where the angle at $O$ is a right angle, and the sides $OA = a$, $OB = b$ are on the $x$ and $y$ axes.   Find also the principal axes passing through $O$.

8.7   Find the equation of the momental ellipsoid of a uniform solid right circular cylinder of radius $a$ and length $b$.   Choose coordinate axes with origin at the center, the $z$ axis being the central axis of the cylinder.   Determine also the ratio of $a$ to $b$ in order that the momental ellipsoid is a sphere.

8.8   Show that the momental ellipsoid of any uniform lamina in the shape of a regular polygon is an ellipsoid of revolution for axes with origin at the center of the polygon.

8.9   A rectangular plate of sides $2a$ and $a$ revolves about a diagonal. Find the magnitude and direction of the torque exerted on the plate by the supporting axis.   Choose the origin to be at the center of the plate.

8.10   A lamina of arbitrary shape rotates freely under zero torque.   Show by means of Euler's equations that the component $\omega_1$ of the angular velocity in the plane of the lamina (the $xy$ plane) is constant in magnitude, although the $z$ component of $\omega$ is not necessarily constant.   [*Hint:* Use the perpendicular axis theorem.]   What kind of lamina gives $\omega_z = $ constant?

8.11 A square plate of side $a$ rotates freely under zero torque. If the axis of rotation makes an angle of 45° with the symmetry axis of the plate, find the period of the precession of the axis of rotation about the symmetry axis and the period of precession of the symmetry axis about the invariable line for two cases (a) a thin plate and (b) a thick plate of thickness $a/4$.

8.12 A rigid body having an axis of symmetry rotates freely about a fixed point under no torques. If $\alpha$ is the angle between the axis of symmetry and the instantaneous axis of rotation, show that the angle between the axis of rotation and the invariable line (the **L** vector) is

$$\tan^{-1}\left[\frac{(I_s - I)\tan\alpha}{I_s + I\tan^2\alpha}\right]$$

where $I_s$ (the moment of inertia about the symmetry axis) is greater than $I$ (the moment of inertia about an axis normal to the symmetry axis). Show that this angle cannot exceed $\tan^{-1}(8^{-1/2})$.

8.13 Find the angle between $\omega$ and **L** for the two cases in Problem 8.11.

8.14 Find the same angle for the earth.

8.15 A rigid body rotates freely about its center of mass. There are no torques. Show by means of Euler's equations that, if all three principal moments of inertia are different, then the body will rotate stably about either the axis of greatest moment of inertia or the axis of least moment of inertia, but that rotation about the axis of intermediate moment of inertia is unstable. (This can be demonstrated by tossing a book into the air. Put an elastic band around the book.)

8.16 A space platform in the form of a thin circular disc of radius $a$ and mass $m$ is rotating with angular speed $\omega$ about its symmetry axis. A meteorite strikes the platform at the edge, inparting an impulse $\hat{\mathbf{P}}$ to the platform. The direction of $\hat{\mathbf{P}}$ is parallel to the axis of the platform. Find the resulting motion of the platform.

8.17 A rigid body having an axis of symmetry rotates with angular velocity $\omega$ in three-dimensional motion about its center of mass. There is a frictional torque $-c\omega$ exerted on the body, such as might be produced by air drag. (a) Show that the component of $\omega$ in the direction of the symmetry axis decreases exponentially with time. (b) Show also that the angle between the angular velocity $\omega$ and the symmetry axis steadily decreases if the moment of inertia about the symmetry axis is the largest principal moment.

8.18 A simple gyroscope consists of a heavy circular disc of mass $m$ and radius $a$ mounted at the center of a thin rod of mass $m/2$ and length $a$. If the gyroscope is set spinning at a given rate $S$, and with the axis at an angle of 45° with the vertical, show that there are two possible values of the precession rate $\dot{\varphi}$ such that the gyroscope precesses steadily at a constant value of $\theta = 45°$. Find the two numerical values of $\dot{\varphi}$ when $S = 900$ rpm and $a = 10$ cm.

8.19 If, instead of steady precession at constant $\theta$, the gyroscope in the above problem is started by releasing it at an angle of $\theta_1 = 45°$ and $\dot{\varphi} = 0$ and with the same spin, set up the energy equation and find the other limit $\theta_2$ which the gyroscope axis makes with the vertical in its nutation.

8.20  The axis of a spinning gyroscope is constrained to remain in a horizontal plane on the surface of the earth, but is free to point in any direction in that plane.  Show that the earth's rotation produces a torque which tends to cause the gyroscope's axis to point in a north-south line.  This is the principle of the gyrocompass.

8.21  A pencil is set spinning in an upright position.  How fast must the spin be in order that the pencil will remain in the upright position?  Assume that the pencil is a uniform cylinder of length $a$ and diameter $b$.  Find the value of the spin in revolutions per minute for $a = 20$ cm and $b = 1$ cm.

# 9. Lagrangian Mechanics

The direct application of Newton's laws to the motion of simple systems will now be supplemented by a general, more sophisticated approach—a very elegant and useful method for finding the equations of motion for all dynamical systems, invented by the French mathematician Joseph Louis Lagrange.

## 9.1. Generalized Coordinates

We have seen that the position of a particle in space can be specified by three coordinates. These may be Cartesian, spherical, cylindrical, or, in fact, any three suitably chosen parameters. If the particle is constrained to move in a plane or on a fixed surface, only two coordinates are needed to specify the particle's position, whereas if the particle moves on a straight line or on a fixed curve, then one coordinate is sufficient.

In the case of a system of $N$ particles we need, in general, $3N$ coordinates to specify completely the simultaneous positions of all the particles—the *configuration* of the system. If there are constraints imposed on the system, however, the number of coordinates actually needed to specify the configuration is less than $3N$. For instance, if the system is a rigid body, we need give only the position of some convenient reference point of the body (for example, the center of mass) and the orientation of the body in space in order to specify the configuration. In this case only six coordinates are needed—three for the reference point and three more (say the Eulerian angles) for the orientation.

In general, a certain minimum number $n$ of coordinates is required to specify the configuration of a given system. We shall designate these coordinates by the symbols

$$q_1, q_2, \ldots, q_n$$

called *generalized coordinates*. A given coordinate $q_k$ may be either an angle or a distance. If, in addition to specifying the configuration of the system, each coordinate can vary independently of the others, the system is said to be *holonomic*. The number of coordinates $n$ in this case is also the number of *degrees of freedom* of the system.

In a nonholonomic system the coordinates cannot all vary independently; that is, the number of degrees of freedom is less than the minimum number of coordinates needed to specify the configuration. An example of a nonholonomic system is a sphere constrained to roll on a perfectly rough plane. Five coordinates are required to specify the configuration—two for the position of the center of the sphere and three for its orientation. But the coordinates cannot all vary independently, for, if the sphere rolls, at least two coordinates must change. For the present, we shall consider only holonomic systems.

If the system is a single particle, the Cartesian coordinates are expressible as functions of the generalized coordinates:

$$x = x(q) \qquad \} \qquad \textit{(one degree of freedom—motion on a curve)}$$

$$\left.\begin{array}{l} x = x(q_1,q_2) \\ y = y(q_1,q_2) \end{array}\right\} \qquad \textit{(two degrees of freedom—motion on a surface)}$$

$$\left.\begin{array}{l} x = x(q_1,q_2,q_3) \\ y = y(q_1,q_2,q_3) \\ z = z(q_1,q_2,q_3) \end{array}\right\} \qquad \textit{(three degrees of freedom—motion in space)}$$

Suppose that the $q$'s change from initial values $(q_1,q_2, \ldots)$ to the neighboring values $(q_1 + \delta q_1, q_2 + \delta q_2, \ldots)$. The corresponding changes in the Cartesian coordinates are given by

$$\delta x = \frac{\partial x}{\partial q_1} \delta q_1 + \frac{\partial x}{\partial q_2} \delta q_2 + \cdots$$

$$\delta y = \frac{\partial y}{\partial q_1} \delta q_1 + \frac{\partial y}{\partial q_2} \delta q_2 + \cdots$$

and so on. The partial derivatives $\partial x/\partial q_1$, and so on, are functions of the $q$'s. As a specific example, consider the motion of a particle in a plane. Let us choose polar coordinates

$$q_1 = r \qquad q_2 = \theta$$

Then

$$x = x(r,\theta) = r \cos \theta$$
$$y = y(r,\theta) = r \sin \theta$$

and

$$\delta x = \frac{\partial x}{\partial r} \delta r + \frac{\partial x}{\partial \theta} \delta \theta = \cos \theta \, \delta r - r \sin \theta \, \delta \theta$$

$$\delta y = \frac{\partial y}{\partial r} \delta r + \frac{\partial y}{\partial \theta} \delta \theta = \sin \theta \, \delta r + r \cos \theta \, \delta \theta$$

giving the changes in $x$ and $y$ that correspond to small changes in $r$ and $\theta$.

Consider now a system consisting of a large number of particles. Let the system have $n$ degrees of freedom and generalized coordinates

$$q_1, q_2, \ldots, q_n$$

Then, in a change from the configuration $(q_1, q_2, \ldots, q_n)$, to the neighboring configuration $(q_1 + \delta q_1, \ldots, q_n + \delta q_n)$, a representative particle $i$ moves from the point $(x_i, y_i, z_i,)$ to the neighboring point $(x_i + \delta x_i, y_i + \delta y_i, z_i + \delta z)$ where

$$\delta x_i = \sum_{k=1}^{n} \frac{\partial x_i}{\partial q_k} \delta q_k$$

$$\delta y_i = \sum_{k=1}^{n} \frac{\partial y_i}{\partial q_k} \delta q_k$$

$$\delta z_i = \sum_{k=1}^{n} \frac{\partial z_i}{\partial q_k} \delta q_k$$

The partial derivatives are again functions of the $q$'s. We shall adopt the convention of letting the subscript $i$ refer to the rectangular coordinates, and the subscript $k$ refer to the generalized coordinates. Let us further adopt the convenient notation of letting the symbol $x_i$ refer to *any* rectangular coordinate. Thus, for a system of $N$ particles, $i$ would take on values between 1 and $3N$.

## 9.2. Generalized Forces

If a particle undergoes a displacement $\delta \mathbf{r}$ under the action of a force $\mathbf{F}$, then we know that the work $\delta W$ done by the force is given by

$$\delta W = \mathbf{F} \cdot \delta \mathbf{r} = F_x \, \delta x + F_y \, \delta y + F_z \, \delta z$$

In our newly adopted notation, the expression for the work is given by

$$\delta W = \sum_i F_i \, \delta x_i \tag{9.1}$$

It is clear that the above formula holds not only for a single particle, but also for a system of many particles. For one particle, $i$ goes from 1 to 3. For $N$ particles, $i$ ranges from 1 to $3N$.

Now let us express the increments $\delta x_i$ in terms of the generalized coordinates. Then

$$\delta W = \sum_i \left( F_i \sum_k \frac{\partial x_i}{\partial q_k} \delta q_k \right)$$

$$= \sum_i \left( \sum_k F_i \frac{\partial x_i}{\partial q_k} \delta q_k \right)$$

By reversing the order of summation, we have

$$\delta W = \sum_k \left( \sum_i F_i \frac{\partial x_i}{\partial q_k} \right) \delta q_k$$

This can be written

$$\delta W = \sum_k Q_k \, \delta q_k \qquad (9.2)$$

where

$$Q_k = \sum_i \left( F_i \frac{\partial x_i}{\partial q_k} \right) \qquad (9.3)$$

The quantity $Q_k$ defined by the above equation is called the *generalized force* associated with the coordinate $q_k$. Since the product $Q_k \, \delta q_k$ has the dimensions of work, then $Q_k$ has the dimensions of force if $q_k$ is a distance, and the dimensions of torque if $q_k$ is an angle.

It is usually unnecessary, and even impractical, to use Equation (9.3) to calculate the actual value of $Q_k$; rather, each generalized force $Q_k$ can be found directly from the fact that $Q_k \, \delta q_k$ is the work done on the system by the external forces when the coordinate $q_k$ changes by the amount $\delta q_k$ (the other generalized coordinates remaining constant). For example, if the system is a rigid body, the work done by the external forces when the body turns through an angle $\delta\theta$ about a given axis is $L_\theta \, \delta\theta$, where $L_\theta$ is the magnitude of the total moment of all the forces about the axis. In this case $L_\theta$ is the generalized force associated with the coordinate $\theta$.

### Generalized Forces for Conservative Systems

We have seen in Chapter 3 that the rectangular components of the force acting on a particle in a conservative field of force are given by

$$F_i = -\frac{\partial V}{\partial x_i}$$

where $V$ is the potential energy function. Accordingly, our formula for the generalized force becomes

$$Q_k = -\left(\sum_i \frac{\partial V}{\partial x_i} \frac{\partial x_i}{\partial q_k}\right)$$

Now the expression in parentheses is just the partial derivative of the function $V$ with respect to $q_k$.   Hence

$$Q_k = -\frac{\partial V}{\partial q_k} \qquad (9.4)$$

For example, if we use polar coordinates $q_1 = r$, $q_2 = \theta$, then the generalized forces are $Q_r = -\partial V/\partial r$; $Q_\theta = -\partial V/\partial \theta$. If $V$ is a function of $r$ alone (central force), then $Q_\theta = 0$.

## 9.3.   Lagrange's Equations

In order to find the differential equations of motion in terms of the generalized coordinates, we could start with the equation

$$F_i = m_i \ddot{x}_i$$

and try to write it directly in terms of the $q$'s.   It turns out, however, to be simpler to use a different approach based on energy considerations.   We shall first calculate the kinetic energy $T$ in terms of Cartesian coordinates and shall then express it as a function of the generalized coordinates and their time derivatives.   Thus, the kinetic energy $T$ of a system of $N$ particles, which we have previously expressed as

$$T = \sum_{i=1}^{N} [\tfrac{1}{2} m_i(\dot{x}_i^2 + \dot{y}_i^2 + \dot{z}_i^2)]$$

will now be written simply

$$T = \sum_{i=1}^{3N} \tfrac{1}{2} m_i \dot{x}_i^2 \qquad (9.5)$$

The Cartesian coordinates $x_i$ are functions of the generalized coordinates $q_k$. For generality, we shall also include the possibility that the functional relationship between the $x$'s and the $q$'s may also involve the time $t$ explicitly.   This would be the case if there were moving constraints, such as a particle constrained to move on a surface which itself is moving in some prescribed manner.   We can write

$$x_i = x_i(q_1, q_2, \ \ldots \ , q_n, t)$$

Thus

$$\dot{x}_i = \sum_k \frac{\partial x_i}{\partial q_k} \dot{q}_k + \frac{\partial x_i}{\partial t} \qquad (9.6)$$

In the above equation and in all that follows, unless stated to the contrary, we shall assume that the range of $i$ is 1, 2, . . . , $3N$, where $N$ is the number of particles in the system, and the range of $k$ is 1, 2, . . . , $n$, where $n$ is the number of generalized coordinates (degrees of freedom) of the system.    In view of the above equation, we see that we can regard $T$ as a function of the generalized coordinates, their time derivatives, and possibly the time.

From the expression for $\dot{x}_i$, it is clear that

$$\frac{\partial \dot{x}_i}{\partial \dot{q}_k} = \frac{\partial x_i}{\partial q_k} \tag{9.7}$$

Now let us multiply by $\dot{x}_i$ and differentiate with respect to $t$.    We then have

$$\frac{d}{dt}\left(\dot{x}_i \frac{\partial \dot{x}_i}{\partial \dot{q}_k}\right) = \frac{d}{dt}\left(\dot{x}_i \frac{\partial x_i}{\partial q_k}\right)$$

$$= \ddot{x}_i \frac{\partial x_i}{\partial q_k} + \dot{x}_i \frac{\partial \dot{x}_i}{\partial q_k}$$

or

$$\frac{d}{dt}\left(\frac{\partial}{\partial \dot{q}_k} \frac{\dot{x}_i^2}{2}\right) = \ddot{x}_i \frac{\partial x_i}{\partial q_k} + \frac{\partial}{\partial q_k}\left(\frac{\dot{x}_i^2}{2}\right)$$

The last step follows from the fact that the order of differentiation with respect to $t$ and $q_k$ or $\dot{q}_k$ can be reversed.    If we next multiply by $m_i$ and set $m_i\ddot{x}_i = F_i$, we can write

$$\frac{d}{dt}\frac{\partial}{\partial \dot{q}_k}\left(\frac{m_i\dot{x}_i^2}{2}\right) = F_i \frac{\partial x_i}{\partial q_k} + \frac{\partial}{\partial q_k}\left(\frac{m_i\dot{x}_i^2}{2}\right)$$

Hence, by summing over $i$, we find

$$\frac{d}{dt}\frac{\partial T}{\partial \dot{q}_k} = \sum_i \left(F_i \frac{\partial x_i}{\partial q_k}\right) + \frac{\partial T}{\partial q_k} \tag{9.8}$$

Finally from the definition of the generalized force $Q_k$, we obtain the result

$$\frac{d}{dt}\frac{\partial T}{\partial \dot{q}_k} = Q_k + \frac{\partial T}{\partial q_k} \tag{9.9}$$

These are the differential equations of motion in the generalized coordinates. They are known as *Lagrange's equations of motion*.

In case the motion is conservative so that the $Q$'s are given by Equation (9.4), then Lagrange's equations can be written

$$\frac{d}{dt}\frac{\partial T}{\partial \dot{q}_k} = \frac{\partial T}{\partial q_k} - \frac{\partial V}{\partial q_k} \tag{9.10}$$

The equations can be written even more compactly by defining a function $L$, known as the *Lagrangian function*, such that

$$L = T - V$$

where it is understood that $T$ and $V$ are expressed in terms of the generalized coordinates.    Thus, since $V = V(q)$ and $\partial V/\partial \dot{q} = 0$, we have

$$\frac{\partial L}{\partial \dot{q}_k} = \frac{\partial T}{\partial \dot{q}_k} \quad \text{and} \quad \frac{\partial L}{\partial q_k} = \frac{\partial T}{\partial q_k} - \frac{\partial V}{\partial q_k}$$

Lagrange's equations can then be written

$$\frac{d}{dt}\frac{\partial L}{\partial \dot{q}_k} = \frac{\partial L}{\partial q_k} \tag{9.11}$$

Thus the differential equations of motion for a conservative system are readily obtained if we know the Lagrangian function in terms of an appropriate set of coordinates.

If part of the generalized forces are not conservative, say $Q_k'$, and part are derivable from a potential function $V$, we can write

$$Q_k = Q_k' - \frac{\partial V}{\partial q_k} \tag{9.12}$$

We can then also define a Lagrangian function $L = T - V$, and write the differential equations of motion in the form

$$\frac{d}{dt}\frac{\partial L}{\partial \dot{q}_k} = Q_k' + \frac{\partial L}{\partial q_k} \tag{9.13}$$

The above form is a convenient one to use, for example, when frictional forces are present.

### 9.4.   Some Applications of Lagrange's Equations

In this section we shall illustrate the remarkable versatility of Lagrange's equations by applying them to a number of specific cases.   The general procedure for finding the differential equations of motion for a system is as follows:

(1) Select a suitable set of coordinates to represent the configuration of the system.

(2) Obtain the kinetic energy $T$ as a function of these coordinates and their time derivatives.

(3) If the system is conservative, find the potential energy $V$ as a function of the coordinates, or, if the system is not conservative, find the generalized forces $Q_k$.

(4) The differential equations of motion are then given by Equations (9.9), (9.11), or (9.13).

*Harmonic Oscillator*

Consider the case of a one-dimensional harmonic oscillator, and suppose that there is a damping force which is proportional to the velocity.   The system is thus nonconservative.   If $x$ is the displacement coordinate, then the Lagrangian function is

$$L = T - V = \tfrac{1}{2}m\dot{x}^2 - \tfrac{1}{2}kx^2$$

in which $m$ is the mass and $k$ is the usual stiffness parameter.   Hence

$$\frac{\partial L}{\partial \dot{x}} = m\dot{x} \qquad \frac{\partial L}{\partial x} = -kx$$

Now since there is a nonconservative force present, Lagrange's equations in the form of Equation (9.13) can be used.   Thus $Q' = -c\dot{x}$, and the equation of motion reads

$$\frac{d}{dt}(m\dot{x}) = -c\dot{x} + (-kx)$$

or

$$m\ddot{x} + c\dot{x} + kx = 0$$

This is the familiar equation of the damped harmonic oscillator that we studied earlier.

### Single Particle in a Central Field

Let us find Lagrange's equations of motion for a particle moving in a plane under a central force.   We shall choose polar coordinates $q_1 = r$, $q_2 = \theta$. Then

$$T = \tfrac{1}{2}mv^2 = \tfrac{1}{2}m(\dot{r}^2 + r^2\dot{\theta}^2)$$
$$V = V(r)$$
$$L = \tfrac{1}{2}m(\dot{r}^2 + r^2\dot{\theta}^2) - V(r)$$

The relevant partial derivatives are as follows:

$$\frac{\partial L}{\partial \dot{r}} = m\dot{r} \qquad \frac{\partial L}{\partial r} = mr\dot{\theta}^2 - \frac{\partial V}{\partial r} = mr\dot{\theta}^2 + F_r$$

$$\frac{\partial L}{\partial \dot{\theta}} = 0 \qquad \frac{\partial L}{\partial \dot{\theta}} = mr^2\dot{\theta}$$

The equations of motion Equation (9.11), are therefore

$$m\ddot{r} = mr\dot{\theta}^2 + F_r \qquad \frac{d}{dt}(mr^2\dot{\theta}) = 0$$

These are identical to the equations found in Section 5.7 for the motion of a particle in a central field.

### Atwood's Machine

A mechanical system known as Atwood's machine consists of two weights of mass $m_1$ and $m_2$, respectively, connected by a light inextensible cord of length $l$ which passes over a pulley (Figure 9.1).   The system has one degree of freedom.   We shall let the variable $x$ represent the configuration of the system, where $x$ is the vertical distance from the pulley to $m_1$, as shown.   The

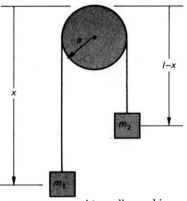

**FIGURE 9.1**   Atwood's machine.

angular speed of the pulley is clearly $\dot{x}/a$, where $a$ is the radius.   The kinetic energy of the system is therefore given by

$$T = \frac{1}{2} m_1 \dot{x}^2 + \frac{1}{2} m_2 \dot{x}^2 + \frac{1}{2} I \frac{\dot{x}^2}{a^2}$$

where $I$ is the moment of inertia of the pulley.   The potential energy is given by

$$V = -m_1 g x - m_2 g (l - x)$$

Assuming that there is no friction, we have the Lagrangian function

$$L = \frac{1}{2} \left( m_1 + m_2 + \frac{I}{a^2} \right) \dot{x}^2 + g(m_1 - m_2)x + m_2 g l$$

and Lagrange's equation

$$\frac{d}{dt} \frac{\partial L}{\partial \dot{x}} = \frac{\partial L}{\partial x}$$

then reads

$$\left( m_1 + m_2 + \frac{I}{a^2} \right) \ddot{x} = g(m_1 - m_2)$$

or

$$\ddot{x} = g \frac{m_1 - m_2}{m_1 + m_2 + I/a^2}$$

giving the acceleration of the system.   We see that if $m_1 > m_2$, then $m_1$ descends with constant acceleration, whereas if $m_1 < m_2$, then $m_1$ ascends with constant acceleration.   The inertial effect of the pulley shows up in the term $I/a^2$ in the denominator.

### The Double Atwood Machine

Consider the system shown in Figure 9.2. Here we have replaced one of the weights in the simple Atwood machine by another pulley supporting two weights connected by another cord. The system now has two degrees of freedom. We shall specify its configuration by the two coordinates $x$ and $x'$,

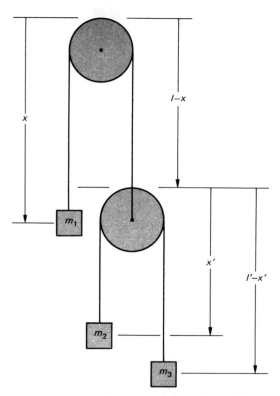

FIGURE 9.2    A compound Atwood machine.

as shown. For simplicity, let us neglect the masses of the pulleys in this case. We have

$$T = \tfrac{1}{2}m_1\dot{x}^2 + \tfrac{1}{2}m_2(-\dot{x} + \dot{x}')^2 + \tfrac{1}{2}m_3(-\dot{x} - \dot{x}')^2$$
$$V = -m_1gx - m_2g(l - x + x') - m_3g(l - x + l' - x')$$

where $m_1$, $m_2$, and $m_3$ are the three masses, and $l$ and $l'$ are the lengths of the two connecting cords. Then

$$L = \tfrac{1}{2}m_1 x_1^2 + \tfrac{1}{2}m_2(-\dot{x} + \dot{x}')^2 + \tfrac{1}{2}m_3(\dot{x} + \dot{x}')^2$$
$$+ g(m_1 - m_2 - m_3)x + g(m_2 - m_3)x' + \text{constant}$$

The equations of motion

$$\frac{d}{dt}\frac{\partial L}{\partial \dot{x}} = \frac{\partial L}{\partial x} \qquad \frac{d}{dt}\frac{\partial L}{\partial \dot{x}'} = \frac{\partial L}{\partial x'}$$

read

$$m_1\ddot{x} + m_2(\ddot{x} - \ddot{x}') + m_3(\ddot{x} + \ddot{x}') = g(m_1 - m_2 - m_3)$$
$$m_2(-\ddot{x} + \ddot{x}') + m_3(\ddot{x} + \ddot{x}') = g(m_2 - m_3)$$

from which the accelerations $\ddot{x}$ and $\ddot{x}'$ are found by simple algebra.

### Particle Sliding on a Movable Inclined Plane

Let us consider the case of a particle sliding on a smooth inclined plane which, itself, is free to slide on a smooth horizontal surface, as shown in Figure 9.3.   In this problem there are two degrees of freedom, so we need two coordinates to specify the configuration.   We shall choose the coordinates $x$ and $x'$, the horizontal displacement of the plane from some reference point and the

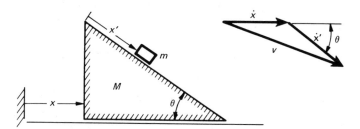

FIGURE 9.3    Block sliding down a movable inclined plane.

displacement of the particle from some reference point on the plane, respectively, as shown.

From a study of the velocity diagram, shown to the right of the figure, we see that the square of the speed of the particle is given by the law of cosines

$$v^2 = \dot{x}^2 + \dot{x}'^2 + 2\dot{x}\dot{x}' \cos \theta$$

Hence the kinetic energy $T$ of the system is given by

$$T = \tfrac{1}{2}mv^2 + \tfrac{1}{2}M\dot{x}^2 = \tfrac{1}{2}m(\dot{x}^2 + \dot{x}'^2 + 2\dot{x}\dot{x}' \cos \theta) + \tfrac{1}{2}M\dot{x}^2$$

where $M$ is the mass of the inclined plane, $\theta$ is the wedge angle as shown, and $m$ is the mass of the particle.   The potential energy of the system does not involve $x$, since the plane is on a horizontal surface.   Hence we can write

$$V = -mgx' \sin \theta + \text{constant}$$

and

$$L = \tfrac{1}{2}m(\dot{x}^2 + \dot{x}'^2 + 2\dot{x}\dot{x}' \cos \theta) + \tfrac{1}{2}M\dot{x}^2 + mgx' \sin \theta + \text{constant}$$

The equations of motion

$$\frac{d}{dt}\frac{\partial L}{\partial \dot{x}} = \frac{\partial L}{\partial x} \qquad \frac{d}{dt}\frac{\partial L}{\partial \dot{x}'} = \frac{\partial L}{\partial x'}$$

then become

$$m(\ddot{x} + \ddot{x}' \cos \theta) + M\ddot{x} = 0 \qquad m(\ddot{x}' + \ddot{x} \cos \theta) = mg \sin \theta$$

Solving for $\ddot{x}$ and $\ddot{x}'$ we find

$$\ddot{x} = \frac{-g \sin \theta \cos \theta}{\dfrac{m+M}{m} - \cos^2 \theta} \qquad \ddot{x}' = \frac{g \sin \theta}{1 - \dfrac{m \cos^2 \theta}{m+M}}$$

The above result can be obtained by analyzing the forces and reactions involved, but that method is much more tedious than the above method of using Lagrange's equations.

### Derivation of Euler's Equations for the Free Rotation of a Rigid Body

Lagrange's method can be used to derive Euler's equations for the motion of a rigid body. In this section we shall consider the case of a rigid body rotating under no torques.

We have seen that the kinetic energy of a rigid body is given by

$$T = \tfrac{1}{2}(I_{xx}\omega_x{}^2 + I_{yy}\omega_y{}^2 + I_{zz}\omega_z{}^2)$$

where the $\omega$'s are referred to principal axes of the body. Let us refer to Figure 8.13 which shows the Eulerian angles $\varphi$, $\psi$, and $\theta$. From a study of the figure we see that the relations between the $\omega$'s and the Eulerian angles and their time derivatives are as follows:

$$\begin{aligned}
\omega_x &= \dot{\theta} \cos \psi + \dot{\varphi} \sin \theta \sin \psi \\
\omega_y &= -\dot{\theta} \sin \psi + \dot{\varphi} \sin \theta \cos \psi \\
\omega_z &= \dot{\psi} + \dot{\varphi} \cos \theta
\end{aligned} \qquad (9.14)$$

Regarding the Eulerian angles as the generalized coordinates, the equations of motion are

$$\frac{d}{dt}\frac{\partial T}{\partial \dot{\theta}} = \frac{\partial T}{\partial \theta}$$

$$\frac{d}{dt}\frac{\partial T}{\partial \dot{\varphi}} = \frac{\partial T}{\partial \varphi}$$

$$\frac{d}{dt}\frac{\partial T}{\partial \dot{\psi}} = \frac{\partial T}{\partial \psi}$$

because the $Q$'s (the generalized forces) are all zero. Now, by the chain rule,

$$\frac{\partial T}{\partial \dot{\psi}} = \frac{\partial T}{\partial \omega_z}\frac{\partial \omega_z}{\partial \dot{\psi}} = I_{zz}\omega_z$$

so

$$\frac{d}{dt}\frac{\partial T}{\partial \dot{\psi}} = I_{zz}\dot{\omega}_z \qquad (9.15)$$

Similarly

$$\frac{\partial T}{\partial \psi} = I_{zz}\omega_z \frac{\partial \omega_x}{\partial \psi} + I_{yy}\omega_y \frac{\partial \omega_y}{\partial \psi}$$

$$= I_{zz}\omega_z(-\dot{\theta}\sin\psi + \dot{\varphi}\sin\theta\cos\psi) + I_{yy}\omega_y(-\dot{\theta}\cos\psi - \dot{\varphi}\sin\theta\sin\psi)$$

$$= I_{zz}\omega_z\omega_y - I_{yy}\omega_y\omega_x \qquad (9.16)$$

From Equations (9.15) and (9.16), the $\psi$ equation becomes

$$I_{zz}\dot{\omega}_z + \omega_x\omega_y(I_{yy} - I_{xx}) = 0$$

which, as we have previously shown (Section 8.6), is one of Euler's equations for the motion of a rigid body under no torques.   The other two equations can be obtained by cyclic permutation of $x$, $y$, and $z$.   This is valid, because we have not designated any special Cartesian coordinate as preferred.

## 9.5.   Generalized Momenta.   Ignorable Coordinates

Consider the motion of a single particle moving in a straight line (rectilinear motion).   The kinetic energy is

$$T = \tfrac{1}{2}m\dot{x}^2$$

where $m$ is the mass of the particle, and $x$ is its positional coordinate.   Now rather than define the momentum $p$ of the particle as the product $m\dot{x}$, we could define $p$ as the quantity $\partial T/\partial \dot{x}$, namely,

$$p = \frac{\partial T}{\partial \dot{x}} = m\dot{x}$$

In the case of a system described by generalized coordinates $q_1$, $q_2$, . . . $q_k$, . . . $q_n$, the quantities $p_k$ defined by

$$p_k = \frac{\partial L}{\partial \dot{q}_k} \qquad (9.17)$$

are called the *generalized momenta*.[1]   Lagrange's equations for a conservative system can then be written

$$\dot{p}_k = \frac{\partial L}{\partial q_k} \qquad (9.18)$$

Suppose, in particular, that one of the coordinates, say $q_\lambda$, is not explicitly contained in $L$.   Then

---

[1] If the potential energy function $V$ does not explicitly involve the $\dot{q}$'s, then $p_k = \partial L/\partial \dot{q}_k = \partial T/\partial \dot{q}_k$.

$$\dot{p}_\lambda = \frac{\partial L}{\partial q_\lambda} = 0 \tag{9.19}$$

and

$$p_\lambda = \text{constant} = c_\lambda \tag{9.20}$$

The coordinate $q_\lambda$ is said to be *ignorable* in this case.   The generalized momentum associated with an ignorable coordinate is therefore a constant of the motion of the system.

For example, in the problem of the particle sliding on the smooth inclined plane (treated in the previous section), we found that the coordinate $x$, the position of the plane, was not contained in the Lagrangian function $L$. Thus $x$ is an ignorable coordinate in this case, and

$$p_x = \frac{\partial L}{\partial \dot{x}} = (M + m)\dot{x} + m\dot{x}' \cos \theta = \text{constant}$$

We can see, as a matter of fact, that $p_x$ is the total horizontal component of the linear momentum of the system, and, since there is no external horizontal force acting on the system, the horizontal component of the linear momentum must be constant.

Another example of an ignorable coordinate is found in the case of the motion of a particle in a central field.   In polar coordinates

$$L = \tfrac{1}{2}m(\dot{r}^2 + r^2\dot{\theta}^2) - V(r)$$

as shown in the example in Section 9.4.   In this case $\theta$ is an ignorable coordinate, and

$$p_\theta = \frac{\partial L}{\partial \dot{\theta}} = mr^2\dot{\theta} = \text{constant}$$

Here $p_\theta$ is just the magnitude of the angular momentum.

### 9.6.  Lagrange's Equations for Impulsive Forces

Suppose we have a dynamic system, described by generalized coordinates $q_k$, in which all the acting generalized forces $Q_k$ are zero except for a short interval of time $\tau$.   We can integrate Lagrange's equations as follows:

$$\frac{d}{dt}\frac{\partial T}{\partial \dot{q}_k} = \frac{\partial T}{\partial q_k} + Q_k$$

$$\int_0^\tau d\left(\frac{\partial T}{\partial \dot{q}_k}\right) = \int_0^\tau \frac{\partial T}{\partial q_k}\, dt + \int_0^\tau Q_k\, dt$$

Now if $Q_k$ tends to infinity in such way that the *generalized impulse*

$$\lim_{\tau \to 0} \int_0^\tau Q_k\, dt = \hat{P}_k \tag{9.21}$$

exists and is finite, then the integral $\int_0^\tau (\partial T/\partial q_k)\, dt$ tends to zero, because the quantity $\partial T/\partial q_k$ remains finite.   We can therefore write

$$\Delta\left(\frac{\partial T}{\partial \dot{q}_k}\right) = \hat{P}_k \tag{9.22}$$

for the changes in the .quantities $\partial T/\partial \dot{q}_k$ following the application of a generalized impulse $\hat{P}_k$ to the system.   For systems in which the potential function $V$ does not involve the $\dot{q}$'s explicitly, so that $\partial T/\partial \dot{q}_k = \partial L/\partial \dot{q}_k = p_k$, we can write Equation (9.22) as

$$\Delta p_k = \hat{P}_k \tag{9.23}$$

where $p_k$ is the generalized momentum associated with the generalized coordinate $q_k$.

The generalized impulses $\hat{P}_k$ are most easily found by calculating the *impulsive work* $\delta\hat{W}$ which is given by

$$\delta\hat{W} = \hat{\mathbf{P}}_a\cdot\delta\mathbf{s}_a + \cdots = \hat{P}_1\delta q_1 + \hat{P}_2\delta q_2 + \cdots = \sum_k \hat{P}_k\delta q_k \tag{9.24}$$

where $\hat{\mathbf{P}}_a \cdots$ are the applied impulses, and $\delta\mathbf{s}_a \cdots$ are arbitrary small displacements through which the applied impulsive forces act (subject to the constraints of the system).

## EXAMPLE

Two rods $AB$ and $BC$, each of length $2a$ and mass $m$, are smoothly joined at $B$ and lie at rest on a smooth horizontal table, the points $A$, $B$, and $C$ being colinear.   Find the motion immediately after an impulse $\hat{\mathbf{P}}$ is applied at point $A$, as shown in Figure 9.4.

Let us choose generalized coordinates $x$, $y$, $\theta_1$, and $\theta_2$ where $x$ and $y$ are the positional coordinates of the joint $B$, and $\theta_1$ and $\theta_2$ are the respective angles which the two rods make with the initial line $ABC$.   The kinetic energy $T$, for the initial motion, is given by

$$T = \tfrac{1}{2}m(\dot{x} + a\dot{\theta})^2 + \tfrac{1}{2}I_{cm}\dot{\theta}_1^2 + \tfrac{1}{2}m(\dot{x} + a\dot{\theta}_2)^2 + \tfrac{1}{2}I_{cm}\dot{\theta}_2^2 + m\dot{y}^2$$

where $I_{cm}$ is the moment of inertia of either rod about its center of mass. Now the impulsive work is equal to $\hat{P}\delta s$ where

$$\delta s = \delta x + 2a\delta\theta_1$$

Thus

$$\delta\hat{W} = \hat{P}\delta s = \hat{P}(\delta x + 2a\delta\theta_1)$$

But for a general displacement of the system we have

$$\delta\hat{W} = \hat{P}_x\delta x + \hat{P}_y\delta y + \hat{P}_{\theta_1}\delta\theta_1 + \hat{P}_{\theta_2}\delta\theta_2$$

Therefore, in our case,

$$\hat{P}_x = \hat{P} \qquad \hat{P}_y = 0 \qquad \hat{P}_{\theta_1} = 2a\hat{P} \qquad \hat{P}_{\theta_2} = 0$$

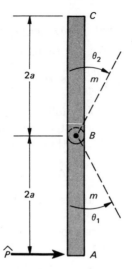

FIGURE 9.4    Impulse applied to one end of a rod that is joined to another rod.

The initial motion of the system is given by Equations (9.22):

$$\Delta\left(\frac{\partial T}{\partial \dot{x}}\right) = \hat{P}_x: \ m(\dot{x} + a\dot{\theta}_1) + m(\dot{x} + a\dot{\theta}_2) = \hat{P}$$

$$\Delta\left(\frac{\partial T}{\partial \dot{\theta}_1}\right) = \hat{P}_\theta: \ ma(\dot{x} + a\dot{\theta}_1) + I_{cm}\dot{\theta}_1 = 2a\hat{P}$$

$$\Delta\left(\frac{\partial T}{\partial \dot{\theta}_2}\right) = \hat{P}_{\theta_2}: ma(\dot{x} + a\dot{\theta}_2) + I_{cm}\dot{\theta}_2 = 0$$

$$\Delta\left(\frac{\partial T}{\partial \dot{y}_2}\right) = \hat{P}_y: \ m\dot{y} = 0$$

Putting $I_{cm} = \frac{1}{3}ma^2$ and solving for the velocities, we finally obtain

$$\dot{x} = -\frac{\hat{P}}{m} \qquad \dot{y} = 0$$

$$\dot{\theta}_1 = \frac{9}{4}\frac{\hat{P}}{am} \qquad \dot{\theta}_2 = \frac{3}{4}\frac{\hat{P}}{am}$$

The reader should verify that the above result gives $\mathbf{v}_{cm} = \hat{\mathbf{P}}/m$, where $\mathbf{v}_{cm}$ is the velocity of the center of mass of the system.

### 9.7.  Hamilton's Variational Principle

Thus far, our study of mechanics has been based largely on the Newtonian laws of motion.  In fact, in the first part of this chapter when we derived Lagrange's equations we used Newton's second law in one of the steps:

Equation (9.8).   In this section we shall investigate an alternative way of deriving Lagrange's equations.   This method is based on a postulate that has proved to be most far-reaching in its consequences—Hamilton's variational principle.

This principle, first announced in 1834 by the Scottish mathematician Sir William R. Hamilton, states that the motion of any system takes place in such a way that the integral

$$\int_{t_1}^{t_2} L \, dt$$

always assumes an extreme value, where $L = T - V$ is the Lagrangian function of the system.   Stated in other words, Hamilton's principle declares that out of all possible ways a system can change in a given finite time interval $t_2 - t_1$, that particular motion which will occur is the one for which the above integral is either a maximum or a minimum.   The statement can be expressed in mathematical form as

$$\delta \int_{t_1}^{t_2} L \, dt = 0 \qquad (9.25)$$

in which $\delta$ denotes a small variation.   This variation results from taking different paths of integration by varying the generalized coordinates and generalized velocities as functions of $t$, Figure 9.5.

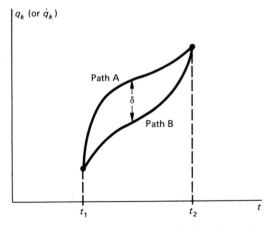

FIGURE 9.5   Illustrating the variation of $q_k$ or $\dot{q}_k$.

To show that the above equation leads directly to Lagrange's equations of motion, let us compute the variation explicitly, assuming that $L$ is a known function of the generalized coordinates $q_k$ and their time derivatives $\dot{q}_k$.   We have

$$\delta \int_{t_1}^{t_2} L \, dt = \int_{t_1}^{t_2} \delta L \, dt = \int_{t_1}^{t_2} \sum_k \left( \frac{\partial L}{\partial q_k} \delta q_k + \frac{\partial L}{\partial \dot{q}_k} \delta \dot{q}_k \right) dt = 0$$

Now $\delta q_k$ is equal to the difference between two slightly different functions of the time $t$. Therefore,

$$\delta \dot{q}_k = \frac{d}{dt} \delta q_k$$

Hence, upon integrating the last term in the integrand by parts, we find

$$\int_{t_1}^{t_2} \sum_k \frac{\partial L}{\partial \dot{q}_k} \delta \dot{q}_k \, dt = \left[ \sum_k \frac{\partial L}{\partial \dot{q}_k} \delta q_k \right]_{t_1}^{t_2} - \int_{t_1}^{t_2} \sum_k \frac{d}{dt} \frac{\partial L}{\partial \dot{q}_k} \delta q_k \, dt$$

But, for fixed values of the limits $t_1$ and $t_2$, the variation $\delta q_k = 0$ at $t_1$ and $t_2$, hence the integrated term vanishes. It follows that

$$\delta \int_{t_1}^{t_2} L \, dt = \int_{t_1}^{t_2} \sum_k \left[ \frac{\partial L}{\partial q_k} - \frac{d}{dt} \frac{\partial L}{\partial \dot{q}_k} \right] \delta q_k \, dt = 0 \qquad (9.26)$$

Now if the generalized coordinates $q_k$ are all independent, then their variations $\delta q_k$ are also independent. Therefore each term in brackets in the integrand must vanish in order that the integral itself vanish. Thus

$$\frac{\partial L}{\partial q_k} - \frac{d}{dt} \frac{\partial L}{\partial \dot{q}_k} = 0 \qquad (k = 1, 2, \ldots, n)$$

These are precisely Lagrange's equations of motion that we found earlier.

In the above derivation it has been assumed that a potential function exists, that is, that the system under consideration is a conservative one. The method of variations can be made to include nonconservative systems by replacing $L$ in the variational integral by the quantity $T + W$ in which $W$ is the work done by *all* forces, whether conservative or nonconservative. The generalized force $Q_k$ is then introduced as defined earlier, Equation (9.2), and the same procedure as above leads to the general form of Lagrange's equations, Equation (9.9).

### 9.8. The Hamiltonian Function. Hamilton's Equations

Consider the following function of the generalized coordinates:

$$H = \sum_k \dot{q}_k p_k - L$$

For simple dynamic systems the kinetic energy $T$ is a homogeneous quadratic function of the $q$'s, and the potential energy $V$ is a function of the $q$'s alone, so that

$$L = T(q_k, \dot{q}_k) - V(q_k)$$

Now, from Euler's theorem for homogeneous functions,[2] we have

$$\sum_k \dot{q}_k p_k = \sum_k \dot{q}_k \frac{\partial L}{\partial \dot{q}_k} = \sum_k \dot{q}_k \frac{\partial T}{\partial \dot{q}_k} = 2T$$

Therefore

$$H = \sum_k \dot{q}_k p_k - L = 2T - (T - V) = T + V \qquad (9.27)$$

That is, the function $H$ is equal to the total energy for the type of systems we are considering.

Suppose we regard the $n$ equations

$$p_k = \frac{\partial L}{\partial \dot{q}_k} \qquad (k = 1, 2, \ldots, n)$$

as solved for the $\dot{q}$'s in terms of the $p$'s and the $q$'s:

$$\dot{q}_k = \dot{q}_k(p_k, q_k)$$

With these equations we can then express $H$ as a function of the $p$'s and the $q$'s:

$$H(p_k, q_k) = \sum_k p_k \dot{q}_k(p_k, q_k) - L \qquad (9.28)$$

Let us calculate the variation of the function $H$ corresponding to a variation $\delta p_k$, $\delta q_k$. We have

$$\delta H = \sum_k \left[ p_k \, \delta \dot{q}_k + \dot{q}_k \, \delta p_k - \frac{\partial L}{\partial \dot{q}_k} \delta \dot{q}_k - \frac{\partial L}{\partial q_k} \delta q_k \right]$$

The first and third terms in the brackets cancel, because $p_k = \partial L / \partial \dot{q}_k$ by definition. Also, since Lagrange's equations can be written as $\dot{p}_k = \partial L / \partial q_k$, we can write

$$\delta H = \sum_k [\dot{q}_k \, \delta p_k - \dot{p}_k \, \delta q_k]$$

Now the variation of $H$ must be given by the equation

$$\delta H = \sum_k \left[ \frac{\partial H}{\partial p_k} \delta p_k + \frac{\partial H}{\partial q_k} \delta q_k \right]$$

It follows that

[2] Euler's theorem states that for a homogeneous function $f$ of degree $n$ in the variables $x_1, x_2, \ldots, x_r$

$$x_1 \frac{\partial f}{\partial x_1} + x_2 \frac{\partial f}{\partial x_2} + \cdots + x_r \frac{\partial f}{\partial x_r} = nf$$

$$\frac{\partial H}{\partial p_k} = \dot{q}_k$$

$$\frac{\partial H}{\partial q_k} = -\dot{p}_k \tag{9.29}$$

These are known as *Hamilton's canonical equations of motion*. They consist of $2n$ first-order differential equations, whereas Lagrange's equations consist of $n$ second-order equations. We have derived Hamilton's equations for simple conservative systems. It can be shown that Equations (9.29) also hold for more general systems, for example, nonconservative systems, systems in which the potential-energy function involves the $\dot{q}$'s, and for systems in which $L$ involves the time explicitly, but in these cases the total energy is no longer necessarily equal to $H$.

Hamilton's equations will be encountered by the student when he studies quantum mechanics (the fundamental theory of atomic phenomena). Hamilton's equations also find application in celestial mechanics.

## EXAMPLES

1. Obtain Hamilton's equations of motion for a one-dimensional harmonic oscillator. We have

$$T = \tfrac{1}{2}m\dot{x}^2 \qquad\qquad V = \tfrac{1}{2}kx^2$$

$$p = \frac{\partial T}{\partial \dot{x}} = m\dot{x} \qquad\qquad \dot{x} = \frac{p}{m}$$

Hence

$$H = T + V = \frac{1}{2m}\,p^2 + \frac{k}{2}\,x^2$$

The equations of motion

$$\frac{\partial H}{\partial p} = \dot{x} \qquad\qquad \frac{\partial H}{\partial x} = -\dot{p}$$

then read

$$\frac{p}{m} = \dot{x} \qquad\qquad kx = -\dot{p}$$

The first equation merely amounts to a restatement of the momentum-velocity relationship in this case. Using the first equation, the second can be written

$$kx = -\frac{d}{dt}\,(m\dot{x})$$

or, upon rearranging terms,

$$m\ddot{x} + kx = 0$$

which is the familiar equation of the harmonic oscillator.

2.   Find the Hamiltonian equations of motion for a particle in a central field.   Here we have

$$T = \frac{m}{2}(\dot{r}^2 + r^2\dot{\theta}^2)$$
$$V = V(r)$$

in polar coordinates.   Hence

$$p_r = \frac{\partial T}{\partial \dot{r}} = m\dot{r} \qquad \dot{r} = \frac{p_r}{m}$$

$$p_\theta = \frac{\partial T}{\partial \dot{\theta}} = mr^2\dot{\theta} \qquad \dot{\theta} = \frac{p_\theta}{mr^2}$$

Consequently

$$H = \frac{1}{2m}\left(p_r{}^2 + \frac{p_\theta{}^2}{r^2}\right) + V(r)$$

The Hamiltonian equations

$$\frac{\partial H}{\partial p_r} = \dot{r} \qquad \frac{\partial H}{\partial r} = -\dot{p}_r \qquad \frac{\partial H}{\partial p_\theta} = \dot{\theta} \qquad \frac{\partial H}{\partial \theta} = -\dot{p}_\theta$$

then read

$$\frac{p_r}{m} = \dot{r}$$

$$\frac{\partial V(r)}{\partial r} - \frac{p_\theta{}^2}{mr^3} = -\dot{p}_r$$

$$\frac{p_\theta}{mr^2} = \dot{\theta}$$

$$0 = -\dot{p}_\theta$$

The last two equations yield the constancy of angular momentum:

$$p_\theta = \text{constant} = mr^2\dot{\theta} = h$$

from which the first two give

$$m\ddot{r} = \dot{p}_r = \frac{h^2}{mr^3} + F_r$$

for the radial equation of motion, where $F_r = -\partial V(r)/\partial r$.

## 9.9.   Lagrange's Equations of Motion with Constraints

It is sometimes convenient to express the differential equations of motion of a constrained system in terms of more coordinates than are actually needed. The differential equations must then also be compatible with the equation, or equations, of the constraint which may be in the form of conditional equations of the type

$$g(q_1, q_2, \ldots q_n) = 0 \tag{9.30}$$

By differentiating, we have the differential form of the condition of constraint

$$\sum_k \frac{\partial g}{\partial q_k} \, \delta q_k = 0 \tag{9.31}$$

There are also certain types of constraints for which a differential relation of the type

$$\sum_k h_k \, \delta q_k = 0 \tag{9.32}$$

can be found but these equations cannot be integrated to give a conditional equation of the type $f(q_1, q_2, \ldots, q_n) = 0$. Such constraints are said to be *nonholonomic*, whereas if the constraint is of the form of Equation (9.30), it is called *holonomic*.

In any case, whether the constraints are holonomic or nonholonomic, it is possible to obtain the differential equations of motion in Lagrangian form by employing the method of *undetermined multipliers*. In this application it is convenient to use the Hamiltonian variational principle.

Let us multiply the differential equation of the constraint, Equation (9.32), by a parameter $\lambda$. This is the undetermined multiplier whose value is, as yet, unspecified. If the resulting expression is added to the integrand of the variational integral of Equation (9.26), there is clearly no change in the result that the integral vanishes:

$$\int_{t_1}^{t_2} \sum_k \left( \frac{\partial L}{\partial q_k} - \frac{d}{dt} \frac{\partial L}{\partial \dot{q}_k} + \lambda h_k \, \delta q_k \right) dt = 0$$

Because of the constraint, only $n - 1$ of the $n$ quantities $\delta q_k$ can be regarded as independent. We now choose $\lambda$ to have a value such that one of the bracketed terms in the summation vanishes, say the first. Then the remaining $n - 1$ can be regarded as independent. Consequently, the remaining terms in brackets must also vanish. Thus we can write

$$\frac{\partial L}{\partial q_k} - \frac{d}{dt} \frac{\partial L}{\partial \dot{q}_k} + \lambda h_k = 0 \qquad (k = 1, 2, \ldots, n) \tag{9.33}$$

$$\sum_k h_k \dot{q}_k = 0 \tag{9.34}$$

The last equation is obtained by dividing the differential equation of the constraint, Equation (9.32), by $\delta t$. There are now $n + 1$ differential equations in all. Hence the $n + 1$ quantities $q_1, q_2, \ldots, q_n, \lambda$ can be found.

The method can be extended to include more than just one equation of constraint by adding on more undetermined multipliers with their corresponding $h$'s to the Lagrangian equations. It can be shown that the equations of motion as given above also apply when there are moving constraints. For a more complete treatment of this method the reader should consult an advanced treatise.[3]

## DRILL EXERCISES

Lagrange's method should be used in all of the following, unless stated otherwise.

9.1 Find the differential equations of motion of a projectile in a uniform gravitational field without air resistance.

9.2 Find the acceleration of a solid uniform sphere rolling down a perfectly rough fixed inclined plane. Compare with the result derived earlier in Section 7.8.

9.3 Two blocks of equal mass $m$ are connected by a flexible cord. One block is placed on a smooth horizontal table, the other block hangs over the edge. Find the acceleration of the system assuming (a) the mass of the cord is negligible, and (b) the cord is heavy, of mass $m'$.

9.4 Find the general differential equations of motion for a particle in cylindrical coordinates: $R$, $\varphi$, $z$. Use the relation

$$v^2 = v_R{}^2 + v_\varphi{}^2 + v_z{}^2$$
$$= \dot{R}^2 + R^2\dot{\varphi}^2 + \dot{z}^2$$

9.5 Find the general differential equations of motion for a particle in spherical coordinates: $r$, $\theta$, $\varphi$. Use the relation

$$v^2 = v_r{}^2 + v_\theta{}^2 + v_\varphi{}^2$$
$$= \dot{r}^2 + r^2\dot{\theta}^2 + r^2 \sin^2 \theta \dot{\varphi}^2$$

[*Note:* Compare your results with the result derived in Chapter 1, Section 1.25.]

## PROBLEMS

9.6 Set up the equations of motion of a "double-double" Atwood machine consisting of one Atwood machine (with masses $m_1$ and $m_2$) connected by means of a light cord passing over a pulley to a second Atwood machine with masses $m_3$ and $m_4$. Neglect the masses of all pulleys. Find the actual accelerations for the case $m_1 = m$, $m_2 = 4m$, $m_3 = 2m$, and $m_4 = 3m$.

9.7 A ball of mass $m$ rolls down a movable wedge of mass $M$. The angle of the wedge is $\theta$, and it is free to slide on a smooth horizontal surface. The contact between the ball and the wedge is perfectly rough. Find the acceleration of the wedge.

9.8 A particle slides on a smooth inclined plane whose inclination $\theta$ is increasing at a constant rate $\omega$. If $\theta = 0$ at time $t = 0$, at which time the particle starts from rest, find the subsequent motion of the particle.

9.9   Show that Lagrange's method automatically yields the correct equations of motion for a particle moving in a plane in a rotating coordinate system $Oxy$. [*Hint:* $T = \frac{1}{2}m\mathbf{v}\cdot\mathbf{v}$, where $\mathbf{v} = \mathbf{i}(\dot{x} - \omega y) + \mathbf{j}(\dot{y} + \omega x)$, and $F_x = -\partial V/\partial x$, $F_y = -\partial V/\partial y$.]

9.10   Repeat the above problem for motion in three dimensions.

9.11   Find the differential equations of motion for an "elastic pendulum": a particle of mass $m$ attached to an elastic string of stiffness $k$ and unstretched length $l_0$. Assume that the motion takes place in a vertical plane.

9.12   The point of support of a simple pendulum is being elevated at a constant acceleration $a$, so that the height of the support is $\frac{1}{2}at^2$, and its vertical velocity is $at$. Find the differential equation of motion for small oscillations of the pendulum by Lagrange's method. Show that the period of the pendulum is $2\pi[l/(g + a)]^{1/2}$ where $l$ is the length of the pendulum.

9.13   The point of support of a simple pendulum is moved in a horizontal direction with constant acceleration $a$. Find the equation of motion and the period for small oscillations.

9.14   Use Lagrange's method to find the differential equations of motion for the spherical pendulum in spherical coordinates.

9.15   Find the differential equations of motion for an elastic spherical pendulum, as in Problem 9.11.

9.16   Find the differential equations of motion for a particle constrained to move on a smooth right-circular cone, the axis of the cone being vertical.

9.17   In the above problem, show that the particle, given an initial motion, will oscillate between two horizontal circles on the cone. [*Hint:* Use spherical coordinates with $\theta = constant$.] Show that $\dot{r}^2 = f(r)$ where $f(r) = 0$ has two roots that define the limits between which the particle must remain.

9.18   Two identical rods $AB$ and $BC$, each of mass $m$ and length $2a$, are joined smoothly at $B$. The rods lie at rest on a smooth horizontal table and are initially at right angles to each other. An impulse $P$ is applied at $A$ lengthwise to the rod $AB$. Find the motion of the system immediately after the application of the impulse.

9.19   Write down the Hamiltonian function and Hamilton's canonical equations for the following:

    (a) A simple Atwood machine
    (b) A simple pendulum
    (c) A projectile in a uniform gravitational field
    (d) A spherical pendulum

9.20   Show that the Lagrangian function

$$L = \tfrac{1}{2}mv^2 - q\varphi + q\mathbf{v}\cdot\mathbf{A}$$

yields the correct equation of motion for a particle in an electromagnetic field, namely,

$$m\ddot{\mathbf{r}} = q(\mathbf{E} + \mathbf{v} \times \mathbf{B})$$

where

$$\mathbf{E} = -\nabla\varphi \qquad \text{and} \qquad \mathbf{B} = \nabla \times \mathbf{A}$$

(The vector quantity $\mathbf{A}$ is called the *vector potential*, and the scalar quantity $\varphi$ is called the *scalar potential*.)

9.21 Find (a) the generalized momenta, and (b) the Hamiltonian function $H$ for the Lagrangian function given in Problem 9.20.

# 10. Dynamics of Oscillating Systems

Simple cases of systems that can undergo oscillations about a configuration of equilibrium include a simple pendulum, a particle suspended on an elastic spring, a physical pendulum, and so on, all being cases of one degree of freedom characterized by a single frequency of oscillation. When we consider more complicated systems—systems with several degrees of freedom—we shall find that not one but several different frequencies of oscillation are possible. In our analysis of oscillating systems, we shall find it very convenient to use generalized coordinates and to employ Lagrange's method for finding the equations of motion in terms of those coordinates.

## 10.1. Potential Energy and Equilibrium. Stability

Before we take up the study of the motion of a system about an equilibrium configuration, let us examine briefly the equilibrium itself. Consider a system with $n$ degrees of freedom, and let the generalized coordinates $q_1$, $q_2$, . . . , $q_n$ specify the configuration. We shall assume that the system is conservative and that the potential energy $V$ is a function of the $q$'s alone:

$$V = V(q_1, q_2, \ . \ . \ . \ , q_n)$$

Now we have shown that the generalized forces $Q_k$ are given by

$$Q_k = -\frac{\partial V}{\partial q_k} \qquad (k = 1, 2, \ . \ . \ . \ , n) \tag{10.1}$$

An equilibrium configuration is defined as a configuration for which all of the generalized forces vanish,

$$Q_k = -\frac{\partial V}{\partial q_k} = 0 \tag{10.2}$$

These equations constitute a necessary condition for the system to remain at rest if, initially, it is at rest.   If the system is given a small displacement, however, it may or may not return to equilibrium.   If a system always tends to return to equilibrium, given a sufficiently small displacement, the equilibrium is *stable;* otherwise, the equilibrium is *unstable.*   (If the system has no tendency to move either toward or away from equilibrium, the equilibrium is said to be *neutral.*)

A ball placed (1) at the bottom of a spherical bowl, (2) on top of a spherical cap, and (3) on a plane horizontal surface are examples of stable, unstable, and neutral equilibrium, respectively.

Let us see how the potential-energy function $V$ enters the picture.   Consider the motion of a system subsequent to the application of a small impulse that leaves the system in motion while it is at an equilibrium configuration. Since the total energy is constant, we can write

$$T + V = T_0 + V_0$$

or

$$T - T_0 = -(V - V_0) \tag{10.3}$$

where $T_0$ is the kinetic energy the system has at the equilibrium configuration (as a result of the impulse), and $V_0$ is the potential energy at the equilibrium configuration.   Now if the potential energy is maximum at equilibrium, then $V - V_0$ is negative, and, consequently $T - T_0$ is positive; that is, $T$ increases as the system moves away from equilibrium.   This is clearly an unstable situation.   On the other hand, if the equilibrium configuration is one of minimum potential energy, then $V - V_0$ is positive, and $T - T_0$ is negative; that is, $T$ decreases.   But $T$ can never be negative, consequently $T$ decreases to zero at some limiting configuration close to equilibrium, provided, of course, that $T_0$ is small enough.   The equilibrium is stable in this case.   Thus the criterion for stable equilibrium is that the potential energy is a *minimum.*

For a system with one degree of freedom, we have

$$V = V(q) \tag{10.4}$$

and, at equilibrium

$$\frac{dV}{dq} = 0 \tag{10.5}$$

The stability is then expressed as follows:

$$\frac{d^2V}{dq^2} > 0 \text{ (stable)} \tag{10.6}$$

$$\frac{d^2V}{dq^2} < 0 \text{ (unstable)} \tag{10.7}$$

If $d^2V/dq^2 = 0$, we must examine the higher-order derivatives.   (This is discussed in the next section.)   In Figure 10.1 is shown a graph of a hypothetical potential function.   The point $A$ corresponds to a position of stable equilibrium, and points $B$ and $C$ correspond to positions of unstable equilibrium.

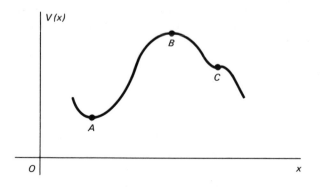

FIGURE 10.1   Potential energy function $V(x)$.   The point $A$ is one of stable
equilibrium.   Points $B$ and $C$ are unstable.

## EXAMPLE

Let us examine the equilibrium of a body having a rounded (spherical or cylindrical) base which is balanced on a plane horizontal surface.   Let $a$ be the radius of curvature of the base, and let the center of mass $CM$ be a distance $b$ from the initial point of contact, as shown in Figure 10.2(a).   In Fig-

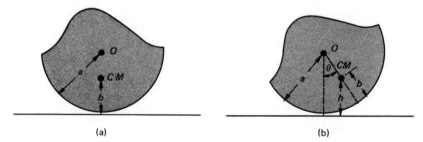

(a)                                                       (b)

FIGURE 10.2   Coordinates for analyzing the stability of equilibrium of a
round-bottomed object.

ure 10.2(b) the body is shown in a displaced position, where $\theta$ is the angle between the vertical and the line $OCM$ ($O$ being the center of curvature), as shown.   Let $h$ denote the distance from the plane to the center of mass. Then the potential energy is given by

$$V = mgh = mg[a - (a - b) \cos \theta]$$

where $m$ is the mass of the body.   We have

$$\frac{dV}{d\theta} = mg(a - b) \sin \theta$$

so

$$\frac{dV}{d\theta} = 0 \qquad \text{for} \qquad \theta = 0$$

Thus $\theta = 0$ is a position of equilibrium.   Furthermore

$$\frac{d^2V}{d\theta^2} = mg(a - b) \cos \theta$$

and

$$\frac{d^2V}{d\theta^2} = mg(a - b) \qquad \text{for} \qquad \theta = 0$$

Hence the equilibrium is stable if $a > b$, that is, if the center of mass lies below the center of curvature.

## 10.2.   Expansion of the Potential-Energy Function in a Power Series

Let us consider first a system having one degree of freedom.   Suppose we expand the potential-energy function $V(q)$ as a power series about the point $q = a$.   We have

$$V(q) = \kappa_0 + \kappa_1(q - a) + \frac{1}{2!} \kappa_2(q - a)^2 + \cdots + \frac{1}{n!} \kappa_n(q - a)^n + \cdots$$

where

$$\kappa_n = \left( \frac{d^n V}{dq^n} \right)_{q=a}$$

Now if the point $q = a$ is a position of equilibrium, then $\kappa_1 = (dV/dq)_{q=a} = 0$. This eliminates the linear term in the expansion, so that

$$V(q) = \kappa_0 + \frac{1}{2!} \kappa_2(q - a)^2 + \cdots \tag{10.8}$$

The stability of the equilibrium at the point $q = a$ depends on the first nonvanishing term, after the first one $\kappa_0$, in the above expansion.   If this is a term of even power $n$, then the equilibrium is stable if the derivative $d^n V/dq^n$ is positive.   If the derivative is negative, or if $n$ is odd, the equilibrium is unstable.   To see why this is so, let $n$ denote the order of the first nonvanishing term.   Then for small departures from the equilibrium point we have

$$F = -\frac{\partial V}{\partial q} \approx -\kappa_n (q - a)^{n-1}$$

Now for stable equilibrium the direction of $F$ must be toward $a$, that is, negative if $q > a$ and positive if $q < a$. This can be the case only if $\kappa_n$ is positive and $n$ is even.

In most cases of physical importance $n = 2$, that is, the potential energy is a quadratic function of the displacement and the force is a linear function. Thus if we transform the origin to the point $q = a$, and arbitrarily set $V(0) = 0$, then we can write

$$V(q) = \tfrac{1}{2}\kappa_2 q^2 \tag{10.9}$$

if we neglect higher powers of $q$.

Similarly, for the case of a system with several degrees of freedom, we can effect a linear transformation so that $q_1 = q_2 = \cdots = q_n = 0$ is a configuration of equilibrium, if an equilibrium configuration exists. The potential-energy function can then be expanded in the form

$$V(q_1,q_2, \ldots ,q_n) = \tfrac{1}{2}(\kappa_{11}q_1{}^2 + 2\kappa_{12}q_1q_2 + \kappa_{22}q_2{}^2 + \cdots) \tag{10.10}$$

where

$$\kappa_{11} = \left(\frac{\partial^2 V}{\partial q_1{}^2}\right)_{q_1=q_2=\cdots=q_n=0}$$

$$\kappa_{12} = \left(\frac{\partial^2 V}{\partial q_1 \partial q_2}\right)_{q_1=q_2=\cdots=q_n=0}$$

and so on. We have arbitrarily set $V(0,0, \ldots ,0) = 0$. The linear terms in the expansion are absent because the expansion is about an equilibrium configuration.

The expression in parentheses in Equation (10.10) is known as a *quadratic form*. If this quadratic form is positive definite,[1] that is, either zero or positive for all values of the $q$'s then the equilibrium configuration $q_1 = q_2 = \cdots = q_n = 0$ is stable.

## 10.3. Oscillations of a System with One Degree of Freedom

If a system has one degree of freedom, the kinetic energy $T$ may be expressed as

$$T = \tfrac{1}{2}\mu\dot{q}^2 \tag{10.11}$$

---

[1] The necessary and sufficient conditions that the quadratic form in Equation (10.10) be positive definite are

$$\kappa_{11} > 0 \qquad \begin{vmatrix} \kappa_{11} & \kappa_{12} \\ \kappa_{21} & \kappa_{22} \end{vmatrix} > 0 \qquad \begin{vmatrix} \kappa_{11} & \kappa_{12} & \kappa_{13} \\ \kappa_{21} & \kappa_{22} & \kappa_{23} \\ \kappa_{31} & \kappa_{32} & \kappa_{33} \end{vmatrix} > 0 \quad \text{and so on}$$

Here the coefficient $\mu$ may be a constant, or it may be a function of the generalized coordinate $q$. In any case we can expand $\mu$ as a power series in $q$ and write

$$\mu = \mu(0) + \left(\frac{d\mu}{dq}\right)_{q=0} q + \cdots \tag{10.12}$$

If $q = 0$ is a position of equilibrium, we shall consider $q$ as small enough so that

$$\mu = \mu(0) = \text{constant} \tag{10.13}$$

is a valid approximation. From Equation (10.9) we see that the Lagrangian function $L$ can be expressed as follows:

$$L = T - V = \tfrac{1}{2}\mu\dot{q}^2 - \tfrac{1}{2}\kappa q^2 \tag{10.14}$$

where $\kappa = \kappa_2 = (d^2V/dq^2)_{q=0}$. Lagrange's equation of motion

$$\frac{d}{dt}\frac{\partial L}{\partial \dot{q}} = \frac{\partial L}{\partial q}$$

is then

$$\mu\ddot{q} + \kappa q = 0 \tag{10.15}$$

Thus if $q = 0$ is a position of stable equilibrium, that is, if $\kappa > 0$, then $q$ oscillates harmonically about the equilibrium position with angular frequency

$$\omega = \sqrt{\frac{\kappa}{\mu}} \tag{10.16}$$

and

$$q = q_0 \cos{(\omega t + \epsilon)} \tag{10.17}$$

where $q_0$ is the amplitude of the oscillation, and $\epsilon$ is a phase angle. The values of these constants of integration are determined, of course, from the initial conditions.


## EXAMPLE

Consider the motion of the round-bottomed object discussed in the example of the preceding section (Figure 10.2). If the contact is perfectly rough, we have pure rolling, and the speed of the center of mass is approximately $b\dot{\theta}$ for small $\theta$. The kinetic energy $T$ is accordingly given by

$$T = \tfrac{1}{2}m(b\dot{\theta})^2 + \tfrac{1}{2}I_{cm}\dot{\theta}^2$$

where $I_{cm}$ is the moment of inertia about the center of mass. Also, we can express the potential-energy function $V$ as follows:

$$V(\theta) = mg[a - (a - b)\cos\theta]$$

$$= mg\left[a - (a - b)\left(1 - \frac{\theta^2}{2!} + \frac{\theta^4}{4!} - \cdots\right)\right]$$

$$= \tfrac{1}{2}mg(a - b)\theta^2 + \text{constant} + \text{higher terms}$$

We can then write

$$L = \tfrac{1}{2}(mb^2 + I_{cm})\dot{\theta}^2 - \tfrac{1}{2}mg(a - b)\theta^2$$

neglecting constants and higher terms. Comparing with Equations (10.14) and (10.15), we see that

$$\mu = mb^2 + I_{cm}$$
$$\kappa = mg(a - b)$$

The motion about the equilibrium position $\theta = 0$ is therefore approximately simple harmonic with angular frequency

$$\omega = \sqrt{\frac{mg(a - b)}{mb^2 + I_{cm}}} \tag{10.18}$$

### 10.4.  Coupled Harmonic Oscillators

Prior to developing the general theory of oscillating systems with any number of degrees of freedom, we shall study a simple specific example, namely, a system consisting of two harmonic oscillators that are coupled together.

For definiteness we use a model comprised of particles attached to elastic springs, although any type of oscillator could be used. For simplicity we assume that the oscillators are identical and are restricted to move in a straight line, Figure 10.3. The coupling is represented by a spring of stiffness $k'$ as shown. The system has two degrees of freedom. We shall choose

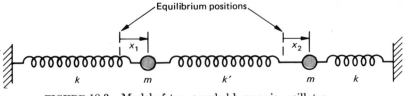

FIGURE 10.3   Model of two coupled harmonic oscillators.

coordinates $x_1$ and $x_2$, the displacements of the particles from their respective equilibrium positions, to represent the configuration of the system.

The kinetic energy of the system is

$$T = \tfrac{1}{2}m\dot{x}_1^2 + \tfrac{1}{2}m\dot{x}_2^2 \tag{10.19}$$

and the potential energy is

$$V = \tfrac{1}{2}kx_1^2 + \tfrac{1}{2}k'(x_2 - x_1)^2 + \tfrac{1}{2}kx_2^2 \tag{10.20}$$

Hence the Lagrangian function $L$ is given by

$$L = \tfrac{1}{2}m\dot{x}_1^2 + \tfrac{1}{2}m\dot{x}_2^2 - \tfrac{1}{2}kx_1^2 - \tfrac{1}{2}k'(x_2 - x_1)^2 - \tfrac{1}{2}kx_2^2 \tag{10.21}$$

The differential equations of motion

$$\frac{d}{dt}\frac{\partial L}{\partial \dot{x}_1} = \frac{\partial L}{\partial x_1} \qquad \frac{d}{dt}\frac{\partial L}{\partial \dot{x}_2} = \frac{\partial L}{\partial x_2}$$

then read

$$\begin{aligned} m\ddot{x}_1 &= -kx_1 + k'(x_2 - x_1) \\ m\ddot{x}_2 &= -kx_2 - k'(x_2 - x_1) \end{aligned} \qquad (10.22)$$

Now if there were no coupling ($k' = 0$), the two equations would be separated and would represent the differential equations of two independent harmonic oscillators of frequency $\sqrt{k/m}$. In the actual system under study it is reasonable to use a trial solution which is also harmonic but of a different and, as yet, unknown frequency $\omega$. Thus our trial solution is of the form $e^{i\omega t}$. Taking the second time derivative, we have

$$\begin{aligned} \ddot{x}_1 &= -\omega^2 x_1 \\ \ddot{x}_2 &= -\omega^2 x_2 \end{aligned} \qquad (10.23)$$

The equations of motion (10.22) then become

$$\frac{k + k'}{m}x_1 - \frac{k'}{m}x_2 = \omega^2 x_1$$

$$\frac{-k'}{m}x_1 + \frac{k + k'}{m}x_2 = \omega^2 x_2 \qquad (10.24)$$

after substitution and rearrangement of terms.

At this point it would be possible simply to eliminate $x_1$ and $x_2$ algebraically between the two equations and thereby obtain an equation giving the unknown frequency $\omega$ in terms of the given parameters $k$, $k'$, and $m$. However, it is more instructive to introduce here the use of matrices to handle problems in coupled harmonic motion. Thus Equations (10.24) are written in matrix notation as

$$\begin{bmatrix} a & -b \\ -b & a \end{bmatrix} \begin{bmatrix} x_1 \\ x_2 \end{bmatrix} = \omega^2 \begin{bmatrix} x_1 \\ x_2 \end{bmatrix} \qquad (10.25)$$

in which we have introduced the abbreviations $a = (k + k')/m$ and $b = k'/m$. This is the standard form of what is known in mathematics as the *eigenvalue* problem: The product of a square matrix by a column vector is equal to a constant times that vector. (We have previously met a similar situation in Chapter 8 when we discussed the problem of finding the principal axes of a rigid body.)

To find the eigenvalues of the constant $\omega^2$ we rewrite the matrix equation (10.25) in the following equivalent form

$$\begin{bmatrix} a - \omega^2 & -b \\ -b & a - \omega^2 \end{bmatrix} \begin{bmatrix} x_1 \\ x_2 \end{bmatrix} = 0 \qquad (10.26)$$

In order that a nontrivial solution exists, namely, one in which $x_1$ and $x_2$ are not both zero, the determinant of the matrix must vanish. That is

$$\begin{vmatrix} a - \omega^2 & -b \\ -b & a - \omega^2 \end{vmatrix} = 0 \tag{10.27}$$

This is known as the *secular* equation. It is an algebraic equation of the $n$th degree in the unknown $\omega^2$, $n$ being the order of the matrix. In our problem it is a quadratic $(a - \omega^2)^2 - b^2 = 0$ whose roots are easily seen to be $\omega^2 = a \pm b$, or

$$\omega_1 = \sqrt{\frac{k}{m}} \qquad \omega_2 = \sqrt{\frac{k + 2k'}{m}} \tag{10.28}$$

These are called the *eigenfrequencies* or *normal* frequencies of the system. Clearly there are two fundamental solutions:

(a) $x_1 = A_1 \cos (\omega_1 t + \varphi_1)$ $\qquad$ $x_2 = A_2 \cos (\omega_1 t + \varphi_2)$
(b) $x_1 = B_1 \cos (\omega_2 t + \varphi_1')$ $\qquad$ $x_2 = B_2 \cos (\omega_2 t + \varphi_2')$

Here the $A$'s, $B$'s, and $\varphi$'s are constants of integration to be determined from the initial conditions. Actually, the amplitudes are not independent, for if we insert the eigenvalues of $\omega$ back into the equations of motion (10.24) we find the following results:

(a) If $\omega = \omega_1 = (k/m)^{1/2}$, then we find that

$$\frac{k + k'}{m} x_1 - \frac{k'}{m} x_2 = \frac{k}{m} x_1$$

which reduces to

$$x_1 = x_2$$

(b) If $\omega = \omega_2 = (k + 2k'/m)^{1/2}$, we obtain

$$\frac{k + k'}{m} x_1 - \frac{k'}{m} = \frac{k + 2k'}{m} x_1$$

which gives

$$x_1 = -x_2$$

The two types of motion represented by the above cases are called the *normal modes* of the system. The first case $x_1 = x_2$ is known as the *symmetric mode*, and the second $x_1 = -x_2$ is called the *antisymmetric mode*. See Figure 10.4. In the language of matrices, the modes correspond to the *eigenvectors* of the matrix in Equation (10.25). In our example they can be expressed as

(a) $\qquad \begin{bmatrix} x_1 \\ x_2 \end{bmatrix} = \begin{bmatrix} x_1 \\ x_1 \end{bmatrix} = \begin{bmatrix} 1 \\ 1 \end{bmatrix} A \cos (\omega_1 t + \varphi)$

(b) $\qquad \begin{bmatrix} x_1 \\ x_2 \end{bmatrix} = \begin{bmatrix} x_1 \\ -x_1 \end{bmatrix} = \begin{bmatrix} 1 \\ -1 \end{bmatrix} B \cos(\omega_2 t + \varphi')$

*The Complete Solution*

Since the original differential equations of motion of the system are linear, we know that any two solutions may be added together to yield another solution. This fact allows us to combine the normal-mode solutions to obtain the most general form of the solution. This is conveniently written in terms of the eigenvectors as follows:

$$\begin{bmatrix} x_1 \\ x_2 \end{bmatrix} = \begin{bmatrix} 1 \\ 1 \end{bmatrix} A \cos(\omega_1 t + \varphi) + \begin{bmatrix} 1 \\ -1 \end{bmatrix} B \cos(\omega_2 t + \varphi') \qquad (10.29)$$

The amplitudes $A$ and $B$ and the phases $\varphi$ and $\varphi'$ are determined from the initial conditions of the system. If the initial condition is such that one of the amplitudes, say $B$, is zero, then it will remain zero, and the system will oscillate permanently in the symmetric mode. Likewise, if $A$ is initially zero, the system remains in the antisymmetric mode. In the general case, the system oscillates as a combination of the two modes.

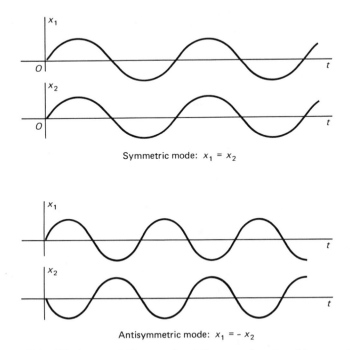

Symmetric mode: $x_1 = x_2$

Antisymmetric mode: $x_1 = -x_2$

FIGURE 10.4   Displacement-time graphs of the normal modes of two coupled harmonic oscillators.

### 10.5.  General Theory of Vibrating Systems

Turning now to a general system with $n$ degrees of freedom, we have shown in the last chapter (Section 9.3) that the kinetic energy $T$ is a homogeneous quadratic function of the generalized velocities, namely,

$$T = \tfrac{1}{2}\mu_{11}\dot{q}_1{}^2 + \mu_{12}\dot{q}_1\dot{q}_2 + \tfrac{1}{2}\mu_{22}\dot{q}_2{}^2 + \cdots = \sum_j \sum_k \tfrac{1}{2}\mu_{jk}\dot{q}_j\dot{q}_k \qquad (10.30)$$

provided there are no moving constraints.  Since we are concerned with motion about an equilibrium configuration, we shall assume, as in Section 10.3, Equation (10.13), that the $\mu$'s are constant and equal to their values at the equilibrium configuration.  We shall further assume that a linear transformation has been introduced so that the equilibrium configuration is given by

$$q_1 = q_2 = \cdots = q_n = 0$$

Accordingly, the potential energy $V$, from Equation (10.10), is given by

$$V = \tfrac{1}{2}\kappa_{11}q_1{}^2 + \kappa_{12}q_1q_2 + \tfrac{1}{2}\kappa_{22}q_2{}^2 + \cdots = \sum_j \sum_k \tfrac{1}{2}\kappa_{jk}q_jq_k \qquad (10.31)$$

The Lagrangian function then assumes the form

$$L = \sum_j \sum_k \tfrac{1}{2}(\mu_{jk}\dot{q}_j\dot{q}_k - \kappa_{jk}q_jq_k) \qquad (10.32)$$

and the equations of motion

$$\frac{d}{dt}\frac{\partial L}{\partial \dot{q}_k} = \frac{\partial L}{\partial q_k}$$

then read

$$\sum_j \mu_{jk}\ddot{q}_j = -\sum_j \kappa_{jk}q_j \qquad (k = 1, 2, \ldots, n) \qquad (10.33)$$

The above set of $n$ differential equations is expressed in matrix notation as

$$\mathbf{M}\ddot{\mathbf{q}} = -\mathbf{Kq} \qquad (10.34)$$

in which

$$\mathbf{M} = \begin{bmatrix} \mu_{11} & \mu_{12} & \cdots \\ \mu_{21} & \mu_{22} & \cdots \\ \cdot & \cdot & \cdots \end{bmatrix} = [\mu_{jk}] \qquad (10.35)$$

$$\mathbf{K} = \begin{bmatrix} \kappa_{11} & \kappa_{12} & \cdots \\ \kappa_{21} & \kappa_{22} & \cdots \\ \cdot & \cdot & \cdots \end{bmatrix} = [\kappa_{jk}] \qquad (10.36)$$

The generalized displacement vector is

$$\mathbf{q} = \begin{bmatrix} q_1 \\ q_2 \\ \cdot \\ \cdot \\ \cdot \\ q_n \end{bmatrix} \tag{10.37}$$

We note that both **M** and **K** are symmetric matrices, according to the energy equations (10.30) and (10.31).

If a harmonic solution of the form

$$q_k = A_k \cos (\omega t + \varphi_k)$$

exists, then $\ddot{q}_k = -\omega^2 q_k$ that is

$$\ddot{\mathbf{q}} = -\omega^2 \mathbf{q}$$

Consequently, the matrix equation (10.34) becomes

$$(\mathbf{K} - \mathbf{M}\omega^2)\mathbf{q} = 0 \tag{10.38}$$

A nontrivial solution requires the secular determinant to vanish

$$\det (\mathbf{K} - \mathbf{M}\omega^2) = 0 \tag{10.39}$$

or

$$\left| \kappa_{ij} - \mu_{ij}\omega^2 \right| = 0 \tag{10.40}$$

The roots give the normal frequencies and the associated eigenvectors define the normal modes.

It is possible to express the problem in different ways by means of matrix algebra. For example, we can multiply both sides of Equation (10.34) by the inverse $\mathbf{M}^{-1}$ of the matrix **M**, assuming of course, that the inverse exists. We then have

$$(\mathbf{M}^{-1}\mathbf{K} - \mathbf{I}\omega^2)\mathbf{q} = 0 \tag{10.41}$$

in which **I** is the identity matrix. This is the standard form of the eigenvalue problem. See Appendix V. The associated secular equation is

$$\det (\mathbf{M}^{-1}\mathbf{K} - \mathbf{I}\omega^2) = 0 \tag{10.42}$$

Now it is easy to prove that the eigenvalues in both Equations (10.39) and (10.42) are the same. This follows from the mathematical theorem that the determinant of the product of two matrices is equal to the product of the respective determinants. Thus

$$\det (\mathbf{M}^{-1}) \det (\mathbf{K} - M\omega^2) = \det (\mathbf{M}^{-1}\mathbf{K} - \mathbf{I}\omega^2)$$

so that if $\det (\mathbf{M}^{-1})$ does not vanish, the roots of the two secular equations in question are identical. The importance of the above result is that we can find the normal frequencies and normal modes without reducing the matrix equation (10.38) to the standard form (10.41).

## EXAMPLE

Let us consider the motion of the so-called "double pendulum" consisting of a light inextensible cord of length $2l$ with one end fixed, the other supporting a particle of mass $m$, and with a second particle of mass $m$ at the center, Figure 10.5. Assuming that the system stays in a single plane, we can specify the configuration by two angles, $\theta$ and $\varphi$ as shown. For small oscillations about the equilibrium position, the speeds of the two particles are approximately $l\dot{\theta}$ and $l(\dot{\theta} + \dot{\varphi})$, and their potential energies are $-mgl \cos \theta$ and $-mgl(\cos \theta + \cos \varphi)$. The Lagrangian function for the system is then

$$L = \frac{m}{2} l^2 \dot{\theta}^2 + \frac{m}{2} l^2 (\dot{\theta} + \dot{\varphi})^2 + 2mgl \cos \theta + mgl \cos \varphi$$

Lagrange's equations of motion

$$\frac{d}{dt}\frac{\partial L}{\partial \dot{\theta}} = \frac{\partial L}{\partial \theta} \qquad \frac{d}{dt}\frac{\partial L}{\partial \dot{\varphi}} = \frac{\partial L}{\partial \varphi}$$

then read

$$ml^2\ddot{\theta} + ml^2(\ddot{\theta} + \ddot{\varphi}) = -2mgl \sin \theta$$
$$ml^2(\ddot{\theta} + \ddot{\varphi}) = -mgl \sin \varphi$$

Assuming that $\sin \theta \approx \theta$, $\sin \varphi \approx \varphi$, and rearranging terms, we find

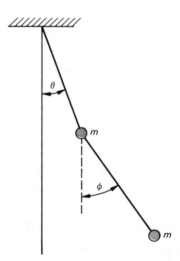

FIGURE 10.5   The double pendulum.

$$2\ddot{\theta} + \frac{2g}{l}\,\theta + \ddot{\varphi} = 0$$

$$\ddot{\theta} + \ddot{\varphi} + \frac{g}{l}\,\varphi = 0 \qquad (10.43)$$

The secular determinant for the system is then

$$\begin{vmatrix} -2\omega^2 + \dfrac{2g}{l} & -\omega^2 \\[2mm] -\omega^2 & -\omega^2 + \dfrac{g}{l} \end{vmatrix} = 0$$

or

$$\omega^4 - 4\omega^2 \left(\frac{g}{l}\right) + 2\left(\frac{g}{l}\right)^2 = 0$$

The two normal frequencies are thus

$$\omega_1 = \left[\frac{g}{l}\,(2 - \sqrt{2})\right]^{1/2}$$

$$\omega_2 = \left[\frac{g}{l}\,(2 + \sqrt{2})\right]^{1/2} \qquad (10.44)$$

If the system is oscillating at either one of its normal frequencies, then the first of Equations (10.43) yields

$$\left(-2\omega^2 + 2\,\frac{g}{l}\right)\theta = \omega^2 \varphi$$

Upon substituting the two values of $\omega$ from Equations (10.44) into the above equation, we find the following relations between $\varphi$ and $\theta$ for the normal modes:

$$\varphi = +\sqrt{2}\,\theta, \quad \omega = \omega_1 \qquad (symmetric\ mode)$$
$$\varphi = -\sqrt{2}\,\theta, \quad \omega = \omega_2 \qquad (antisymmetric\ mode)$$

The corresponding eigenvectors are

$$\begin{bmatrix} \theta \\ \varphi \end{bmatrix} = \begin{bmatrix} 1 \\ \pm 2^{1/2} \end{bmatrix}$$

and the general solution may be written

$$\begin{bmatrix} \theta \\ \varphi \end{bmatrix} = \begin{bmatrix} 1 \\ 2^{1/2} \end{bmatrix} A \cos(\omega_1 t + \alpha_1) + \begin{bmatrix} 1 \\ -2^{1/2} \end{bmatrix} B \cos(\omega_2 t + \alpha_2)$$

## 10.6.   Normal Coordinates

In the language of matrices, the kinetic and potential energy functions defined by Equations (10.30) and (10.31) take the compact forms

$$T = \tfrac{1}{2}\dot{\mathbf{q}}^T\mathbf{M}\dot{\mathbf{q}} \qquad V = \tfrac{1}{2}\mathbf{q}^T\mathbf{K}\mathbf{q} \tag{10.45}$$

Now a considerable simplification results if these quantities are expressed in terms of coordinates in which the matrices $\mathbf{M}$ and $\mathbf{K}$ are both diagonal. In this case the expressions reduce to sums of squares, and consequently the secular determinant is also diagonal. Hence the roots giving the normal frequencies are obtained simply by equating to zero each diagonal term.

It turns out that it is indeed generally possible to find coordinates such that the matrices in question are diagonal. These are known as *normal coordinates*. In order to find the normal coordinates, we consider a transformation defined by a certain matrix $\mathbf{A}$ which transforms between the original displacement vector $\mathbf{q}$ and a new vector $\mathbf{q}'$ according to the usual rule

$$\mathbf{q} = \mathbf{A}\mathbf{q}' \qquad \mathbf{q}' = \mathbf{A}^{-1}\mathbf{q}$$

The corresponding transformed expressions for $T$ and $V$ are

$$T' = \tfrac{1}{2}\dot{\mathbf{q}}'^T\mathbf{A}^T\mathbf{M}\mathbf{A}\dot{\mathbf{q}}' \qquad V' = \tfrac{1}{2}\mathbf{q}'^T\mathbf{A}^T\mathbf{K}\mathbf{A}\mathbf{q}'$$

In order for these to be sums of squares the transformed matrices must be diagonal, namely,

$$\mathbf{A}^T\mathbf{M}\mathbf{A} = \mathbf{M}' = \begin{bmatrix} \mu_1' & 0 & \cdot \\ 0 & \mu_2' & \cdot \\ \cdot & \cdot & \cdot \end{bmatrix}$$

$$\mathbf{A}^T\mathbf{K}\mathbf{A} = \mathbf{K}' = \begin{bmatrix} k_1' & 0 & \cdot \\ 0 & k_2' & \cdot \\ \cdot & \cdot & \cdot \end{bmatrix} \tag{10.46}$$

There are several ways of arriving at the required transformation. One method consists of finding a set of vectors that are simultaneously eigenvectors of both the matrices $\mathbf{M}$ and $\mathbf{K}$. Such eigenvectors, denoted by $\mathbf{a}_k$, must then satisfy the two equations

$$\begin{aligned} \mathbf{M}\mathbf{a}_k &= \mu_k'\mathbf{a}_k \\ \mathbf{K}\mathbf{a}_k &= \kappa_k'\mathbf{a}_k \end{aligned} \qquad (k = 1, 2, \ldots, n) \tag{10.47}$$

Let us multiply the first equation by $\kappa_k'/\mu_k'$ and subtract it from the second. The result is

$$(\mathbf{K} - \mathbf{M}\lambda_k)\mathbf{a}_k = 0 \tag{10.48}$$

in which we have set $\lambda_k = \kappa_k'/\mu_k'$. The above matrix equation is formally identical to Equation (10.38) found earlier. It must therefore have the same eigenvalues, namely, $\lambda_k = \omega_k^2$.

Returning now to Equations (10.46), suppose we multiply by $(\mathbf{A}^T)^{-1}$. This gives $\mathbf{M}\mathbf{A} = (\mathbf{A}^T)^{-1}\mathbf{M}'$ and $\mathbf{K}\mathbf{A} = (\mathbf{A}^T)^{-1}\mathbf{K}'$. By the usual rules of matrix algebra, these will reduce to

$$\mathbf{MA} = \mathbf{M'A}$$
$$\mathbf{KA} = \mathbf{K'A} \qquad (10.49)$$

provided $\mathbf{A}^T = \mathbf{A}^{-1}$. This means that $\mathbf{A}$ must be what is known as an *orthogonal matrix*.

At this point a little reflection will show that the two sets of equations (10.47) and (10.49) are actually equivalent to one another. This equivalence may be stated in the following way. The columns of the matrix $\mathbf{A}$ are equal to the vectors $\mathbf{a}_k$, to within a constant multiplicative factor. Thus, if we write

$$\mathbf{a}_k = \begin{bmatrix} a_{1k} \\ a_{2k} \\ \cdot \\ \cdot \\ \cdot \end{bmatrix} \qquad (10.50)$$

then $(\mathbf{A})_{jk} \propto a_{jk}$. Hence, to construct the matrix $\mathbf{A}$, we solve for the roots of the secular equation $\det(\mathbf{K} - \mathbf{M}\omega^2) = 0$. For each root $\omega_k$ we construct the corresponding eigenvector $\mathbf{a}_k$ by solving the system of equations $(\mathbf{K} - \mathbf{M}\omega_k^2)\mathbf{a}_k = 0$. Each equation is equivalent to the $n$ algebraic equations

$$\sum_j (k_{ij} - \mu_{ij}\omega_k^2)a_{jk} = 0 \qquad (i = 1, 2, \ldots, n) \qquad (10.51)$$

Since there are $n$ different values of $\omega_k$, except in the case of repeated roots, there are $n^2$ equations to be solved. For large $n$ this can lead to a considerable amount of computational labor. In practical problems, the use of high-speed electronic computers is clearly advantageous.

## EXAMPLE

For the problem of the two coupled harmonic oscillators treated in Section 10.4, we showed that the eigenvectors were proportional to $\begin{bmatrix} 1 \\ 1 \end{bmatrix}$ and $\begin{bmatrix} 1 \\ -1 \end{bmatrix}$. Thus the transformation matrix for normal coordinates is

$$\mathbf{A} = 2^{-1/2} \begin{bmatrix} 1 & 1 \\ 1 & -1 \end{bmatrix}$$

The factor $2^{-1/2}$ is inserted in order that $|\det \mathbf{A}| = 1$ and thus $\mathbf{A}^T = \mathbf{A}^{-1} = \mathbf{A}$. Hence, in this problem, $\mathbf{A}$ is its own inverse. Denoting the normal coordinates by $x_1'$ and $x_2'$, we have $\mathbf{q}' = \mathbf{A}^{-1}\mathbf{q}$, or

$$\begin{bmatrix} x_1' \\ x_2' \end{bmatrix} = 2^{-1/2} \begin{bmatrix} 1 & 1 \\ 1 & -1 \end{bmatrix} \begin{bmatrix} x_1 \\ x_2 \end{bmatrix}$$

That is

$$x_1' = 2^{-1/2}(x_1 + x_2)$$
$$x_2' = 2^{-1/2}(x_1 - x_2)$$

and since $\mathbf{A} = \mathbf{A}^{-1}$, the primes can be interchanged.

Let us express the Lagrangian function in terms of the normal coordinates. We then find

$$T = \frac{m}{2}\frac{1}{2}(\dot{x}_1' + \dot{x}_2')^2 + \frac{m}{2}\frac{1}{2}(\dot{x}_1' - \dot{x}_2')^2 = \frac{m}{2}(\dot{x}_1'^2 + \dot{x}_2'^2)$$

$$V = \frac{k}{2}\frac{1}{2}(x_1' + x_2')^2 + \frac{k}{2}\frac{1}{2}(x_1' - x_2')^2 + \frac{k}{2}x_2'^2 = \frac{k}{2}x_1'^2 + \frac{k + 2k'}{2}x_2'^2$$

and so

$$L = \frac{m}{2}\dot{x}_1'^2 - \frac{k}{2}x_1'^2 + \frac{m}{2}\dot{x}_2'^2 - \frac{k + 2k'}{2}x_2'^2$$

Thus $L$ consists entirely of squares, that is, there are no product terms like $x_1'x_2'$, and the equations of motion $(d/dt)(\partial L/\partial \dot{x}_i') = \partial L/\partial x_i'$ then read simply

$$m\ddot{x}_1' = -kx_1' \qquad m\ddot{x}_2' = -(k + 2k')x_2'$$

The equations are thus in separated form, and the solutions are

$$x_1' = 2^{-1/2}(x_1 + x_2) = A \cos(\omega_1 t + \varphi)$$
$$x_2' = 2^{-1/2}(x_1 - x_2) = B \cos(\omega_2 t + \varphi')$$

where $\omega_1$ and $\omega_2$ are the same as those found previously. For any motion of the system the coordinate $x_1'$ always oscillates at the frequency $\omega_1$, and $x_2'$ always oscillates at the frequency $\omega_2$. If the initial conditions are such that one of the two constants $A$ or $B$ is zero, but not both, then the system oscillates in one of its normal modes.

### Orthogonality of Eigenvectors.    Repeated Roots

It is a well-known theorem, discussed in Appendix V, that the eigenvectors of a symmetric matrix are orthogonal to one another if they belong to different eigenvalues. Hence the eigenvectors defining the normal modes of an oscillating system are, as a rule, mutually orthogonal. Exceptions to this rule occur for mechanical systems in which there exist solutions of the matrix equation $(\mathbf{K} - \mathbf{M}\omega^2)\mathbf{q} = 0$ but for which the eigenvectors are not simultaneous eigenvectors of the matrices $\mathbf{K}$ and $\mathbf{M}$, as was assumed in the previous discussion. An example of such a case is given below.

Another type of exception to the orthogonality rule arises when the secular equation $\det(\mathbf{K} - \mathbf{M}\omega^2) = 0$ has repeated roots, so that two or more eigenvectors can belong to the same eigenvalue. However, in this case it is possible to construct a set of mutually orthogonal eigenvectors. The procedure is as follows. Suppose the multiplicity of a given root is two. Then there will be two different eigenvectors associated with the same eigenvalue or normal frequency. Let $\mathbf{q}_\alpha$ and $\mathbf{q}_\beta$ denote these two eigenvectors, assumed to be linearly independent. Clearly any linear combination is also a solution of the given matrix equation and belongs to the same eigenvalue, that is, $c\mathbf{q}_\alpha + \mathbf{q}_\beta$ has the same eigenvalue as $\mathbf{q}_\alpha$ and $\mathbf{q}_\beta$ where $c$ is a constant. Now consider the expression

$$\mathbf{q}_\alpha{}^T(c\mathbf{q}_\alpha + \mathbf{q}_\beta) = c\mathbf{q}_\alpha{}^T\mathbf{q}_\alpha + \mathbf{q}_\alpha{}^T\mathbf{q}_\beta$$

Hence if we choose $c$ such that the expression vanishes, that is,

$$c = -\frac{\mathbf{q}_\alpha{}^T\mathbf{q}_\beta}{\mathbf{q}_\alpha{}^T\mathbf{q}_\alpha} \tag{10.52}$$

then the eigenvector $\mathbf{q}_\alpha$ is orthogonal to $c\mathbf{q}_\alpha + \mathbf{q}_\beta$. A similar procedure can be used if the multiplicity is higher than two.

It may happen that an eigenvalue of $\omega$ is zero. In this case there is no oscillation, rather, there is merely uniform motion of the normal coordinate in question.

## EXAMPLES

1. In the problem of the double pendulum discussed earlier the two eigenvectors were found to be $\begin{bmatrix} 1 \\ 2^{1/2} \end{bmatrix}$ and $\begin{bmatrix} 1 \\ -2^{1/2} \end{bmatrix}$. These are not orthogonal, but they are eigenvectors of the system, and they have different eigenvalues as shown above. They satisfy the matrix equation $(\mathbf{K} - \mathbf{M}\omega^2)\mathbf{q} = 0$ in which

$$\mathbf{K} = \frac{g}{l}\begin{bmatrix} 2 & 0 \\ 0 & 1 \end{bmatrix} \qquad \mathbf{M} = \begin{bmatrix} 2 & 1 \\ 1 & 1 \end{bmatrix}$$

However, the two eigenvectors in question are not eigenvectors of the matrices $\mathbf{K}$ and $\mathbf{M}$. The nonorthogonality here is related to the fact that $\mathbf{K}$ and $\mathbf{M}$ do not commute. To show this, we refer to Eq. (10.41) and to Appendix V in which it is shown that different eigenvectors of a symmetric matrix are orthogonal. Now if $\mathbf{K}$ and $\mathbf{M}$ commute, then the product $\mathbf{M}^{-1}\mathbf{K}$ is symmetric. (The student should verify this.) Hence different eigenvectors must be orthogonal. On the other hand, if the product in question is not symmetric, the eigenvectors are not necessarily orthogonal.

2. Three particles of the same mass $m$ are constrained to move in a common circular path. They are connected by three identical springs of stiffness $k$, as shown in Figure 10.6. Find the normal frequencies and normal modes.

We choose $x_1$, $x_2$, and $x_3$ to specify the configuration of the system. The $x$'s are measured from the equilibrium positions, as shown. Then

$$T = \tfrac{1}{2}m(\dot{x}_1{}^2 + \dot{x}_2{}^2 + \dot{x}_3{}^2) \qquad V = \tfrac{1}{2}k[(x_1 - x_2)^2 + (x_2 - x_3)^2 + (x_3 - x_1)^2]$$

The energy matrices are then

$$\mathbf{M} = \begin{bmatrix} 1 & 0 & 0 \\ 0 & 1 & 0 \\ 0 & 0 & 1 \end{bmatrix} m \qquad \mathbf{K} = \begin{bmatrix} 2 & -1 & -1 \\ -1 & 2 & -1 \\ -1 & -1 & 2 \end{bmatrix} k$$

The secular determinant is

$$\begin{vmatrix} 2 - \lambda & -1 & -1 \\ -1 & 2 - \lambda & -1 \\ -1 & -1 & 2 - \lambda \end{vmatrix} = 0$$

or

$$\lambda^3 - 6\lambda^2 + 9\lambda = 0$$

where $\lambda = \omega^2 m/k$. The three roots are $\lambda = 0$, $\lambda = 3$, $\lambda = 3$. Thus we have two repeated roots.

The associated matrix equations for determining the eigenvectors are

$$(\lambda = 0): \begin{bmatrix} 2 & -1 & -1 \\ -1 & 2 & -1 \\ -1 & -1 & 2 \end{bmatrix} \begin{bmatrix} x_1 \\ x_2 \\ x_3 \end{bmatrix} = 0 \qquad (\lambda = 3): \begin{bmatrix} -1 & -1 & -1 \\ -1 & -1 & -1 \\ -1 & -1 & -1 \end{bmatrix} \begin{bmatrix} x_1 \\ x_2 \\ x_3 \end{bmatrix} = 0$$

The first matrix equation gives $2x_1 = x_2 + x_3$ together with two more obtained by cyclic permutation of the subscripts. Solving them, we find $x_1 = x_2 = x_3$. Thus the eigenvector for this mode ($\omega = 0$) is $\begin{bmatrix} 1 \\ 1 \\ 1 \end{bmatrix}$. The second matrix equation gives $x_1 + x_2 + x_3 = 0$ for all three component equations. This type of degeneracy occurs when there are repeated roots. In this case we are free to choose the value of one component of the associated eigenvector (two if it is a triple root, and so on). Let us then set $x_3 = 0$, so that $x_2 = -x_1$, and the corresponding eigenvector can be expressed $\begin{bmatrix} 1 \\ -1 \\ 0 \end{bmatrix}$. Clearly, a similar procedure with $x_2 = 0$ gives another eigenvector $\begin{bmatrix} 1 \\ 0 \\ -1 \end{bmatrix}$. A third would be $\begin{bmatrix} 0 \\ -1 \\ 1 \end{bmatrix}$ with $x_1 = 0$. However, these are not linearly independent, for the last is the difference of the first two. There are only two linearly independent eigenvectors associated with the root $\lambda = 3$. Inspection shows that the above eigenvectors are not orthogonal, however. Using the theory leading to Equation (10.52), we know that the vector $c \begin{bmatrix} 1 \\ -1 \\ 0 \end{bmatrix} + \begin{bmatrix} 1 \\ 0 \\ -1 \end{bmatrix}$ is orthogonal

to the vector $\begin{bmatrix} 1 \\ -1 \\ 0 \end{bmatrix}$ if we choose $c = -\frac{1}{2}[1, -1, 0] \begin{bmatrix} 1 \\ 0 \\ -1 \end{bmatrix} = -\frac{1}{2}$. This gives a

vector $\frac{1}{2} \begin{bmatrix} 1 \\ 1 \\ -2 \end{bmatrix}$ which is orthogonal to $\begin{bmatrix} 1 \\ -1 \\ 0 \end{bmatrix}$. Finally, we can write the

general solution as

$$\begin{bmatrix} x_1 \\ x_2 \\ x_3 \end{bmatrix} = \begin{bmatrix} 1 \\ 1 \\ 1 \end{bmatrix} (a + bt) + \begin{bmatrix} 1 \\ -1 \\ 0 \end{bmatrix} A \cos(\omega t + \alpha) + \begin{bmatrix} 1 \\ 1 \\ -2 \end{bmatrix} B \cos(\omega t + \beta)$$

The first mode ($\omega = 0$) is pure rotation, and the other two modes are oscillation of the system at the frequency $\omega = \sqrt{3k/m}$.

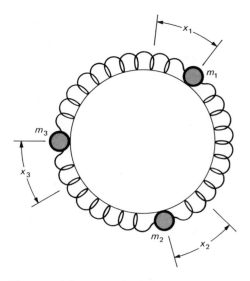

FIGURE 10.6   Three particles constrained to move in a common circular path and connected by elastic springs. The model is an approximation to a symmetrical planar triatomic molecule.

*Motion of a General System When Damping Forces*
*and External Driving Forces Are Present*

In the foregoing analysis of the oscillation of a general system, we neglected the presence of any frictional forces. If the system is subject to viscous damping forces proportional to the first powers of the velocities of the particles, we can write Lagrange's equations in the form

$$\frac{d}{dt} \frac{\partial L}{\partial \dot{q}_k} = \frac{\partial L}{\partial q_k} + R_k$$

where the generalized damping force $R_k$ is given by

$$R_k = -c_{1k}\dot{q}_1 - c_{2k}\dot{q}_2 - \cdots - c_{nk}\dot{q}_n$$

The resulting differential equations of motion are similar to the undamped case, except that terms involving the $\dot{q}$'s are present. It is often (but not always) possible in this case to find a normal coordinate transformation such that the resulting differential equations are of the form

$$\mu_k'\ddot{q}_k' + c_k'\dot{q}_k' + \kappa_k'q_k' = 0$$

so that

$$q_k' = A'_k e^{-\lambda_k t} \cos(\omega_k t + \epsilon_k)$$

The amplitudes of the normal modes thus die out exponentially with time. There is also the possibility of a nonoscillatory situation analogous to the critically damped or overdamped one-dimensional case.

Finally, for the motion of a system which, in addition to linear restoring forces and dissipative forces, is subject to external driving forces that vary harmonically with time, we can express the situation analytically by including terms of the form $Q_{k\text{ext}} \cos \omega t$ (or $Q_{k\text{ext}}e^{i\omega t}$) in each equation of motion. The resulting equations of motion in the normalized coordinates assume the form

$$\mu_k'\ddot{q}_k' + c_k\dot{q}_k + \kappa_k'q_k' = Q_k e^{i\omega t}$$

Thus, for example, if the system is subject to a single driving force varying harmonically at a frequency equal to one of the normal frequencies of the system, then the corresponding normal mode is the one that assumes the largest amplitude in the steady-state condition. In fact, if the damping constants are vanishingly small, then the normal mode whose frequency is equal to the driving frequency is the only one that is excited.

### 10.7.  Vibration of a Loaded String

In this section we consider the motion of a simple mechanical system consisting of a light elastic string that is clamped at both ends and loaded with a given number $n$ of particles equally spaced along the length of the string, each being of equal mass $m$. The problem illustrates the general theory of vibrations and also leads naturally into the theory of wave motion, briefly treated in the next section.

Let us label the displacements of the various particles from their equilibrium positions by the coordinates $q_1, q_2, \ldots, q_n$. Actually, there are two types of displacement that can occur, namely a longitudinal displacement in which the particle moves along the direction of the string, and a transverse displacement in which the particle moves at right angles to the length of the string.   These are illustrated in Figure 10.7.   For simplicity we shall assume that the motion is either purely longitudinal or purely transverse, although in the actual physical situation a combination of the two could occur.   The kinetic energy of the system is then given by

$$T = \frac{m}{2} (\dot{q}_1{}^2 + \dot{q}_2{}^2 + \cdots + \dot{q}_n{}^2)$$

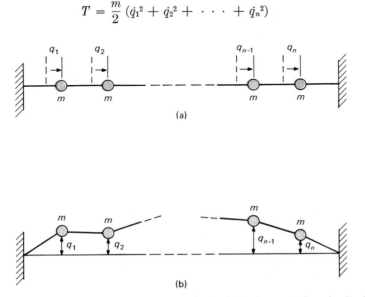

(a)

(b)

FIGURE 10.7   Linear array of particles or the loaded string.   (a) Longitudinal motion.   (b) Transverse motion.

If we use the letter $\nu$ to denote any given particle, then, in the case of longitudinal motion, the stretch of the section of string between particle $\nu$ and particle $\nu +1$ is

$$q_{\nu+1} - q_\nu$$

Hence the potential energy of this section of the string is

$$\tfrac{1}{2}K(q_{\nu+1} - q_\nu)^2$$

in which $K$ is the elastic stiffness coefficient of the section of string connecting the two adjacent particles.

For the case of transverse motion, the distance between particle $\nu$ and $\nu + 1$ is

$$[h^2 + (q_{\nu+1} - q_\nu)^2]^{1/2} = h + \frac{1}{2h} (q_{\nu+1} - q_\nu)^2 + \cdots$$

in which $h$ is the equilibrium distance between two adjacent particles.   The stretch of the section of string connecting the two particles is then approximately

$$\Delta l = \frac{1}{2h} (q_{\nu+1} - q_\nu)^2$$

Thus, if $S$ is the tension in the string, the potential energy of the section under consideration is given by

$$S\Delta l = \frac{S}{2h} (q_{\nu+1} - q_\nu)^2$$

It follows that the total potential energy of the system in either the longitudinal or the transverse type of motion is expressible as a quadratic function of the form

$$V = \frac{k}{2} [q_1^2 + (q_2 - q_1)^2 + \cdot \cdot \cdot + (q_n - q_{n-1})^2 + q_n^2] \qquad (11.66)$$

in which

$$k = \frac{S}{h} \qquad (\textit{transverse motion})$$

or

$$k = K \qquad (\textit{longitudinal motion})$$

The Lagrangian function of the loaded string is thus given by

$$L = \tfrac{1}{2} \sum_\nu [m\dot{q}_\nu^2 - k(q_{\nu+1} - q_\nu)^2] \qquad (10.53)$$

The Lagrangian equations of motion

$$\frac{d}{dt} \frac{\partial L}{\partial \dot{q}_\nu} = \frac{\partial L}{\partial q_\nu}$$

then become

$$m\ddot{q}_\nu = -k(q_\nu - q_{\nu-1}) + k(q_{\nu+1} - q_\nu) \qquad (10.54)$$

where $\nu = 1, 2, \ldots, n$.

To solve the above system of $n$ equations, we use a trial solution in which the $q$'s are assumed to vary harmonically with time. It is convenient to use the exponential form

$$q_\nu = a_\nu e^{i\omega t}$$

where $a_\nu$ is the amplitude of vibration of the $\nu$th particle. Substitution of the above trial solution into the differential equations (10.54) yields the following recursion formula for the amplitudes:

$$-m\omega^2 a_\nu = k(a_{\nu-1} - 2a_\nu + a_{\nu+1}) \qquad (10.55)$$

This formula will include the end points of the string if we set

$$a_0 = a_{n+1} = 0$$

The secular determinant is thus

$$\begin{vmatrix} -m\omega^2 + 2k & -k & 0 & \cdots & 0 \\ -k & -m\omega^2 + 2k & -k & \cdots & 0 \\ 0 & -k & -m\omega^2 + 2k & \cdots & 0 \\ \cdots & \cdots & \cdots & \cdots & \cdots \\ 0 & 0 & 0 & \cdots & -m\omega^2 + 2k \end{vmatrix} = 0$$

The determinant is of the $n$th order and there are thus $n$ values of $\omega$ that satisfy the equation. However, rather than find these $n$ roots by algebra, it turns out that we can find them by working directly with the recursion relation, Equation (10.55).

To this end, we define a quantity $\varphi$ related to the amplitudes $a_\nu$ by the following equation

$$a_\nu = A \sin (\nu\varphi) \tag{10.56}$$

Direct substitution into the recursion formula then yields

$$-m\omega^2 A \sin (\nu\varphi) = kA[\sin (\nu\varphi - \varphi) - 2 \sin (\nu\varphi) + \sin (\nu\varphi + \varphi)]$$

which easily reduces to

$$m\omega^2 = k(2 - 2 \cos \varphi) = 4k \sin^2 \frac{\varphi}{2}$$

or

$$\omega = 2\omega_0 \sin \frac{\varphi}{2} \tag{10.57}$$

in which

$$\omega_0 = \left(\frac{k}{m}\right)^{1/2} \tag{10.58}$$

Equation (10.57) gives the normal frequencies in terms of the quantity $\varphi$ which we have not, as yet determined. Now, as a matter of fact, the same relation would have been obtained by any of the following substitutions for the amplitude $a_\nu$: $A \cos (\nu\varphi)$, $A e^{i\nu\varphi}$, $A e^{-i\nu\varphi}$, or any linear combination of these. However, only the substitution $a_\nu = A \sin (\nu\varphi)$ satisfies the end condition $a_0 = 0$. In order to determine the actual value of the parameter $\varphi$, and thus find the normal frequencies of the vibrating string, we use the other end condition, namely $a_{n+1} = 0$. This condition will be met if we set

$$(n + 1)\varphi = N\pi \tag{10.59}$$

in which $N$ is an integer, because we then have

$$a_{n+1} = A \sin (N\pi) = 0$$

Having found $\varphi$, we can now calculate the normal frequencies. They are given by

$$\omega_N = 2\omega_0 \sin \left(\frac{N\pi}{2n + 2}\right) \tag{10.60}$$

Furthermore, from Equations (10.56) and (10.59) we see that the amplitudes for the normal modes are given by

$$a_\nu = A \sin \left( \frac{N\pi\nu}{n+1} \right) \tag{10.61}$$

Here the value of $\nu = 1, 2, \ldots, n$ denotes a particular particle in the linear array, and the value of $N = 1, 2, \ldots, n$ refers to the normal mode in which the system is oscillating.

The different normal modes are illustrated graphically by plotting the amplitudes as given by Equation (10.61). These fall on a sine curve as shown in Figure 10.8 which shows the case of three particles $n = 3$. The actual motion of the system, when it is vibrating in a single pure mode is given by the equation

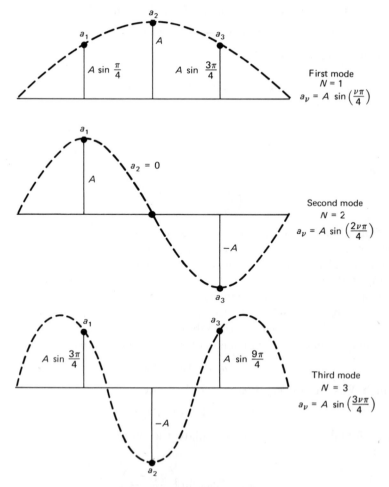

FIGURE 10.8   The normal modes of a three-particle system.

$$q_\nu = a_\nu \cos(\omega_N t) = A \sin\left(\frac{\pi N \nu}{n+1}\right) \cos(\omega_N t)$$

The general type of motion is a linear combination of all the normal modes. This can be expressed as

$$q_\nu = \sum_{N=1}^{n} A_N \sin\left(\frac{N\pi\nu}{n+1}\right) \cos(\omega_N t + \varphi_N) \qquad (10.62)$$

in which the values of $A_N$ and $\varphi_N$ are determined from the initial conditions.

In the event that the number $n$ of particles is large compared to the mode number $N$, so that the ratio $N\pi/(2n+2)$ is small, we can replace the sine term in Equation (10.60) by the argument.   Thus we have approximately

$$\omega_N \approx N\left(\frac{\pi\omega_0}{n+1}\right)$$

This means that the normal frequencies are approximately integral multiples of the lowest frequency $\pi\omega_0/(n+1)$.   In other words, we can regard the different normal frequencies as the fundamental, the second harmonic, the third harmonic, and so on.   The accuracy of this integral harmonic relationship is improved as the number of particles is made larger.

### 10.8.   Vibration of a Continuous System.
### The Wave Equation

Let us consider the motion of a linear array of connected particles in which the number of particles is made indefinitely large and the distance between adjacent particles indefinitely small.   In other words, we have a continuous heavy cord or rod.   To analyze this type of system it is convenient to rewrite the differential equations of motion of a finite system, Equation (10.54), in the following form:

$$m\ddot{q} = kh\left[\left(\frac{q_{\nu+1} - q_\nu}{h}\right) - \left(\frac{q_\nu - q_{\nu-1}}{h}\right)\right]$$

in which $h$ is the distance between the equilibrium positions of any two adjacent particles.   Now if the variable $x$ represents general distances in the longitudinal direction, and if the number $n$ of particles is very large so that $h$ is small compared to the total length, then we can write

$$\frac{q_{\nu+1} - q_\nu}{h} \approx \left(\frac{\partial q}{\partial x}\right)_{x=\nu h + h/2}$$

$$\frac{q_\nu - q_{\nu-1}}{h} \approx \left(\frac{\partial q}{\partial x}\right)_{x=\nu h - h/2}$$

Consequently the difference between the above two expressions is equal to the second derivative multiplied by $h$, namely,

$$\frac{q_{\nu+1} - q_\nu}{h} - \frac{q_\nu - q_{\nu-1}}{h} \approx h \left(\frac{\partial^2 q}{\partial x^2}\right)_{x = \nu h}$$

The equation of motion can therefore be written

$$\frac{\partial^2 q}{\partial t^2} = \frac{kh^2}{m} \frac{\partial^2 q}{\partial x^2}$$

or

$$\frac{\partial^2 q}{\partial t^2} = v^2 \frac{\partial^2 q}{\partial x^2} \tag{10.63}$$

in which we have introduced the abbreviation

$$v^2 = \frac{kh^2}{m} \tag{10.64}$$

Equation (10.62) is a well-known differential equation of mathematical physics. It is called the *one-dimensional wave equation*. It is encountered in many different places. Solutions of the wave equation represent traveling disturbances of some sort. It is easy to verify that a very general type of solution of the wave equation is given by

$$q = f(x + vt)$$

or

$$q = f(x - vt)$$

where $f$ is *any* differentiable function of the argument $x \pm vt$. The first solution represents a disturbance that is propagating in the negative $x$ direction with speed $v$, and the second equation represents a disturbance moving with speed $v$ in the positive $x$ direction. In our particular problem, the disturbance $q$ is a *displacement* of a small portion of the system from its equilibrium configuration, Figure 10.9. For the cord, this displacement could be a kink that travels along the cord; and for a solid rod, it could be a region of compression or of rarefaction moving along the length of the rod.

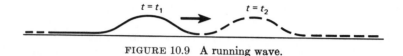

FIGURE 10.9   A running wave.

*Evaluation of the Wave Speed*

In the preceding section we found that the constant $k$, for transverse motion of a loaded string, is equal to the ratio $S/h$ where $S$ is the tension in the string. For the continuous string this ratio would, of course, become infinite as $h$ approaches zero. However, if we introduce the linear density or mass per unit length $\rho$, we have

$$\rho = \frac{m}{h}$$

Consequently, the expression for $v^2$, Equation (10.64), can be written

$$v^2 = \frac{(S/h)h^2}{\rho h} = \frac{S}{\rho}$$

so that $h$ cancels out.   The speed of propagation for transverse waves in a continuous string is then

$$v = \left(\frac{S}{\rho}\right)^{1/2} \tag{10.65}$$

For the case of longitudinal vibrations, we introduce the elastic modulus $Y$ which is defined as the ratio of the force to the elongation per unit length. Thus $k$, the stiffness of a small section of length $h$, is given by

$$k = \frac{Y}{h}$$

Consequently, Equation (10.64) can be written as

$$v^2 = \frac{(Y/h)h^2}{\rho h} = \frac{Y}{\rho}$$

and again we see that $h$ cancels out.   Hence the speed of propagation of longitudinal waves in an elastic rod is

$$v = \left(\frac{Y}{\rho}\right)^{1/2} \tag{10.66}$$

## 10.9.   Sinusoidal Waves

In the study of wave motion, those particular solutions of the wave equation

$$\frac{\partial^2 q}{\partial t^2} = v^2 \frac{\partial^2 q}{\partial x^2}$$

in which $q$ is a sinusoidal function of $x$ and $t$, namely

$$q = A \frac{\sin}{\cos} \left[ \frac{2\pi}{\lambda} (x + vt) \right] \tag{10.67}$$

$$q = A \frac{\sin}{\cos} \left[ \frac{2\pi}{\lambda} (x - vt) \right] \tag{10.68}$$

are of fundamental importance.   These solutions represent traveling disturbances in which the displacement, at a given point $x$, varies harmonically in time.   The amplitude of this motion is the constant $A$, and the *frequency f* is given by

$$f = \frac{\omega}{2\pi} = \frac{v}{\lambda}$$

Furthermore, at a given value of the time $t$, say $t = 0$, the displacement varies sinusoidally with the distance $x$. The distance between two successive maxima, or minima, of the displacement is the constant $\lambda$, called the *wavelength*. The waves represented by Equation (10.67) propagate in the negative $x$ direction, and those represented by Equation (10.68) propagate in the positive $x$ direction, as shown in Figure 10.10. They are special cases of the general type of solution mentioned earlier.

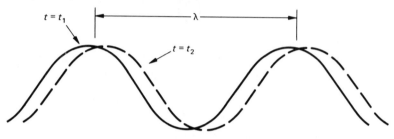

FIGURE 10.10   A sinusoidal wave.

*Standing Waves*

Since the wave equation, Equation (10.63), is linear, we can build up any number of solutions by making linear combinations of known solutions. One possible linear combination which is of particular significance is obtained by adding together two waves of equal amplitude that are traveling in opposite directions. In our notation, such a solution is given by

$$q = \frac{1}{2} A \sin \left[ \frac{2\pi}{\lambda} (x + vt) \right] + \frac{1}{2} A \sin \left[ \frac{2\pi}{\lambda} (x - vt) \right]$$

By using the appropriate trigonometric identity and collecting terms, we find that the equation reduces to

$$q = A \sin \left( \frac{2\pi}{\lambda} x \right) \cos (\omega t) \qquad (10.69)$$

in which $\omega = 2\pi v/\lambda$. The above equation represents what are known as *standing waves*. Here we see that the amplitude of the displacement is no longer constant, but varies with the value of $x$. Thus, for $x = 0$, $\lambda/2$, $\lambda$, $3\lambda/2$, . . . , the displacement is always zero since the sine term vanishes at these points. The points of zero displacement are called *nodes*. On the other hand, at those values of $x$ for which the absolute value of the sine term is unity, namely, $x = \lambda/4$, $3\lambda/4$, $5\lambda/4$, . . . , the amplitude of the harmonic oscillation has its maximum value of $A$. These points are called *antinodes*. The distance between two successive nodes, or two successive antinodes, is just one half of the wavelength. The above facts are illustrated in Figure 10.11.

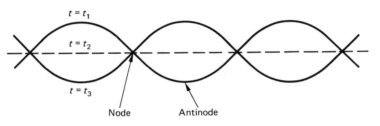

FIGURE 10.11   Standing waves.

### *Interpretation of the Motion of a Loaded String in Terms of Standing Waves*

If we compare the equation for a standing wave, Equation (10.69), with our previous solution for the motion of a loaded string, Equation (10.62), we observe that the two expressions are very similar.   The similarity can be brought out even more by noting that the standing wave solution will satisfy the boundary conditions of our original problem, namely

$$q = 0: x = 0 \quad \text{and} \quad x = l$$

provided that the end points of the string correspond to nodes of the standing wave.   This condition is met if the length $l$ of the string is an integral number $N$ of half wavelengths, that is

$$l = (n + 1)h = N \frac{\lambda}{2}$$

Solving for $\lambda$ and substituting in Equation (10.69) yields

$$q = A \sin \left[ \frac{\pi N x}{(n + 1)h} \right] \cos (\omega t)$$

This agrees with our previous solution, Equation (10.62), since at the positions of the various particles we have

$$x_\nu = \nu h \quad (\nu = 1, 2, \ldots n)$$

Thus the vibration of a loaded string can be regarded as a standing wave. Each normal mode contains a certain integral number of nodes in the standing-wave pattern.

### DRILL EXERCISES

10.1   Determine the equilibrium positions and the conditions for stability for a particle of mass $m$ moving in the following potentials:

(a) $V(x) = \dfrac{A}{x^3} - \dfrac{B}{x^2}$

(b) $V(x) = -kxe^{-bx}$

(c) $V(x) = k(x^4 - 2a^2x^2)$

10.2    Find the equilibrium configuration and stability of a system whose potential energy function is of the form

$$V(q_1, q_2, \ldots, q_n) = \sum_{i=1}^{n} \tfrac{1}{2}k_i(q_i - b_i)^2$$

10.3    Find the frequencies of oscillation of the particle in Exercise 10.1 for small oscillations about the respective equilibrium positions.

## PROBLEMS

10.4    A light spring of length $2l$ and stiffness $k$ is held with the ends fixed a distance $2l$ apart in a horizontal position.    A block of mass $m$ is then fastened to the midpoint of the spring.    Show that the potential energy of the system is given by the expression

$$V(y) = 2k[y^2 - 2l(y^2 + l^2)^{1/2}] - mgy$$

where $y$ is the vertical sag of the center of the spring.    From this show that the equilibrium position is given by a root of the equation

$$u^4 - 2au^3 + a^2u^2 - 2au + a^2 = 0$$

where $u = y/l$ and $a = mg/4kl$.

10.5    A uniform cubical block of mass $m$ and sides $2a$ is balanced on top of a rough sphere of radius $b$.    Show that the potential energy function can be expressed as

$$V = mg[(a + b)\cos\theta + b\theta\sin\theta]$$

where $\theta$ is the angle between the line of contact and the vertical, measured from the center of the sphere.    From this, show that the equilibrium is stable, or unstable, depending on whether $a$ is less than or greater than $b$, respectively. Investigate the stability for the case $a = b$.

10.6    A solid homogeneous hemisphere of radius $a$ rests on top of a rough hemispherical cap of radius $b$, the curved faces being in contact.    Show that the equilibrium is stable if $a$ is less than $3b/5$.

10.7    Determine the frequency of vertical oscillations about the equilibrium position in Problem 10.4.

10.8    Determine the period of oscillation of the block in Problem 10.5.

10.9    Determine the period of oscillation of the hemisphere in Problem 10.6.

10.10    A small steel ball rolls back and forth about its equilibrium position in a rough spherical bowl.    Show that the period of oscillation is

$2\pi[7a^2/5g(b - a)]^{1/2}$ where $a$ is the radius of the ball and $b$ is the radius of the bowl.

**10.11** A heavy elastic spring of uniform stiffness and density supports a particle of mass $m$. If $m'$ is the mass of the spring and $k$ its stiffness, show that the period of vertical oscillations of the particle is

$$2\pi \sqrt{\frac{m + (m'/3)}{k}}$$

This problem shows the effect of the mass of the spring on the period of oscillation. [*Hint:* To set up the Lagrangian function for the system, assume that velocity of any part of the spring is proportional to its distance from the point of suspension.]

**10.12** Find the normal frequencies of the coupled harmonic oscillator system, Figure 10.3, for the general case in which the two particles have unequal mass, and the springs have different stiffness. In particular, find the frequencies for the case $m_1 = m$, $m_2 = 2m$, $k_1 = k$, $k_2 = 2k$, $k' = k/2$. Express the result in terms of the quantity $\omega_0 = (k/m)^{1/2}$.

**10.13** Write down the complete solution of the symmetric coupled harmonic oscillator system, Equation (10.29), for the following initial conditions: $t = 0$, $x_1 = A_0$, $x_2 = 0$, $\dot{x}_1 = \dot{x}_2 = 0$. Show that the amplitude $A$ of the symmetric component is equal to the amplitude $B$ of the antisymmetric component in this case.

**11.14** A light elastic spring of stiffness $k$ is clamped at its upper end and supports a particle of mass $m$ at its lower end. A second spring of stiffness $k$ is fastened to the particle and, in turn, supports a particle of mass $2m$ at its lower end. Find the normal frequencies of the system for vertical oscillations about the equilibrium configuration. Find also the normal coordinates.

**10.15** Find the eigenvectors and normal coordinates for (a) Problem 10.12, second part, and (b) Problem 10.14. (c) Are the eigenvectors orthogonal in both cases?

**10.16** Consider the case of a double pendulum, Figure 10.5, in which the two sections are of different length, the upper one being of length $l_1$, and the lower of length $l_2$. Both particles are of equal mass $m$. Find the normal frequencies of the system, the corresponding eigenvectors, and the normal coordinates.

**10.17** Set up the secular equation for the case of three coupled particles in a linear array and show that the normal frequencies are the same as those given by Equation (10.61).

**10.18** Two identical simple pendulums are coupled together by a very weak force of attraction that varies as the inverse square of the distance between the two particles. (This force might be the gravitational attraction between the two particles, for instance.) Show that, for small departures from the equilibrium configuration, the Lagrangian can be reduced to the same form as that of the two coupled harmonic oscillators. Show further that if one pendulum is started oscillating with the other at rest, then eventually the second pendulum will be moving and the first one will be at rest, and so on.

10.19  A linear triatomic molecule ($CO_2$, for example) consists of a central atom of mass $m$ and two other atoms, each of mass $m'$, the three atoms being in a straight line. Set up the Lagrangian function for such a molecule, assuming that the motion takes place along a single straight line (the $x$ axis) and find the normal modes and normal frequencies. Assume that the forces between adjacent atoms can be represented by a spring of stiffness $k$.

10.20  Illustrate the normal modes for the case of four particles in a linear array. Find the numerical values of the ratios of the 2nd, 3rd, and 4th normal frequencies to the lowest or first normal frequency.

10.21  A light elastic cord of natural length $l$ and stiffness $k$ is stretched out to a length $l + \Delta l$ and loaded with a number $n$ of particles evenly spaced along its length. If $m$ is the total mass of all $n$ particles, find the speed of transverse and of longitudinal waves in the cord.

10.22  Work the above problem for the case in which, instead of being loaded, the cord is heavy with linear density $\rho$.

# Selected References

## Mechanics

Becker, R. A., *Introduction to Theoretical Mechanics*, McGraw-Hill, New York, 1954.

Lindsay, R. B., *Physical Mechanics*, Van Nostrand, Princeton, N. J., 1961.

Rutherford, D. E., *Classical Mechanics*, Interscience, New York, 1951.

Slater, J. C., and Frank, N. H., *Mechanics*, McGraw-Hill, New York, 1947.

Synge, J. L., and Griffith, B. A., *Principles of Mechanics*, McGraw-Hill, New York, 1959.

## Advanced Mechanics

Corbin, H. C., and Stehle, P., *Classical Mechanics*, Wiley, New York, 1950.

Goldstein, H., *Classical Mechanics*, Addison-Wesley, Reading, Mass., 1950.

Landau, L. D., and Lifshitz, E. M., *Mechanics*, Addison-Wesley, Reading, Mass., 1960.

Wells, D. A., *Lagrangian Dynamics*, Shaum, New York, 1967.

Whittaker, E. T., *Advanced Dynamics*, Cambridge University Press, London and New York, 1937.

## Mathematical Methods

Jeffreys, H., and Jeffreys, B. S., *Methods of Mathematical Physics*, Cambridge University Press, London and New York, 1946.

Kaplan, W., *Advanced Calculus*, Addison-Wesley, Reading, Mass., 1952.

Mathews, J., and Walker, R. L., *Methods of Mathematical Physics*, W. A. Benjamin, New York, 1964.

Margenau, J., and Murphy, G. M., *The Mathematics of Physics and Chemistry*, Van Nostrand, New York, 1943.

Wylie, C. R., Jr., *Advanced Engineering Mathematics*, McGraw-Hill, New York, 1951.

## Tables

Dwight, H. B., *Mathematical Tables*, Dover, New York, 1958.

Pierce, B. O., *A Short Table of Integrals*, Ginn, Boston, 1929.

*Handbook of Chemistry and Physics, Mathematical Tables*, Chemical Rubber Co., Cleveland, Ohio, 1962 or after.

# Appendixes

# Appendix I

## Complex Numbers

The quantity

$$z = x + iy$$

is said to be a *complex number* if $x$ and $y$ are real and $i = \sqrt{-1}$.  The *complex conjugate* is defined as

$$z^* = x - iy$$

The *absolute value* $|z|$ is given by

$$|z|^2 = zz^* = x^2 + y^2$$

The following are true

$$z + z^* = 2x = 2 \operatorname{Re} z$$
$$z - z^* = 2y = 2 \operatorname{Im} z$$

*Exponential Notation*

$$z = x + iy = |z|e^{i\theta} = |z|(\cos \theta + i \sin \theta)$$
$$z^* = x - iy = |z|e^{-i\theta} = |z|(\cos \theta - i \sin \theta)$$

where

$$\tan \theta = \frac{y}{x}$$

*Circular and Hyperbolic Functions*

The following relations are often useful

$$\cos \theta = \frac{e^{i\theta} + e^{-i\theta}}{2}$$

$$\sin \theta = \frac{e^{i\theta} - e^{-i\theta}}{2}$$

$$\cosh \theta = \frac{e^{\theta} + e^{-\theta}}{2} \qquad \text{(hyperbolic cosine)}$$

$$\sinh \theta = \frac{e^{\theta} - e^{-\theta}}{2} \qquad \text{(hyperbolic sine)}$$

$$\tanh \theta = \frac{\sinh \theta}{\cosh \theta} = \frac{e^{\theta} - e^{-\theta}}{e^{\theta} + e^{-\theta}} \qquad \text{(hyperbolic tangent)}$$

*Relations Between Circular and Hyperbolic Functions*

$$\sin i\theta = i \sinh \theta$$
$$\cos i\theta = \cosh \theta$$
$$\sinh i\theta = i \sin \theta$$
$$\cosh i\theta = \cos \theta$$

*Derivatives*

$$\frac{d}{d\theta} \sinh \theta = \cosh \theta$$

$$\frac{d}{d\theta} \cosh \theta = \sinh \theta$$

*Identities*

$$\cosh^2 \theta - \sinh^2 \theta = 1$$
$$\sinh (\theta + \varphi) = \sinh \theta \cosh \varphi + \cosh \theta \sinh \varphi$$
$$\cosh (\theta + \varphi) = \cosh \theta \cosh \varphi + \sinh \theta \sinh \varphi$$

# Appendix II

*Taylor's Series*

$$f(x + a) = f(a) + xf'(a) + \frac{x^2}{2!}f''(a) + \cdots + \frac{x^n}{n!}f^n(a) + \cdots$$

$$f(x) = f(0) + xf'(0) + \frac{x^2}{2!}f''(0) + \cdots + \frac{x^n}{n!}f^n(0) + \cdots$$

where

$$f^n(a) = \frac{d^n}{dx^n}f(x)\Bigg|_{x=a}$$

*Often-Used Expansions*

$$e^x = 1 + x + \frac{x^2}{2!} + \cdots + \frac{x^n}{n!} + \cdots$$

$$\sin x = x - \frac{x^3}{3!} + \frac{x^5}{5!} - \cdots$$

$$\cos x = 1 - \frac{x^2}{2!} + \frac{x^4}{4!} - \cdots$$

$$\sinh x = x + \frac{x^3}{3!} + \frac{x^5}{5!} + \cdots$$

$$\cosh x = 1 + \frac{x^2}{2!} + \frac{x^4}{4!} + \cdots$$

$$\ln (1 + x) = x - \frac{x^2}{2} + \frac{x^3}{3} - \cdots \qquad |x| < 1$$

$$\tan x = x + \frac{x^3}{3} + \frac{2}{15}x^5 + \cdots \qquad |x| < \frac{\pi}{2}$$

*Binomial Series*

$$(a + x)^n = a^n + na^{n-1}x + \frac{n(n-1)}{2!}a^{n-2}x^2 + \cdots + \binom{n}{m}a^{n-m}x^m + \cdots$$

where the binomial coefficient is

$$\binom{n}{m} = \frac{n!}{(n - m)!m!}$$

The series converges for $|x/a| < 1$.

*Useful Approximations*

For small $x$, the following approximations are often used

$$e^x \approx 1 + x$$
$$\sin x \approx x$$
$$\cos x \approx 1 - \tfrac{1}{2}x^2$$
$$\sqrt{1 + x} \approx 1 + \tfrac{1}{2}x$$

$$\frac{1}{1 + x} \approx 1 - x$$

$$\frac{1}{1 - x} \approx 1 + x$$

The last three are based on the binomial series, and the list can be extended for other values of the exponent.

# Appendix III

**Special Functions**

*Elliptic Integrals*

The elliptic integral of the *first kind* is given by the expressions

$$F(k,\varphi) = \int_0^\varphi \frac{d\varphi}{(1 - k^2 \sin^2 \varphi)^{1/2}}$$

$$= \int_0^x \frac{dx}{(1 - x^2)^{1/2}(1 - k^2 x^2)^{1/2}}$$

and the elliptic integral of the *second kind* by

$$E(k,\varphi) = \int_0^\varphi (1 - k^2 \sin^2 \varphi)^{1/2} \, d\varphi$$

$$= \int_0^x \frac{(1 - k^2 x^2)^{1/2}}{(1 - x^2)^{1/2}} \, dx$$

Both converge for $|k| < 1$.  They are called *incomplete* if $x = \sin \varphi < 1$, and *complete* if $x = \sin \varphi = 1$.  The complete integrals have the following series expansions

$$F(k) = F\left(k, \frac{\pi}{2}\right) = \frac{\pi}{2}\left(1 + \frac{k^2}{4} + \frac{9}{64}k^4 + \cdots\right)$$

$$E(k) = E\left(k, \frac{\pi}{2}\right) = \frac{\pi}{2}\left(1 - \frac{k^2}{4} - \frac{9}{64}k^4 - \cdots\right)$$

*Gamma Function*

The gamma function is defined as

$$\Gamma(n) = \int_0^\infty x^{n-1}\, e^{-x}\, dx$$

$$= \int_0^1 \left[\ln\left(\frac{1}{x}\right)\right]^{-1} dx$$

For any value of $n$

$$n\Gamma(n) = \Gamma(n+1)$$

If $n$ is a positive integer

$$\Gamma(n) = (n-1)!$$

Special values

$$\Gamma(\tfrac{1}{2}) = \sqrt{\pi}$$
$$\Gamma(1) = 1$$
$$\Gamma(\tfrac{3}{2}) = \tfrac{1}{2}\sqrt{\pi}$$
$$\Gamma(2) = 1$$

Integrals expressible in terms of gamma functions

$$\int_0^1 \frac{dx}{\sqrt{1 - x^n}} = \frac{\sqrt{\pi}}{n}\,\frac{\Gamma(1/n)}{\Gamma[(1/2) + (1/n)]}$$

$$\int_0^1 (1 - x^2)x^m\, dx = \frac{\Gamma(n+1)\Gamma[(m+1)/2]}{2\Gamma[(2n + m + 3)/2]}$$

# Appendix IV

## Curvilinear Coordinates

We consider a general orthogonal system of coordinates $u$, $v$, and $w$ with unit vectors $e_1$, $e_2$, and $e_3$. The volume element is

$$dV = h_1 h_2 h_3 \, du \, dv \, dw$$

and the line element is

$$ds = e_1 h_1 \, du + e_2 h_2 \, dv + e_3 h_3 \, dw$$

The gradient, divergence, and curl are as follows

$$\text{grad } f = \frac{e_1}{h_1} \frac{\partial f}{\partial u} + \frac{e_2}{h_2} \frac{\partial f}{\partial v} + \frac{e_3}{h_3} \frac{\partial f}{\partial w}$$

$$\text{div } \mathbf{Q} = \frac{1}{h_1 h_2 h_3} \left[ \frac{\partial}{\partial u} (h_2 h_3 Q_1) + \frac{\partial}{\partial v} (h_3 h_1 Q_2) + \frac{\partial}{\partial w} (h_1 h_2 Q_3) \right]$$

$$\text{curl } \mathbf{Q} = \frac{1}{h_1 h_2 h_3} \begin{vmatrix} h_1 e_1 & h_2 e_2 & h_3 e_3 \\ \dfrac{\partial}{\partial u} & \dfrac{\partial}{\partial v} & \dfrac{\partial}{\partial w} \\ h_1 Q_1 & h_2 Q_2 & h_3 Q_3 \end{vmatrix}$$

The $h$ functions for some common coordinate systems are listed below.

*Rectangular Coordinates*: $x$, $y$, $z$

$$h_x = 1 \qquad h_y = 1 \qquad h_z = 1$$

*Cylindrical Coordinates*: $r$, $\theta$, $z$

$$x = r \cos \theta \qquad y = r \sin \theta$$
$$h_r = 1 \qquad h_\theta = r \qquad h_z = 1$$

*Spherical Coordinates*: $r$, $\theta$, $\varphi$

$$x = r \sin \theta \cos \varphi \qquad y = r \sin \theta \sin \varphi \qquad z = r \cos \theta$$
$$h_r = 1 \qquad h_\theta = r \qquad h_\varphi = r \sin \theta$$

*Parabolic Coordinates*: $u$, $v$, $\theta$

$$x = uv \cos \theta \qquad y = uv \sin \theta \qquad z = \tfrac{1}{2}(u^2 - v^2)$$
$$h_u = h_v = \sqrt{u^2 + v^2} \qquad h_\theta = uv$$

Example: The curl in spherical coordinates is

$$\text{curl } \mathbf{Q} = \begin{vmatrix} e_r & r e_\theta & r \sin \theta e_\varphi \\ \dfrac{\partial}{\partial r} & \dfrac{\partial}{\partial \theta} & \dfrac{\partial}{\partial \varphi} \\ Q_r & r Q_\theta & r \sin \theta Q_\varphi \end{vmatrix} \frac{1}{r^2 \sin \theta}$$

# Appendix V

## Matrices

A *matrix* **A** is an array of elements $a_{ij}$ arranged thus

$$\mathbf{A} = \begin{bmatrix} a_{11} & a_{12} & \cdots & a_{1j} & \cdots & a_{1m} \\ a_{21} & a_{22} & \cdots & a_{2j} & \cdots & a_{2m} \\ \cdot & \cdot & & \cdot & & \cdot \\ \cdot & \cdot & & \cdot & & \cdot \\ a_{i1} & a_{i2} & \cdots & a_{ij} & \cdots & a_{im} \\ \cdot & \cdot & & \cdot & & \cdot \\ \cdot & \cdot & & \cdot & & \cdot \\ \cdot & \cdot & & \cdot & & \cdot \\ a_{n1} & a_{n2} & \cdots & a_{nj} & \cdots & a_{nm} \end{bmatrix}$$

If $n = m$, it is called a *square* matrix. Unless stated otherwise, we shall consider only square matrices in this Appendix. A *symmetric* matrix is one such that $a_{ij} = a_{ji}$. If $a_{ij} = -a_{ji}$, it is *antisymmetric*.

The sum of two matrices is defined as

$$(\mathbf{A} + \mathbf{B})_{ij} = a_{ij} + b_{ij}$$

The product of two matrices is defined as

$$(\mathbf{AB})_{ij} = a_{i1}b_{1j} + a_{i2}b_{2j} + \cdots = \sum_k a_{ik}b_{kj}$$

The product **AB** is not, in general, equal to **BA**. If **AB** = **BA**, the two matrices are said to *commute*. A *diagonal matrix* is one whose nondiagonal elements are zero, $a_{ij} = 0$ for $i \neq j$. The *identity* matrix[1] is a diagonal matrix with all diagonal elements equal to unity,

$$\mathbf{I} = \begin{bmatrix} 1 & 0 & 0 & \cdots & 0 \\ 0 & 1 & 0 & \cdots & 0 \\ 0 & 0 & 1 & \cdots & 0 \\ \cdot & \cdot & \cdot & \cdots & \\ 0 & 0 & 0 & \cdots & 1 \end{bmatrix}$$

From the definition of the product, it is easily shown that

$$\mathbf{AI} = \mathbf{IA}$$

The *inverse* $\mathbf{A}^{-1}$ of a matrix **A** is defined by

$$\mathbf{AA}^{-1} = \mathbf{I} = \mathbf{A}^{-1}\mathbf{A}$$

---

[1] This should not be confused with the inertia tensor defined in Chapter 8.

316

The *transpose* $\mathbf{A}^T$ of a matrix $\mathbf{A}$ is defined as

$$(\mathbf{A}^T)_{ij} = (\mathbf{A})_{ji}$$

For two matrices $\mathbf{A}$ and $\mathbf{B}$, $(\mathbf{AB})^T = \mathbf{B}^T\mathbf{A}^T$.

The determinant of a matrix is the determinant of its elements,

$$\det \mathbf{A} = \begin{vmatrix} a_{11} & a_{12} & \cdots \\ a_{21} & a_{22} & \cdots \\ \cdot & \cdot & \cdots \end{vmatrix}$$

The determinant of the product of two matrices is equal to the product of the respective determinants,

$$\det \mathbf{AB} = \det \mathbf{A} \det \mathbf{B}$$

It can be shown that the inverse of a matrix $\mathbf{A}$ is given by the formula

$$\mathbf{A}^{-1} = \begin{bmatrix} \dfrac{\det \mathbf{A}_{11}}{\det \mathbf{A}} & \dfrac{\det \mathbf{A}_{21}}{\det \mathbf{A}} & \cdots \\[3mm] \dfrac{\det \mathbf{A}_{12}}{\det \mathbf{A}} & \dfrac{\det \mathbf{A}_{22}}{\det \mathbf{A}} & \cdots \\[3mm] \cdots & \cdots & \cdots \end{bmatrix}$$

where the matrix $\mathbf{A}_{ij}$ is the matrix left after the $i$th row and $j$th column have been removed.

### *Matrix Representation of Vectors*

A matrix with one row, or one column, defines a *row vector*, or *column vector*, respectively. If $\mathbf{a}$ is a column vector, then $\mathbf{a}^T$ is the corresponding row vector,

$$\mathbf{a} = \begin{bmatrix} a_1 \\ a_2 \\ \cdot \\ \cdot \\ \cdot \\ a_n \end{bmatrix} \qquad \mathbf{a}^T = [a_1, a_2, \cdots, a_n]$$

For two column vectors $\mathbf{a}$ and $\mathbf{b}$ with the same number of elements, the product $\mathbf{a}^T\mathbf{b}$ is a scalar, analogous to the dot product,

$$\mathbf{a}^T\mathbf{b} = [a_1, a_2, \ldots] \begin{bmatrix} b_1 \\ b_2 \\ \cdot \\ \cdot \\ \cdot \end{bmatrix} = a_1 b_1 + a_2 b_2 + \cdots$$

Two vectors $\mathbf{a}$ and $\mathbf{b}$ are *orthogonal* if $\mathbf{a}^T\mathbf{b} = 0$.

*Matrix Transformations*

A matrix $\mathbf{Q}$ is said to *transform* a vector $\mathbf{a}$ into another vector $\mathbf{a}'$ according to the rule

$$\mathbf{a}' = \mathbf{Q}\mathbf{a} = \begin{bmatrix} q_{11} & q_{12} & \cdots \\ q_{21} & q_{22} & \cdots \\ \cdot & \cdot & \cdots \\ \cdot & \cdot & \cdots \\ \cdot & \cdot & \end{bmatrix} \begin{bmatrix} a_1 \\ a_2 \\ \cdot \\ \cdot \\ \cdot \end{bmatrix} = \begin{bmatrix} q_{11}a_1 + q_{12}a_2 + \cdots \\ q_{21}a_1 + q_{22}a_2 + \cdots \\ \cdot & \cdot & \cdots \\ \cdot & \cdot & \cdots \end{bmatrix}$$

The transpose of $\mathbf{a}'$ is then

$$\mathbf{a}'^T = \mathbf{a}^T\mathbf{Q}^T = [a_1, a_2, \ldots] \begin{bmatrix} q_{11} & q_{12} & \cdots \\ q_{21} & q_{22} & \cdots \\ \cdot & \cdot & \cdots \end{bmatrix}$$

$$= [q_{11}a_1 + q_{12}a_2 + \ldots, q_{21}a_1 + q_{22}a_2 + \ldots]$$

A matrix $\mathbf{Q}$ is said to be *orthogonal* if $\mathbf{Q}^T = \mathbf{Q}^{-1}$. It defines an *orthogonal transformation*. It leaves $\mathbf{a}^T\mathbf{b}$ unchanged, since $\mathbf{a}'^T\mathbf{b}' = \mathbf{a}^T\mathbf{Q}^T\mathbf{Q}\mathbf{b} = \mathbf{a}^T\mathbf{Q}^{-1}\mathbf{Q}\mathbf{b} = \mathbf{a}^T\mathbf{b}$.

The transformation defined by the matrix product $\mathbf{Q}^{-1}\mathbf{A}\mathbf{Q}$ is called a *similarity transformation*. The transformation defined by the product $\mathbf{Q}^T\mathbf{A}\mathbf{Q}$ is called a *congruent transformation*.

If the elements of $\mathbf{Q}$ are complex, then $\mathbf{Q}$ is called *Hermitian* if $q_{ij}{}^* = q_{ji}$, that is, $\mathbf{Q}^{T*} = \mathbf{Q}$. If $\mathbf{Q}^{T*} = \mathbf{Q}^{-1}$, then $\mathbf{Q}$ is called a *unitary* matrix, and the transformation $\mathbf{Q}^{-1}\mathbf{A}\mathbf{Q}$ is called a *unitary transformation*.

*Eigenvectors of a Matrix*

An *eigenvector* $\mathbf{a}$ of a matrix $\mathbf{Q}$ is a vector such that

$$\mathbf{Q}\mathbf{a} = \lambda\mathbf{a}$$

or

$$(\mathbf{Q} - \mathbf{I}\lambda)\mathbf{a} = 0$$

where $\lambda$ is a scalar, called the *eigenvalue*. The eigenvalues are found by solving the *secular equation*

$$\det(\mathbf{Q} - \mathbf{I}\lambda) = \begin{vmatrix} q_{11} - \lambda & q_{12} & \cdots \\ q_{21} & q_{22} - \lambda & \cdots \\ \cdot & \cdot & \cdots \end{vmatrix} = 0$$

which is an algebraic equation of degree $n$ (the number of rows or columns or order of the matrix.)

If the matrix $\mathbf{Q}$ is diagonal, then the eigenvalues are its elements.

Consider two different eigenvectors $\mathbf{a}_\alpha$ and $\mathbf{a}_\beta$ of a symmetric matrix $\mathbf{Q}$. Then

$$\mathbf{Q}\mathbf{a}_\alpha = \lambda_\alpha \mathbf{a}_\alpha$$
$$\mathbf{Q}\mathbf{a}_\beta = \lambda_\beta \mathbf{a}_\beta$$

where $\lambda_\alpha$ and $\lambda_\beta$ are the eigenvalues. Multiply the first by $\mathbf{a}_\beta{}^T$ and the second, transposed, by $\mathbf{a}_\alpha$. Then

$$\mathbf{a}_\beta{}^T\mathbf{Q}\mathbf{a}_\alpha = \lambda_\alpha \mathbf{a}_\beta{}^T\mathbf{a}_\alpha$$
$$\mathbf{a}_\beta{}^T\mathbf{Q}^T\mathbf{a}_\alpha = \lambda_\beta \mathbf{a}_\beta{}^T\mathbf{a}_\alpha$$

But if $\mathbf{Q}$ is symmetric, then $\mathbf{Q}^T = \mathbf{Q}$, so the two expressions on the left are equal. Hence

$$(\lambda_\beta - \lambda_\alpha)\mathbf{a}_\beta{}^T\mathbf{a}_\alpha = 0$$

If the eigenvalues are different, then the two eigenvectors must be orthogonal.

*Reduction to Diagonal Form*

Given a matrix $\mathbf{Q}$, we seek a matrix $\mathbf{A}$ such that

$$\mathbf{A}^{-1}\mathbf{Q}\mathbf{A} = \mathbf{D}$$

where $\mathbf{D}$ is diagonal. Now

$$\mathbf{D} - \lambda\mathbf{I} = \mathbf{A}^{-1}\mathbf{Q}\mathbf{A} - \lambda\mathbf{I} = \mathbf{A}^{-1}(\mathbf{Q} - \lambda\mathbf{I})\mathbf{A}.$$

Hence the eigenvalues of $\mathbf{Q}$ are the same as those of $\mathbf{D}$, namely, the elements of $\mathbf{D}$. Let $\lambda_k$ be a particular eigenvalue, found by solving the secular equation $\det(\mathbf{Q} - \lambda\mathbf{I}) = 0$. Then the corresponding eigenvector $\mathbf{a}_k$ satisfies the equation

$$\mathbf{Q}\mathbf{a}_k = \lambda_k \mathbf{a}_k$$

which is equivalent to $n$ linear homogeneous algebraic equations

$$\sum_j q_{ij}a_{jk} = \lambda_k a_{ik} \qquad (i = 1, 2, \ldots, n)$$

These may be solved for the ratios of the $a$'s to yield the components of the eigenvector $\mathbf{a}_k$. The same procedure is repeated for each eigenvalue in turn. We then form the matrix $\mathbf{A}$ whose columns are the eigenvectors $\mathbf{a}_k$, that is, $[\mathbf{A}]_{ik} = a_{ik}$. Thus the matrix $\mathbf{A}$ must satisfy

$$\mathbf{Q}\mathbf{A} = \mathbf{A}\begin{bmatrix} \lambda_1 & 0 & \cdots \\ 0 & \lambda_2 & \cdots \\ & & \cdots \\ 0 & 0 & \cdots \lambda_n \end{bmatrix} = \mathbf{A}\mathbf{D}$$

so that $\mathbf{A}^{-1}\mathbf{Q}\mathbf{A} = \mathbf{D}$ as required. The above method can always be done if $\mathbf{A}$ is symmetric and the eigenvalues are all different. In the case of repeated roots, orthogonal eigevectors are constructed by the method of Section 10.6.

# Appendix VI

**Values of $k^2$ of Various Bodies**
**(Moment of Inertia = Mass $\times$ $k^2$)**

| Body | Axis | $k^2$ |
|------|------|-------|
| Thin rod, length $a$ | Normal to rod at its center | $\dfrac{a^2}{12}$ |
| | Normal to rod at one end | $\dfrac{a^2}{3}$ |
| Thin rectangular lamina, sides $a$ and $b$ | Through the center, parallel to side $b$ | $\dfrac{a^2}{12}$ |
| | Through the center, normal to the lamina | $\dfrac{a^2 + b^2}{12}$ |
| Thin circular disc, radius $a$ | Through the center, in the plane of the disc | $\dfrac{a^2}{4}$ |
| | Through the center, normal to the disc | $\dfrac{a^2}{2}$ |
| Thin hoop (or ring) radius $a$ | Through the center, in the plane of the loop | $\dfrac{a^2}{2}$ |
| | Through the center, normal to the plane of the hoop | $a^2$ |
| Thin cylindrical shell, radius $a$, length $b$ | Central longitudinal axis | $a^2$ |
| Uniform solid right circular cylinder radius $a$, length $b$ | Central longitudinal axis | $\dfrac{a^2}{2}$ |
| | Through center, perpendicular to longitudinal axis | $\dfrac{a^2}{4} + \dfrac{b^2}{12}$ |
| Thin spherical shell, radius $a$ | Any diameter | $\dfrac{2}{3}a^2$ |
| Uniform solid sphere, radius $a$ | Any diameter | $\dfrac{2}{5}a^2$ |
| Uniform solid rectangular parallelepiped, sides $a$, $b$, and $c$ | Through center, normal to face $ab$, parallel to edge $c$ | $\dfrac{a^2 + b^2}{12}$ |

320

# Answers to Selected Od-Numbered Problems

## Chapter 1

1.1  (a) $\mathbf{i} + 2\mathbf{j} - \mathbf{k}, 6^{\frac{1}{2}}$
(b) $\mathbf{i} + \mathbf{k}, 2^{\frac{1}{2}}$
(c) 1
(d) $-\mathbf{i} + \mathbf{j} + \mathbf{k}, 3^{\frac{1}{2}}$

1.3  $\cos^{-1} (2/3)^{\frac{1}{2}} = 35°15'$

1.5  $(-1 \pm \sqrt{17})/4$

1.15  $\omega b (1 + 3\cos^2\omega t)^{\frac{1}{2}}$

1.19  $(b^2\omega^2 + 4c^2t^2)^{\frac{1}{2}}, (b^2\omega^4 + 4c^2)^{\frac{1}{2}}$

1.21  $4c^2t(b^2\omega^2 + 4c^2t^2)^{-\frac{1}{2}}, [b^2\omega^4 + 4c^2 - 8c^2t^2(b^2\omega^2 + 4c^2t^2)^{-1}]^{\frac{1}{2}}$

1.25  $(a\upsilon - \upsilon\dot{\upsilon})/|\mathbf{a} \times \mathbf{v}|$

## Chapter 2

2.1  $3F_0 t_0/m$

2.3  (a) $\upsilon = [(2F_0 x + kx^2)m^{-1}]^{\frac{1}{2}}$
(c) $x = [k\upsilon - F_0\ln(1 + k\upsilon/F_0]k^{-2}$

2.7  $\upsilon = (A + Bt)^\alpha$ where $A = \upsilon_0^{1-n}, B = (c/m)(n-1), \alpha = (1-n)^{-1}$
$x = C(\upsilon_0^\beta - \upsilon^\beta)$ where $C = (m/c)(2-n)^{-1}$ and $n \neq 1,2$

2.9  (a) $x = -(m/c)[\upsilon + (mg/c)\ln(1 - c\upsilon/mg)]$
(b) $x = -(m/2c)\ln(1 - c\upsilon^2/mg)$

321

2.11   $F(x) = -mb^2x^{-3}$
2.13   $x = a \tan(bt)$ where $a = (2mv_0/k)^{\frac{1}{2}}$ , $b = (kv_0/2m)^{\frac{1}{2}}$
2.15   $(A_1/A_2)(m_1/m_2)^{\frac{1}{2}}$


## Chapter 3

3.1    (a), (b), and (e) are conservative.
3.3    (a) $\mathbf{F} = \mathbf{i}yz + \mathbf{j}xz + \mathbf{k}xy$
       (c) $\mathbf{F} = kx^\alpha y^\beta z^\gamma (\mathbf{i}x^{-1} + \mathbf{k}z^{-1})$
3.5    $(v_0^2 - 16m^{-1})^{\frac{1}{2}}$
3.7    $F = -mg + mgz/2r_e$
3.11   $m\ddot{x} = -cx\dot{s}, \; m\ddot{y} = -cy\dot{s}, \; m\ddot{z} = -mg - cz\dot{s}$
3.13   $x = 6^{-\frac{1}{2}} \cos(2^{\frac{1}{2}} t), y = 24^{-\frac{1}{2}} \cos(8^{\frac{1}{2}} t), z = 54^{-\frac{1}{2}} \cos(18^{\frac{1}{2}} t)$
3.17   $v = (2gb)^{\frac{1}{2}}$ , $R = 3mg$
3.19   $T = 4(l/g)^{\frac{1}{2}} \; K(15°)$


## Chapter 4

4.1    $\mathbf{A} = \mathbf{i} - y\mathbf{j} + 2\mathbf{k}, \; |\mathbf{A}| = (y^2 + 9)^{\frac{1}{2}}$
4.3    about 41 nt (west)
4.5    $A_{max} = a_0 + (a_0^2 + v^4/b^2)^{\frac{1}{2}}$ where $b$ is the wheel radius and $v$ is the instantaneous forward speed. The maximum occurs at the point defined by $\tan\Theta = a_0b/v^2$ where $\Theta$ is measured upward from the rear of the wheel.
4.9    $d^3\mathbf{R}/dt^3 = \dddot{\mathbf{r}} + 3\dot{\omega} \times \dot{\mathbf{r}} + 3\omega \times \ddot{\mathbf{r}} + \ddot{\omega} \times \mathbf{r} + 2\omega \times (\dot{\omega} \times \mathbf{r}) + \dot{\omega} \times (\omega \times \mathbf{r})$
       $+ 3\omega \times (\omega \times \dot{\mathbf{r}}) + \omega \times [\omega \times (\omega \times \mathbf{r})] + d\mathbf{A}_0/dt$
       Hint: Expand the operator expression $[(\;\dot{}\;) + \omega\mathbf{x} (\;)]^3$.
4.11   For small $\theta_0$, $\sin^2\theta_0 \approx \omega_e \sin\lambda (4/3) (l/g)^{\frac{1}{2}}$ which gives $\theta_0 \approx 0.5°$.


## Chapter 5

5.1    $9.4 \times 10^{-6}$ dynes

5.5    $T = \sqrt{\dfrac{3\pi}{G\rho}} = 2\pi \sqrt{\dfrac{r_e}{g}} \approx 1.4$ hr

5.7    $F = -GMmr^{-2} - (4/3) \pi \rho mGr$
5.9    I: $r = ae^{k\Theta} + be^{-k\Theta}$        $k^2 = 1 - c/mh^2 > 0$
       II: $r = a\cos k\Theta + b\sin k\Theta$        $k^2 = -1 + c/mh^2 > 0$
       III: $r = (a\Theta + b)^{-1}$        $c = mh^2$

5.11 No. $\Theta$ varies as $t^{\frac{1}{3}}$

5.13 Period $= \left(\frac{r_0}{1-e}\right)^{\frac{3}{2}} = 75.6$ yr

$r_{max} = 35.16$ a.u.

Use $r_0 v_0 = r_1 v_1$

5.17 $\pi (1 - ab)^{-\frac{1}{2}}$

5.21 $a > (\epsilon/k)^{\frac{1}{2}}$

5.25 Approximately $0.7°$ for orbits near the earth

## Chapter 6

6.1 (a) $(2\mathbf{i} + \mathbf{j} + 3\mathbf{k})/3$

(c) $3\mathbf{j} + \mathbf{k}$

(d) $-\mathbf{i} + 3\mathbf{k}$

6.3 Fraction lost $= (1 + \gamma)^{-1}$, $s = \dfrac{v_0^2 \gamma^2}{2g\mu(1 + \gamma)^2}$

6.5 $v_0, 2v_0, 4v_0$

6.13 Proton: $v_x' = v_y' \approx 0.6\, v$

Helium: $v_x' \approx 0.09v$, $v_y' \approx 0.15v$

6.15 $\sim 55°$

6.19 $\ddot{z} = g - 3\dot{z}^2/g$

6.21 $m\dot{v} + V\dot{m} + kv = 0$

## Chapter 7

7.1 (a) $\frac{1}{3}a$ from bottom center

(b) $\frac{3}{5}\, b$ from vertex

(c) $\frac{2}{3}\, b$ from vertex

(d) $\frac{3}{4}\, h$ from vertex

7.3 (a) $\frac{7}{36}\, ma^2$ where $m = $ total mass

(d) $\frac{3}{10}\, ma^2$ where $a = $ radius of base

7.7 $\sin\Theta = \dfrac{8\mu(1 + \mu)}{3 (1 + \mu^2)}$

7.11 $2\pi (2a/g)^{\frac{1}{2}}$ , $2\pi (3a/2g)^{\frac{1}{2}}$

7.13 $5g/7$

7.17 When line of centers makes an angle of $\cos^{-1} (4/7)$ with the vertical

7.19 (a) Horizontal: $(3/4)mg \sin\theta(3 \cos\theta - 2)$

Vertical: $(1/4)mg(3 \cos\theta - 1)^2$

(b) Slipping begins when $|3 \sin \theta(3 \cos \theta - 2)| = \mu(3 \cos \theta - 1)^2$
Rod slips backwards if above equation is satisfied for $\theta < \cos^{-1}$
(2/3), otherwise rod slips forward.

7.21  $s = v_0 t - \frac{1}{2}gt^2(\sin \Theta + \mu \cos \Theta)$
Pure rolling begins when
$s = (2v_0^2/g) (\sin\Theta + 6\mu \cos\Theta) (2\sin\Theta + 7\mu\cos\Theta)^{-2}$

7.25  $v_{cm1} = -\hat{P}/4m, \; \omega_1 = -3\hat{P}/2ml$
$v_{cm2} = 5\hat{P}/4m, \; \omega_2 = 9\hat{P}/ml$
$v_B = -\hat{P}/m$

## Chapter 8

8.1 $\begin{bmatrix} 1 & 0 & 0 \\ 0 & 1 & 0 \\ 0 & 0 & 2 \end{bmatrix} \dfrac{ml^2}{12}$

8.3 $\begin{bmatrix} b^2 + c^2 & 0 & 0 \\ 0 & a^2 + c^2 & 0 \\ 0 & 0 & a^2 + b^2 \end{bmatrix} m$

8.5  The inertia tensor is the same expression as in Problem 8.3 with $m/3$
instead of $m$, and where $2a$, $2b$, and $2c$ are the dimensions of the
block.

8.7  $Ax^2 + By^2 + Cz^2 = 1$ where

$$B = A = m \left( \frac{a^2 + b^2}{4} + \frac{b^2}{12} \right), C = \frac{ma^2}{2}$$

$a:b = 1:\sqrt{3}$

8.9  $ma\omega^2/12$ about $z$ axis

8.11  (a) $(2\pi/\omega)2^{1/2}$ , $(2\pi/\omega) (2/5)^{1/2}$

8.13  (a) $45° - \tan^{-2} (1/2) = 18.5°$

8.19  $\Theta_2 \approx 10°$

8.21  $S^2 > (64ga/b^4) (a^2/3 + b^2/16)$
Approximately 1,230,000 rpm
The pencil would be torn apart by centrifugal force.

## Chapter 9

9.1  Use $L = \dfrac{m}{2} (\dot{x}^2 + \dot{y}^2 + \dot{z}^2) - mgz$

9.3  $g/2, g(m + m'z/b) (m + m'/2)^{-1}$ where $z$ is the length of the cord hanging over the table, and $b$ is the total length of the cord.

9.7 $mg \sin \Theta \cos \Theta \, [(7/5) \, (m + M) - m\cos^2\Theta]^{-1}$

9.11 $d^2r/dt^2 = r\dot{\Theta}^2 + g\cos\Theta - (k/m) \, (r - l_0)$
$d(r^2\dot\Theta)/adt = -grsin\Theta$
where $r$ is the instantaneous length of the pendulum

9.13 $T = 2\pi \, l^{\frac{1}{2}} \, (g^2 + a^2)^{-\frac{1}{4}}$

9.15 Use $L = \dfrac{m}{2} \, (\dot{r}^2 + r^2\dot{\Theta}^2 + r^2\dot{\varphi}^2 \sin^2\Theta) + mgr\cos\Theta - \dfrac{k}{2} \, (r - l_0)^2$

9.19 $H = \dfrac{p^2}{2(m_1+m_2)} - m_2gz$

9.21 $p_x = m\dot{x} - qA_x, p_y = m\dot{y} - qA_y, p_z = m\dot{z} - qA_z$

## Chapter 10

10.1 (a) $x = 3A/2B$, stable only if $A$ and $B$ are of opposite sign
(b) $x = b^{-1}$, stable for $k > 0$
(c) $x = \pm a$, stable for $k > 0$

10.3 (a) $(B/3A)^2 \, (2/m)^{\frac{1}{2}}$
(b) $(k/m)^{\frac{1}{2}}$
(c) $2a(2k/m)^{\frac{1}{2}}$

10.7 $(1.93k/m)^{\frac{1}{2}}$

10.9 $2\pi \, (26ab^2/5g)^{\frac{1}{2}} \, [(a+b) \, (3b-5a)]^{-\frac{1}{2}}$

10.13 $\begin{bmatrix} x_1 \\ x_2 \end{bmatrix} = \begin{bmatrix} 1 \\ 1 \end{bmatrix} A_0\cos\omega_1 t + \begin{bmatrix} 1 \\ -1 \end{bmatrix} A_0\cos\omega_2 t$

10.15 (a) $\omega_1 = 1$ (symmetric), $\omega_2 = (7/4)12$ (antisymmetric)

$\begin{bmatrix} 1 \\ 1 \end{bmatrix}$ (symmetric eigenvector)

$\begin{bmatrix} 2 \\ -1 \end{bmatrix}$ (antisymmetric eigenvector)

$q_1 = x_1 + 2x_2$
$\qquad\qquad$ (normal coordinates)
$q_2 = x_1 - x_2$

10.19 $\omega_1 = (k/m')^{\frac{1}{2}}$ , $\omega_2 = [(k/m') + (2k/m)]^{\frac{1}{2}}$

10.21 $v_{\text{long}} = (k/m)^{\frac{1}{2}} \, (l + \Delta l)$
$v_{\text{trans}} = (k/m)^{\frac{1}{2}} \, [(l + \Delta l)\Delta l]^{\frac{1}{2}}$

# INDEX

# Index